Numerical Modeling Strategies for Sustainable Concrete Structures

RILEM BOOKSERIES

Volume 38

RILEM, The International Union of Laboratories and Experts in Construction Materials, Systems and Structures, founded in 1947, is a non-governmental scientific association whose goal is to contribute to progress in the construction sciences, techniques and industries, essentially by means of the communication it fosters between research and practice. RILEM's focus is on construction materials and their use in building and civil engineering structures, covering all phases of the building process from manufacture to use and recycling of materials. More information on RILEM and its previous publications can be found on www.RILEM.net.

Indexed in SCOPUS, Google Scholar and SpringerLink.

More information about this series at https://link.springer.com/bookseries/8781

Pierre Rossi · Jean-Louis Tailhan
Editors

Numerical Modeling Strategies for Sustainable Concrete Structures

SSCS 2022

Springer

Editors
Pierre Rossi
MAST-EMGCU
Université Gustave Eiffel
Marne-la-Vallée, France

Jean-Louis Tailhan
MAST-EMGCU
Université Gustave Eiffel
Marne-la-Vallée, France

ISSN 2211-0844 ISSN 2211-0852 (electronic)
RILEM Bookseries
ISBN 978-3-031-07745-6 ISBN 978-3-031-07746-3 (eBook)
https://doi.org/10.1007/978-3-031-07746-3

This Springer imprint is published by the registered company Springer Nature Switzerland AG
The registered company address is: Gewerbestrasse 11, 6330 Cham, Switzerland

Preface

Cement is the main constituent of concrete, the building material most widely used today, and that will continue to be in the years to come. Its production generates, among other things, CO_2 emissions and, so, greenhouse effects. It is, thus, of primary importance to optimize the utilization of cement in concrete constructions while checking that these constructions have a lifespan compatible with the stakes of sustainable development. It is essential to take up the challenge and use adapted quantification tools, making it possible to rigorously and reliably justify the adopted strategic and technical choices.

Numerical methods (finite elements, finite volumes, and finite differences) are a relevant response to this challenge. They are indeed accurate tools for an optimized design of concrete constructions. They allow us to consider all types of complexities, for example, those linked to rheological, physicochemical, and mechanical properties of concrete, those linked to the geometry of the structures or even to the environmental boundary conditions. This optimization must also respect constraints of time, money, security, energy, CO2 emissions, and, more generally, life cycle more reliably than the codes and analytical approaches currently used.

Numerical methods are, undoubtedly, the best calculation tools at the service of concrete eco-construction.

After Aix-en-Provence (France, 2012), Rio de Janeiro (Brazil, 2015), and Lecco (Italy, 2019), the RILEM International Conference on Numerical Modeling Strategies for Sustainable Concrete Structures (SSCS 2022) is taking place in the city of Marseille (France).

July 2022

Pierre Rossi
Jean-Louis Tailhan

Contents

RILEM Publications

The following list is presenting the global offer of RILEM Publications, sorted by series. Each publication is available in printed version and/or in online version.

RILEM Proceedings (PRO)

PRO 1: Durability of High Performance Concrete (ISBN: 2-912143-03-9; e-ISBN: 2-351580-12-5; e-ISBN: 2351580125); *Ed. H. Sommer*

PRO 2: Chloride Penetration into Concrete (ISBN: 2-912143-00-04; e-ISBN: 2912143454); *Eds. L.-O. Nilsson and J.-P. Ollivier*

PRO 3: Evaluation and Strengthening of Existing Masonry Structures (ISBN: 2-912143-02-0; e-ISBN: 2351580141); *Eds. L. Binda and C. Modena*

PRO 4: Concrete: From Material to Structure (ISBN: 2-912143-04-7; e-ISBN: 2351580206); *Eds. J.-P. Bournazel and Y. Malier*

PRO 5: The Role of Admixtures in High Performance Concrete (ISBN: 2-912143-05-5; e-ISBN: 2351580214); *Eds. J. G. Cabrera and R. Rivera-Villarreal*

PRO 6: High Performance Fiber Reinforced Cement Composites - HPFRCC 3 (ISBN: 2-912143-06-3; e-ISBN: 2351580222); *Eds. H. W. Reinhardt and A. E. Naaman*

PRO 7: 1st International RILEM Symposium on Self-Compacting Concrete (ISBN: 2-912143-09-8; e-ISBN: 2912143721); *Eds. Å. Skarendahl and Ö. Petersson*

PRO 8: International RILEM Symposium on Timber Engineering (ISBN: 2-912143-10-1; e-ISBN: 2351580230); *Ed. L. Boström*

PRO 9: 2nd International RILEM Symposium on Adhesion between Polymers and Concrete ISAP '99 (ISBN: 2-912143-11-X; e-ISBN: 2351580249); *Eds. Y. Ohama and M. Puterman*

PRO 10: 3rd International RILEM Symposium on Durability of Building and Construction Sealants (ISBN: 2-912143-13-6; e-ISBN: 2351580257); *Eds. A. T. Wolf*

PRO 11: 4th International RILEM Conference on Reflective Cracking in Pavements (ISBN: 2-912143-14-4; e-ISBN: 2351580265); *Eds. A. O. Abd El Halim, D. A. Taylor and El H. H. Mohamed*

PRO 12: International RILEM Workshop on Historic Mortars: Characteristics and Tests (ISBN: 2-912143-15-2; e-ISBN: 2351580273); *Eds. P. Bartos, C. Groot and J. J. Hughes*

PRO 13: 2nd International RILEM Symposium on Hydration and Setting (ISBN: 2-912143-16-0; e-ISBN: 2351580281); *Ed. A. Nonat*

PRO 14: Integrated Life-Cycle Design of Materials and Structures - ILCDES 2000 (ISBN: 951-758-408-3; e-ISBN: 235158029X); (ISSN: 0356-9403); *Ed. S. Sarja*

PRO 15: Fifth RILEM Symposium on Fibre-Reinforced Concretes (FRC) - BEFIB'2000 (ISBN: 2-912143-18-7; e-ISBN: 291214373X); *Eds. P. Rossi and G. Chanvillard*

PRO 16: Life Prediction and Management of Concrete Structures (ISBN: 2-912143-19-5; e-ISBN: 2351580303); *Ed. D. Naus*

PRO 17: Shrinkage of Concrete – Shrinkage 2000 (ISBN: 2-912143-20-9; e-ISBN: 2351580311); *Eds. V. Baroghel-Bouny and P.-C. Aïtcin*

PRO 18: Measurement and Interpretation of the On-Site Corrosion Rate (ISBN: 2-912143-21-7; e-ISBN: 235158032X); *Eds. C. Andrade, C. Alonso, J. Fullea, J. Polimon and J. Rodriguez*

PRO 19: Testing and Modelling the Chloride Ingress into Concrete (ISBN: 2-912143-22-5; e-ISBN: 2351580338); *Eds. C. Andrade and J. Kropp*

PRO 20: 1st International RILEM Workshop on Microbial Impacts on Building Materials (CD 02) (e-ISBN 978-2-35158-013-4); *Ed. M. Ribas Silva*

PRO 21: International RILEM Symposium on Connections between Steel and Concrete (ISBN: 2-912143-25-X; e-ISBN: 2351580346); *Ed. R. Eligehausen*

PRO 22: International RILEM Symposium on Joints in Timber Structures (ISBN: 2-912143-28-4; e-ISBN: 2351580354); *Eds. S. Aicher and H.-W. Reinhardt*

PRO 23: International RILEM Conference on Early Age Cracking in Cementitious Systems (ISBN: 2-912143-29-2; e-ISBN: 2351580362); *Eds. K. Kovler and A. Bentur*

PRO 24: 2nd International RILEM Workshop on Frost Resistance of Concrete (ISBN: 2-912143-30-6; e-ISBN: 2351580370); *Eds. M. J. Setzer, R. Auberg and H.-J. Keck*

PRO 25: International RILEM Workshop on Frost Damage in Concrete (ISBN: 2-912143-31-4; e-ISBN: 2351580389); *Eds. D. J. Janssen, M. J. Setzer and M. B. Snyder*

PRO 26: International RILEM Workshop on On-Site Control and Evaluation of Masonry Structures (ISBN: 2-912143-34-9; e-ISBN: 2351580141); *Eds. L. Binda and R. C. de Vekey*

PRO 27: International RILEM Symposium on Building Joint Sealants (CD03; e-ISBN: 235158015X); *Ed. A. T. Wolf*

PRO 28: 6th International RILEM Symposium on Performance Testing and Evaluation of Bituminous Materials - PTEBM'03 (ISBN: 2-912143-35-7; e-ISBN: 978-2-912143-77-8); *Ed. M. N. Partl*

PRO 29: 2nd International RILEM Workshop on Life Prediction and Ageing Management of Concrete Structures (ISBN: 2-912143-36-5; e-ISBN: 2912143780); *Ed. D. J. Naus*

PRO 30: 4th International RILEM Workshop on High Performance Fiber Reinforced Cement Composites - HPFRCC 4 (ISBN: 2-912143-37-3; e-ISBN: 2912143799); *Eds. A. E. Naaman and H. W. Reinhardt*

PRO 31: International RILEM Workshop on Test and Design Methods for Steel Fibre Reinforced Concrete: Background and Experiences (ISBN: 2-912143-38-1; e-ISBN: 2351580168); *Eds. B. Schnütgen and L. Vandewalle*

PRO 32: International Conference on Advances in Concrete and Structures 2 vol. (ISBN (set): 2-912143-41-1; e-ISBN: 2351580176); *Eds. Ying-shu Yuan, Surendra P. Shah and Heng-lin Lü*

PRO 33: 3rd International Symposium on Self-Compacting Concrete (ISBN: 2-912143-42-X; e-ISBN: 2912143713); *Eds. Ó. Wallevik and I. Níelsson*

PRO 34: International RILEM Conference on Microbial Impact on Building Materials (ISBN: 2-912143-43-8; e-ISBN: 2351580184); *Ed. M. Ribas Silva*

PRO 35: International RILEM TC 186-ISA on Internal Sulfate Attack and Delayed Ettringite Formation (ISBN: 2-912143-44-6; e-ISBN: 2912143802); *Eds. K. Scrivener and J. Skalny*

PRO 36: International RILEM Symposium on Concrete Science and Engineering – A Tribute to Arnon Bentur (ISBN: 2-912143-46-2; e-ISBN: 2912143586); *Eds. K. Kovler, J. Marchand, S. Mindess and J. Weiss*

PRO 37: 5th International RILEM Conference on Cracking in Pavements – Mitigation, Risk Assessment and Prevention (ISBN: 2-912143-47-0; e-ISBN: 2912143764); *Eds. C. Petit, I. Al-Qadi and A. Millien*

PRO 38: 3rd International RILEM Workshop on Testing and Modelling the Chloride Ingress into Concrete (ISBN: 2-912143-48-9; e-ISBN: 2912143578); *Eds. C. Andrade and J. Kropp*

PRO 39: 6th International RILEM Symposium on Fibre-Reinforced Concretes - BEFIB 2004 (ISBN: 2-912143-51-9; e-ISBN: 2912143748); *Eds. M. Di Prisco, R. Felicetti and G. A. Plizzari*

PRO 40: International RILEM Conference on the Use of Recycled Materials in Buildings and Structures (ISBN: 2-912143-52-7; e-ISBN: 2912143756); *Eds. E. Vázquez, Ch. F. Hendriks and G. M. T. Janssen*

PRO 41: RILEM International Symposium on Environment-Conscious Materials and Systems for Sustainable Development (ISBN: 2-912143-55-1; e-ISBN: 2912143640); *Eds. N. Kashino and Y. Ohama*

PRO 42: SCC'2005 - China: 1st International Symposium on Design, Performance and Use of Self-Consolidating Concrete (ISBN: 2-912143-61-6; e-ISBN: 2912143624); *Eds. Zhiwu Yu, Caijun Shi, Kamal Henri Khayat and Youjun Xie*

PRO 43: International RILEM Workshop on Bonded Concrete Overlays (e-ISBN: 2-912143-83-7); *Eds. J. L. Granju and J. Silfwerbrand*

PRO 44: 2nd International RILEM Workshop on Microbial Impacts on Building Materials (CD11) (e-ISBN: 2-912143-84-5); *Ed. M. Ribas Silva*

PRO 45: 2nd International Symposium on Nanotechnology in Construction, Bilbao (ISBN: 2-912143-87-X; e-ISBN: 2912143888); *Eds. Peter J. M. Bartos, Yolanda de Miguel and Antonio Porro*

PRO 46: ConcreteLife'06 - International RILEM-JCI Seminar on Concrete Durability and Service Life Planning: Curing, Crack Control, Performance in Harsh Environments (ISBN: 2-912143-89-6; e-ISBN: 291214390X); *Ed. K. Kovler*

PRO 47: International RILEM Workshop on Performance Based Evaluation and Indicators for Concrete Durability (ISBN: 978-2-912143-95-2; e-ISBN: 9782912143969); *Eds. V. Baroghel-Bouny, C. Andrade, R. Torrent and K. Scrivener*

PRO 48: 1st International RILEM Symposium on Advances in Concrete through Science and Engineering (e-ISBN: 2-912143-92-6); *Eds. J. Weiss, K. Kovler, J. Marchand, and S. Mindess*

PRO 49: International RILEM Workshop on High Performance Fiber Reinforced Cementitious Composites in Structural Applications (ISBN: 2-912143-93-4; e-ISBN: 2912143942); *Eds. G. Fischer and V. C. Li*

PRO 50: 1st International RILEM Symposium on Textile Reinforced Concrete (ISBN: 2-912143-97-7; e-ISBN: 2351580087); *Eds. Josef Hegger, Wolfgang Brameshuber and Norbert Will*

PRO 51: 2nd International Symposium on Advances in Concrete through Science and Engineering (ISBN: 2-35158-003-6; e-ISBN: 2-35158-002-8); *Eds. J. Marchand, B. Bissonnette, R. Gagné, M. Jolin and F. Paradis*

PRO 52: Volume Changes of Hardening Concrete: Testing and Mitigation (ISBN: 2-35158-004-4; e-ISBN: 2-35158-005-2); *Eds. O. M. Jensen, P. Lura and K. Kovler*

PRO 53: High Performance Fiber Reinforced Cement Composites - HPFRCC5 (ISBN: 978-2-35158-046-2; e-ISBN: 978-2-35158-089-9); *Eds. H. W. Reinhardt and A. E. Naaman*

PRO 54: 5th International RILEM Symposium on Self-Compacting Concrete (ISBN: 978-2-35158-047-9; e-ISBN: 978-2-35158-088-2); *Eds. G. De Schutter and V. Boel*

PRO 55: International RILEM Symposium Photocatalysis, Environment and Construction Materials (ISBN: 978-2-35158-056-1; e-ISBN: 978-2-35158-057-8); *Eds. P. Baglioni and L. Cassar*

PRO 56: International RILEM Workshop on Integral Service Life Modelling of Concrete Structures (ISBN 978-2-35158-058-5; e-ISBN: 978-2-35158-090-5); *Eds. R. M. Ferreira, J. Gulikers and C. Andrade*

PRO 57: RILEM Workshop on Performance of cement-based materials in aggressive aqueous environments (e-ISBN: 978-2-35158-059-2); *Ed. N. De Belie*

PRO 58: International RILEM Symposium on Concrete Modelling - CONMOD'08 (ISBN: 978-2-35158-060-8; e-ISBN: 978-2-35158-076-9); *Eds. E. Schlangen and G. De Schutter*

PRO 59: International RILEM Conference on On Site Assessment of Concrete, Masonry and Timber Structures - SACoMaTiS 2008 (ISBN set: 978-2-35158-061-5; e-ISBN: 978-2-35158-075-2); *Eds. L. Binda, M. di Prisco and R. Felicetti*

PRO 60: Seventh RILEM International Symposium on Fibre Reinforced Concrete: Design and Applications - BEFIB 2008 (ISBN: 978-2-35158-064-6; e-ISBN: 978-2-35158-086-8); *Ed. R. Gettu*

PRO 61: 1st International Conference on Microstructure Related Durability of Cementitious Composites 2 vol., (ISBN: 978-2-35158-065-3; e-ISBN: 978-2-35158-084-4); *Eds. W. Sun, K. van Breugel, C. Miao, G. Ye and H. Chen*

PRO 62: NSF/ RILEM Workshop: In-situ Evaluation of Historic Wood and Masonry Structures (e-ISBN: 978-2-35158-068-4); *Eds. B. Kasal, R. Anthony and M. Drdácký*

PRO 63: Concrete in Aggressive Aqueous Environments: Performance, Testing and Modelling, 2 vol., (ISBN: 978-2-35158-071-4; e-ISBN: 978-2-35158-082-0); *Eds. M. G. Alexander and A. Bertron*

PRO 64: Long Term Performance of Cementitious Barriers and Reinforced Concrete in Nuclear Power Plants and Waste Management - NUCPERF 2009 (ISBN: 978-2-35158-072-1; e-ISBN: 978-2-35158-087-5); *Eds. V. L'Hostis, R. Gens, C. Gallé*

PRO 65: Design Performance and Use of Self-consolidating Concrete - SCC'2009 (ISBN: 978-2-35158-073-8; e-ISBN: 978-2-35158-093-6); *Eds. C. Shi, Z. Yu, K. H. Khayat and P. Yan*

PRO 66: 2nd International RILEM Workshop on Concrete Durability and Service Life Planning - ConcreteLife'09 (ISBN: 978-2-35158-074-5; ISBN: 978-2-35158-074-5); *Ed. K. Kovler*

PRO 67: Repairs Mortars for Historic Masonry (e-ISBN: 978-2-35158-083-7); *Ed. C. Groot*

PRO 68: Proceedings of the 3rd International RILEM Symposium on 'Rheology of Cement Suspensions such as Fresh Concrete (ISBN 978-2-35158-091-2; e-ISBN: 978-2-35158-092-9); *Eds. O. H. Wallevik, S. Kubens and S. Oesterheld*

PRO 69: 3rd International PhD Student Workshop on 'Modelling the Durability of Reinforced Concrete (ISBN: 978-2-35158-095-0); *Eds. R. M. Ferreira, J. Gulikers and C. Andrade*

PRO 70: 2nd International Conference on 'Service Life Design for Infrastructure' (ISBN set: 978-2-35158-096-7, e-ISBN: 978-2-35158-097-4); *Ed. K. van Breugel, G. Ye and Y. Yuan*

PRO 71: Advances in Civil Engineering Materials - The 50-year Teaching Anniversary of Prof. Sun Wei' (ISBN: 978-2-35158-098-1; e-ISBN: 978-2-35158-099-8); *Eds. C. Miao, G. Ye, and H. Chen*

PRO 72: First International Conference on 'Advances in Chemically-Activated Materials – CAM'2010' (2010), 264 pp, ISBN: 978-2-35158-101-8; e-ISBN: 978-2-35158-115-5, *Eds. Caijun Shi and Xiaodong Shen*

PRO 73: 2nd International Conference on 'Waste Engineering and Management - ICWEM 2010' (2010), 894 pp, ISBN: 978-2-35158-102-5; e-ISBN: 978-2-35158-103-2, *Eds. J. Zh. Xiao, Y. Zhang, M. S. Cheung and R. Chu*

PRO 74: International RILEM Conference on 'Use of Superabsorsorbent Polymers and Other New Addditives in Concrete' (2010) 374 pp., ISBN: 978-2-35158-104-9; e-ISBN: 978-2-35158-105-6; *Eds. O. M. Jensen, M. T. Hasholt, and S. Laustsen*

PRO 75: International Conference on 'Material Science - 2nd ICTRC - Textile Reinforced Concrete - Theme 1' (2010) 436 pp., ISBN: 978-2-35158-106-3; e-ISBN: 978-2-35158-107-0; *Ed. W. Brameshuber*

PRO 76: International Conference on 'Material Science - HetMat - Modelling of Heterogeneous Materials - Theme 2' (2010) 255 pp., ISBN: 978-2-35158-108-7; e-ISBN: 978-2-35158-109-4; *Ed. W. Brameshuber*

PRO 77: International Conference on 'Material Science - AdIPoC - Additions Improving Properties of Concrete - Theme 3' (2010) 459 pp., ISBN: 978-2-35158-110-0; e-ISBN: 978-2-35158-111-7; *Ed. W. Brameshuber*

PRO 78: 2nd Historic Mortars Conference and RILEM TC 203-RHM Final Workshop – HMC2010 (2010) 1416 pp., e-ISBN: 978-2-35158-112-4; *Eds. J. Válek, C. Groot, and J. J. Hughes*

PRO 79: International RILEM Conference on Advances in Construction Materials Through Science and Engineering (2011) 213 pp., ISBN: 978-2-35158-116-2, e-ISBN: 978-2-35158-117-9; *Eds. Christopher Leung and K.T. Wan*

PRO 80: 2nd International RILEM Conference on Concrete Spalling due to Fire Exposure (2011) 453 pp., ISBN: 978-2-35158-118-6, e-ISBN: 978-2-35158-119-3; *Eds. E. A. B. Koenders and F. Dehn*

PRO 81: 2nd International RILEM Conference on Strain Hardening Cementitious Composites (SHCC2-Rio) (2011) 451 pp., ISBN: 978-2-35158-120-9, e-ISBN: 978-2-35158-121-6; *Eds. R.D. Toledo Filho, F. A. Silva, E. A. B. Koenders and E. M. R. Fairbairn*

PRO 82: 2nd International RILEM Conference on Progress of Recycling in the Built Environment (2011) 507 pp., e-ISBN: 978-2-35158-122-3; *Eds. V. M. John, E. Vazquez, S. C. Angulo and C. Ulsen*

PRO 83: 2nd International Conference on Microstructural-related Durability of Cementitious Composites (2012) 250 pp., ISBN: 978-2-35158-129-2; e-ISBN: 978-2-35158-123-0; *Eds. G. Ye, K. van Breugel, W. Sun and C. Miao*

PRO 84: CONSEC13 - Seventh International Conference on Concrete under Severe Conditions – Environment and Loading (2013) 1930 pp., ISBN: 978-2-35158-124-7; e-ISBN: 978-2- 35158-134-6; *Eds. Z. J. Li, W. Sun, C. W. Miao, K. Sakai, O. E. Gjorv & N. Banthia*

PRO 85: RILEM-JCI International Workshop on Crack Control of Mass Concrete and Related issues concerning Early-Age of Concrete Structures – ConCrack 3 – Control of Cracking in Concrete Structures 3 (2012) 237 pp., ISBN: 978-2-35158-125-4; e-ISBN: 978-2-35158-126-1; *Eds. F. Toutlemonde and J.-M. Torrenti*

PRO 86: International Symposium on Life Cycle Assessment and Construction (2012) 414 pp., ISBN: 978-2-35158-127-8, e-ISBN: 978-2-35158-128-5; *Eds. A. Ventura and C. de la Roche*

PRO 87: UHPFRC 2013 – RILEM-fib-AFGC International Symposium on Ultra-High Performance Fibre-Reinforced Concrete (2013), ISBN: 978-2-35158-130-8, e-ISBN: 978-2-35158-131-5; *Eds. F. Toutlemonde*

PRO 88: 8th RILEM International Symposium on Fibre Reinforced Concrete (2012) 344 pp., ISBN: 978-2-35158-132-2, e-ISBN: 978-2-35158-133-9; *Eds. Joaquim A. O. Barros*

PRO 89: RILEM International workshop on performance-based specification and control of concrete durability (2014) 678 pp, ISBN: 978-2-35158-135-3, e-ISBN: 978-2-35158-136-0; *Eds. D. Bjegović, H. Beushausen and M. Serdar*

PRO 90: 7th RILEM International Conference on Self-Compacting Concrete and of the 1st RILEM International Conference on Rheology and Processing of Construction Materials (2013) 396 pp, ISBN: 978-2-35158-137-7, e-ISBN: 978-2-35158-138-4; *Eds. Nicolas Roussel and Hela Bessaies-Bey*

PRO 91: CONMOD 2014 - RILEM International Symposium on Concrete Modelling (2014), ISBN: 978-2-35158-139-1; e-ISBN: 978-2-35158-140-7; *Eds. Kefei Li, Peiyu Yan and Rongwei Yang*

PRO 92: CAM 2014 - 2nd International Conference on advances in chemically-activated materials (2014) 392 pp., ISBN: 978-2-35158-141-4; e-ISBN: 978-2-35158-142-1; *Eds. Caijun Shi and Xiadong Shen*

PRO 93: SCC 2014 - 3rd International Symposium on Design, Performance and Use of Self-Consolidating Concrete (2014) 438 pp., ISBN: 978-2-35158-143-8; e-ISBN: 978-2-35158-144-5; *Eds. Caijun Shi, Zhihua Ou, Kamal H. Khayat*

PRO 94 (online version): HPFRCC-7 - 7th RILEM conference on High performance fiber reinforced cement composites (2015), e-ISBN: 978-2-35158-146-9; *Eds. H. W. Reinhardt, G. J. Parra-Montesinos, H. Garrecht*

PRO 95: International RILEM Conference on Application of superabsorbent polymers and other new admixtures in concrete construction (2014), ISBN: 978-2-35158-147-6; e-ISBN: 978-2-35158-148-3; *Eds. Viktor Mechtcherine, Christof Schroefl*

PRO 96 (online version): XIII DBMC: XIII International Conference on Durability of Building Materials and Components(2015), e-ISBN: 978-2-35158-149-0; *Eds. M. Quattrone, V. M. John*

PRO 97: SHCC3 – 3rd International RILEM Conference on Strain Hardening Cementitious Composites (2014), ISBN: 978-2-35158-150-6; e-ISBN: 978-2-35158-151-3; *Eds. E. Schlangen, M. G. Sierra Beltran, M. Lukovic, G. Ye*

PRO 98: FERRO-11 – 11th International Symposium on Ferrocement and 3rd ICTRC - International Conference on Textile Reinforced Concrete (2015), ISBN: 978-2-35158-152-0; e-ISBN: 978-2-35158-153-7; *Ed. W. Brameshuber*

PRO 99 (online version): ICBBM 2015 - 1st International Conference on Bio-Based Building Materials (2015), e-ISBN: 978-2-35158-154-4; *Eds. S. Amziane, M. Sonebi*

PRO 100: SCC16 - RILEM Self-Consolidating Concrete Conference (2016), ISBN: 978-2-35158-156-8; e-ISBN: 978-2-35158-157-5; *Ed. Kamal H. Kayat*

PRO 101 (online version): III Progress of Recycling in the Built Environment (2015), e-ISBN: 978-2-35158-158-2; *Eds. I. Martins, C. Ulsen and S. C. Angulo*

PRO 102 (online version): RILEM Conference on Microorganisms-Cementitious Materials Interactions (2016), e-ISBN: 978-2-35158-160-5; *Eds. Alexandra Bertron, Henk Jonkers, Virginie Wiktor*

PRO 103 (online version): ACESC'16 - Advances in Civil Engineering and Sustainable Construction (2016), e-ISBN: 978-2-35158-161-2; *Eds. T.Ch. Madhavi, G. Prabhakar, Santhosh Ram and P. M. Rameshwaran*

PRO 104 (online version): SSCS'2015 - Numerical Modeling - Strategies for Sustainable Concrete Structures (2015), e-ISBN: 978-2-35158-162-9

PRO 105: 1st International Conference on UHPC Materials and Structures (2016), ISBN: 978-2-35158-164-3, e-ISBN: 978-2-35158-165-0

PRO 106: AFGC-ACI-fib-RILEM International Conference on Ultra-High-Performance Fibre-Reinforced Concrete – UHPFRC 2017 (2017), ISBN: 978-2-35158-166-7, e-ISBN: 978-2-35158-167-4; *Eds. François Toutlemonde & Jacques Resplendino*

PRO 107 (online version): XIV DBMC – 14th International Conference on Durability of Building Materials and Components (2017), e-ISBN: 978-2-35158-159-9; *Eds. Geert De Schutter, Nele De Belie, Arnold Janssens, Nathan Van Den Bossche*

PRO 108: MSSCE 2016 - Innovation of Teaching in Materials and Structures (2016), ISBN: 978-2-35158-178-0, e-ISBN: 978-2-35158-179-7; *Ed. Per Goltermann*

PRO 109 (2 volumes): MSSCE 2016 - Service Life of Cement-Based Materials and Structures (2016), ISBN Vol. 1: 978-2-35158-170-4, Vol. 2: 978-2-35158-171-4, Set Vol. 1&2: 978-2-35158-172-8, e-ISBN : 978-2-35158-173-5; *Eds. Miguel Azenha, Ivan Gabrijel, Dirk Schlicke, Terje Kanstad and Ole Mejlhede Jensen*

PRO 110: MSSCE 2016 - Historical Masonry (2016), ISBN: 978-2-35158-178-0, e-ISBN: 978-2-35158-179-7; *Eds. Inge Rörig-Dalgaard and Ioannis Ioannou*

PRO 111: MSSCE 2016 - Electrochemistry in Civil Engineering (2016), ISBN: 978-2-35158-176-6, e-ISBN: 978-2-35158-177-3; *Ed. Lisbeth M. Ottosen*

PRO 112: MSSCE 2016 - Moisture in Materials and Structures (2016), ISBN: 978-2-35158-178-0, e-ISBN: 978-2-35158-179-7; *Eds. Kurt Kielsgaard Hansen, Carsten Rode and Lars-Olof Nilsson*

PRO 113: MSSCE 2016 - Concrete with Supplementary Cementitious Materials (2016), ISBN: 978-2-35158-178-0, e-ISBN: 978-2-35158-179-7; *Eds. Ole Mejlhede Jensen, Konstantin Kovler and Nele De Belie*

PRO 114: MSSCE 2016 - Frost Action in Concrete (2016), ISBN: 978-2-35158-182-7, e-ISBN: 978-2-35158-183-4; *Eds. Marianne Tange Hasholt, Katja Fridh and R. Doug Hooton*

PRO 115: MSSCE 2016 - Fresh Concrete (2016), ISBN: 978-2-35158-184-1, e-ISBN: 978-2-35158-185-8; *Eds. Lars N. Thrane, Claus Pade, Oldrich Svec and Nicolas Roussel*

PRO 116: BEFIB 2016 – 9th RILEM International Symposium on Fiber Reinforced Concrete (2016), ISBN: 978-2-35158-187-2, e-ISBN: 978-2-35158-186-5; *Eds. N. Banthia, M. di Prisco and S. Soleimani-Dashtaki*

PRO 117: 3rd International RILEM Conference on Microstructure Related Durability of Cementitious Composites (2016), ISBN: 978-2-35158-188-9, e-ISBN: 978-2-35158-189-6; *Eds. Changwen Miao, Wei Sun, Jiaping Liu, Huisu Chen, Guang Ye and Klaas van Breugel*

PRO 118 (4 volumes): International Conference on Advances in Construction Materials and Systems (2017), ISBN Set: 978-2-35158-190-2, Vol. 1: 978-2-35158-193-3, Vol. 2: 978-2-35158-194-0, Vol. 3: ISBN:978-2-35158-195-7, Vol. 4: ISBN:978-2-35158-196-4, e-ISBN: 978-2-35158-191-9; *Eds. Manu Santhanam, Ravindra Gettu, Radhakrishna G. Pillai and Sunitha K. Nayar*

PRO 119 (online version): ICBBM 2017 - Second International RILEM Conference on Bio-based Building Materials, (2017), e-ISBN: 978-2-35158-192-6; *Ed. Sofiane Amziane*

PRO 120 (2 volumes): EAC-02 - 2nd International RILEM/COST Conference on Early Age Cracking and Serviceability in Cement-based Materials and Structures, (2017), Vol. 1: 978-2-35158-199-5, Vol. 2: 978-2-35158-200-8, Set: 978-2-35158-197-1, e-ISBN: 978-2-35158-198-8; *Eds. Stéphanie Staquet and Dimitrios Aggelis*

PRO 121 (2 volumes): SynerCrete18: Interdisciplinary Approaches for Cement-based Materials and Structural Concrete: Synergizing Expertise and Bridging Scales of Space and Time, (2018), Set: 978-2-35158-202-2, Vol.1: 978-2-35158-211-4, Vol.2: 978-2-35158-212-1, e-ISBN: 978-2-35158-203-9; *Eds. Miguel Azenha, Dirk Schlicke, Farid Benboudjema, Agnieszka Knoppik*

PRO 122: SCC'2018 China - Fourth International Symposium on Design, Performance and Use of Self-Consolidating Concrete, (2018), ISBN: 978-2-35158-204-6, e-ISBN: 978-2-35158-205-3; *Eds. C. Shi, Z. Zhang, K. H. Khayat*

PRO 123: Final Conference of RILEM TC 253-MCI: Microorganisms-Cementitious Materials Interactions (2018), Set: 978-2-35158-207-7, Vol.1: 978-2-35158-209-1, Vol.2: 978-2-35158-210-7, e-ISBN: 978-2-35158-206-0; *Ed. Alexandra Bertron*

PRO 124 (online version): Fourth International Conference Progress of Recycling in the Built Environment (2018), e-ISBN: 978-2-35158-208-4; *Eds. Isabel M. Martins, Carina Ulsen, Yury Villagran*

PRO 125 (online version): SLD4 - 4th International Conference on Service Life Design for Infrastructures (2018), e-ISBN: 978-2-35158-213-8; *Eds. Guang Ye, Yong Yuan, Claudia Romero Rodriguez, Hongzhi Zhang, Branko Savija*

PRO 126: Workshop on Concrete Modelling and Material Behaviour in honor of Professor Klaas van Breugel (2018), ISBN: 978-2-35158-214-5, e-ISBN: 978-2-35158-215-2; *Ed. Guang Ye*

PRO 127 (online version): CONMOD2018 - Symposium on Concrete Modelling (2018), e-ISBN: 978-2-35158-216-9; *Eds. Erik Schlangen, Geert de Schutter, Branko Savija, Hongzhi Zhang, Claudia Romero Rodriguez*

PRO 128: SMSS2019 - International Conference on Sustainable Materials, Systems and Structures (2019), ISBN: 978-2-35158-217-6, e-ISBN: 978-2-35158-218-3

PRO 129: 2nd International Conference on UHPC Materials and Structures (UHPC2018-China), ISBN: 978-2-35158-219-0, e-ISBN: 978-2-35158-220-6;

PRO 130: 5th Historic Mortars Conference (2019), ISBN: 978-2-35158-221-3, e-ISBN: 978-2-35158-222-0; *Eds. José Ignacio Álvarez, José María Fernández, Íñigo Navarro, Adrián Durán, Rafael Sirera*

PRO 131 (online version): 3rd International Conference on Bio-Based Building Materials (ICBBM2019), e-ISBN: 978-2-35158-229-9; *Eds. Mohammed Sonebi, Sofiane Amziane, Jonathan Page*

PRO 132: IRWRMC'18 - International RILEM Workshop on Rheological Measurements of Cement-based Materials (2018), ISBN: 978-2-35158-230-5, e-ISBN: 978-2-35158-231-2; *Eds. Chafika Djelal and Yannick Vanhove*

PRO 133 (online version): CO2STO2019 - International Workshop CO2 Storage in Concrete (2019), e-ISBN: 978-2-35158-232-9; *Eds. Assia Djerbi, Othman Omikrine-Metalssi and Teddy Fen-Chong*

PRO 134: 3rd ACF/HNU International Conference on UHPC Materials and Structures - UHPC'2020, ISBN: 978-2-35158-233-6, e-ISBN: 978-2-35158-234-3; *Eds. Caijun Shi and Jiaping Liu*

RILEM Reports (REP)

Report 19: Considerations for Use in Managing the Aging of Nuclear Power Plant Concrete Structures (ISBN: 2-912143-07-1); *Ed. D. J. Naus*

Report 20: Engineering and Transport Properties of the Interfacial Transition Zone in Cementitious Composites (ISBN: 2-912143-08-X); *Eds. M. G. Alexander, G. Arliguie, G. Ballivy, A. Bentur and J. Marchand*

Report 21: Durability of Building Sealants (ISBN: 2-912143-12-8); *Ed. A. T. Wolf*

Report 22: Sustainable Raw Materials - Construction and Demolition Waste (ISBN: 2-912143-17-9); *Eds. C. F. Hendriks and H. S. Pietersen*

Report 23: Self-Compacting Concrete state-of-the-art report (ISBN: 2-912143-23-3); *Eds. Å. Skarendahl and Ö. Petersson*

Report 24: Workability and Rheology of Fresh Concrete: Compendium of Tests (ISBN: 2-912143-32-2); *Eds. P. J. M. Bartos, M. Sonebi and A. K. Tamimi*

Report 25: Early Age Cracking in Cementitious Systems (ISBN: 2-912143-33-0); *Ed. A. Bentur*

Report 26: Towards Sustainable Roofing (Joint Committee CIB/RILEM) (CD 07) (e-ISBN 978-2-912143-65-5); *Eds. Thomas W. Hutchinson and Keith Roberts*

Report 27: Condition Assessment of Roofs (Joint Committee CIB/RILEM) (CD 08) (e-ISBN 978-2-912143-66-2); *Ed. CIB W 83/RILEM TC166-RMS*

Report 28: Final report of RILEM TC 167-COM 'Characterisation of Old Mortars with Respect to Their Repair (ISBN: 978-2-912143-56-3); *Eds. C. Groot, G. Ashall and J. Hughes*

Report 29: Pavement Performance Prediction and Evaluation (PPPE): Interlaboratory Tests (e-ISBN: 2-912143-68-3); *Eds. M. Partl and H. Piber*

Report 30: Final Report of RILEM TC 198-URM 'Use of Recycled Materials' (ISBN: 2-912143-82-9; e-ISBN: 2-912143-69-1); *Eds. Ch. F. Hendriks, G. M. T. Janssen and E. Vázquez*

Report 31: Final Report of RILEM TC 185-ATC 'Advanced testing of cement-based materials during setting and hardening' (ISBN: 2-912143-81-0; e-ISBN: 2-912143-70-5); *Eds. H. W. Reinhardt and C. U. Grosse*

Report 32: Probabilistic Assessment of Existing Structures. A JCSS publication (ISBN 2-912143-24-1); *Ed. D. Diamantidis*

Report 33: State-of-the-Art Report of RILEM Technical Committee TC 184-IFE 'Industrial Floors' (ISBN 2-35158-006-0); *Ed. P. Seidler*

Report 34: Report of RILEM Technical Committee TC 147-FMB 'Fracture mechanics applications to anchorage and bond' Tension of Reinforced Concrete Prisms – Round Robin Analysis and Tests on Bond (e-ISBN 2-912143-91-8); *Eds. L. Elfgren and K. Noghabai*

Report 35: Final Report of RILEM Technical Committee TC 188-CSC 'Casting of Self Compacting Concrete' (ISBN 2-35158-001-X; e-ISBN: 2-912143-98-5); *Eds. Å. Skarendahl and P. Billberg*

Report 36: State-of-the-Art Report of RILEM Technical Committee TC 201-TRC 'Textile Reinforced Concrete' (ISBN 2-912143-99-3); *Ed. W. Brameshuber*

Report 37: State-of-the-Art Report of RILEM Technical Committee TC 192-ECM 'Environment-conscious construction materials and systems' (ISBN: 978-2-35158-053-0); *Eds. N. Kashino, D. Van Gemert and K. Imamoto*

Report 38: State-of-the-Art Report of RILEM Technical Committee TC 205-DSC 'Durability of Self-Compacting Concrete' (ISBN: 978-2-35158-048-6); *Eds. G. De Schutter and K. Audenaert*

Report 39: Final Report of RILEM Technical Committee TC 187-SOC 'Experimental determination of the stress-crack opening curve for concrete in tension' (ISBN 978-2-35158-049-3); *Ed. J. Planas*

Report 40: State-of-the-Art Report of RILEM Technical Committee TC 189-NEC 'Non-Destructive Evaluation of the Penetrability and Thickness of the Concrete Cover' (ISBN 978-2-35158-054-7); *Eds. R. Torrent and L. Fernández Luco*

Report 41: State-of-the-Art Report of RILEM Technical Committee TC 196-ICC 'Internal Curing of Concrete' (ISBN 978-2-35158-009-7); *Eds. K. Kovler and O. M. Jensen*

Report 42: 'Acoustic Emission and Related Non-destructive Evaluation Techniques for Crack Detection and Damage Evaluation in Concrete' - Final Report of RILEM Technical Committee 212-ACD (e-ISBN: 978-2-35158-100-1); *Ed. M. Ohtsu*

Report 45: Repair Mortars for Historic Masonry - State-of-the-Art Report of RILEM Technical Committee TC 203-RHM (e-ISBN: 978-2-35158-163-6); *Eds. Paul Maurenbrecher and Caspar Groot*

Report 46: Surface delamination of concrete industrial floors and other durability related aspects guide - Report of RILEM Technical Committee TC 268-SIF (e-ISBN: 978-2-35158-201-5); *Ed. Valerie Pollet*

Artificial Neural Network-Based Methodology for Optimization of Low-Cost Green UHPFRC Under Ductility Requirements

Joaquín Abellán-García(✉)

Universidad del Norte, Barranquilla, Atlántico, Colombia
jabellan@uninorte.edu.co

Abstract. Several constructions in earthquake-prone areas in developing countries do not meet current seismic codes, mainly because of the rampant informal construction. These circumstances require effective seismic retrofitting interventions through solutions of an acceptable cost that allow the most extensive application possible. This research focuses on developing a low-cost, low-carbon-footprint material with the required ductility parameters for seismic retrofitting applications. First, a plain UHPC is optimized under compressive strength, cost, and carbon footprint criteria. After that, the second stage of this study determines the binary combination of fibers, among those available in the Colombian market, that permit reaching the necessary ductility parameters for the desired application at a lower cost. The ductility parameters considered are the energy capacity absorption (g) and the strain capacity at maximum tensile strength (ε_{pc}) measured in the direct tensile test. Various statistical and computational tools such as Artificial Neural Networks, Design of Experiments, and Multi-Objective Optimization were utilized to lesser the experimental campaign. The mathematically optimized dosage was experimentally evaluated. Finally, the optimal fiber volume fraction for the necessary UHPFRC ductility parameters for seismic strengthening applications ($g \geq 50$ kJ/m^3 and $\varepsilon_{pc} \geq 0.3\%$) was selected at only 1.7%. This optimal fiber combination was composed of 0.34% of smooth high-strength steel ($l_f/d_f = 65$) fibers, and 1.36% of normal strength hooked end steel fibers ($l_f/d_f = 80$). It is relevant to highlight that this optimized UHPFRC outperforms the ductility parameters obtained by other authors with successful applications in the seismic strengthening field.

Keywords: UHPFRC · ANN · Design of Experiments · Multi-objective optimization · Experimental validation · Direct tensile behavior

1 Introduction

1.1 Motivation

Seismic retrofitting existing structures requests practical, inexpensive, and feasible strategies with a high dependence on the different motivations [1]. E.g., the seismic

P. Rossi and J.-L. Tailhan (Eds.): SSCS 2022, RILEM Bookseries 38, pp. 1–11, 2023.
https://doi.org/10.1007/978-3-031-07746-3_1

hazard map is constantly transforming and evolving, prompting the considered maximum ground acceleration, as defined by seismic regulations currently in force today, to be more demanding than those suggested in the past [1]. Furthermore, many hypotheses of modern seismic design codes as ultimate limit states and capacity design principles were unknown when most structures were originally built [2]. The seismic retrofitting requirement can also be ascribed to the increased importance of a given structure, thereby leading to elevated seismic demand [1].

Colombia, where this research took place, sits at the complex junction of the Caribbean Plate, the Cocos Plate, and the Nazca plate [3]. Thus, as per the National Statistics, an estimated 90% of the population lives under high or moderate seismic hazards [4, 5]. Besides, since 1984 Colombia has had regulations for earthquake-resistant constructions [4, 5]. Thus, buildings constructed before that date have a good chance of being considered vulnerable. The same goes for informal construction [3].

1.2 UHPFRC

Definition. Ultra-high-performance fiber reinforced concrete (UHPFRC) refers to high-tech fiber-reinforced cementitious composites [6–8], which are excellent mechanical and durable compared to other concrete classes. Their properties can be attributed to their low porosity cementitious matrix [3, 8]. ACI 239 [9] defines UHPFRC as a class of advanced cementitious material with a minimum compressive strength of 150 MPa and with specified tensile ductility requirements, for which fibers are generally added.

Ductility Parameters of UHPFRC. The ductility of UHPFRC is usually measured by the direct tensile test [10–13]. Regarding direct tensile tests behavior, UHPFRC can be classified as strain-softening or strain-hardening [14, 15]. The fibers' bridging effect significantly affects performance in the hardening and softening domain [16]. Per Naaman [17], strain hardening happens as the output of the maximum post-cracking tensile strength, σ_{pc}, exceeding the cracking tensile strength, σ_{cc}. Figure 1 shows the three phases of a strain-hardening UHPFRC strain stress curve under uniaxial loading [13, 14]. Zone I shows elastic behavior up to cracking strength σ_{cc} and its associated strain ε_{cc}. Thereafter, the material enters its strain hardening behavior, i.e., Zone II. Here the multiple-cracking pattern develops, and the energy absorption capacity g can be computed as the area under the stress-strain curve up to σ_{pc} [10, 13, 18]. Finally, Zone III is characterized by the fiber slip until the pullout.

Seismic Retrofitting Applications of UHPFRC. A set of particular traits, such as ductility, durability, and ease of application, make UHPFRC particularly attractive for retrofitting non-ductile concrete structures [16]. Furthermore, several authors have demonstrated that, with a slight surface treatment of the structure to be repaired, UHPFRC can create a high bond strength with the old concrete [16, 19, 20]. The jacketing system is maybe the most successful technology used [11, 21]. Table 1 summarizes the ductility of UHPFRC´s parameters reported in successful cases of seismic retrofitting applications.

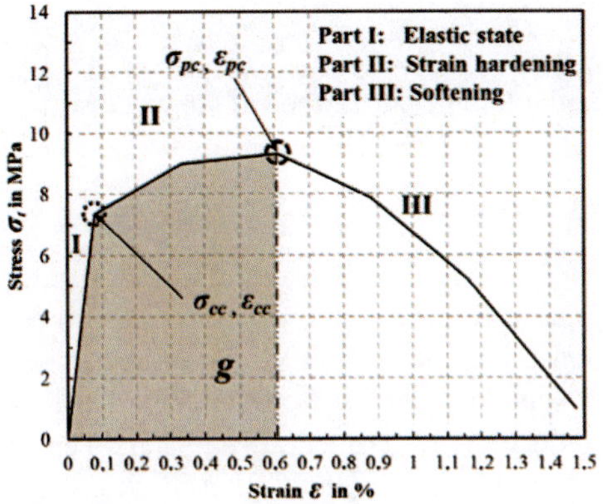

Fig. 1. Idealized strain hardening UHPFRC behavior under direct tensile test.

Table 1. Ductility properties of UHPFRC successfully applied in seismic retrofitting of reinforced concrete structures.

V_{fT} (%)	g (kJ/m^3)	ε_{pc} (%)	Reference
3%	–	0.2	[19]
3%	30	0.2	[11]
2%	30	0.3	[21]

Objectives and Organization. This research applies Artificial Neural Networks (ANN) models to design a high-energy absorbent UHPFRC at a lower cost for its application as a material for seismic retrofitting of non-ductile structures. The study focuses on raw materials, including cement, supplementary cementitious materials, aggregates, superplasticizers, and fibers, locally available in the Colombian market. Furthermore, different by-products of the local industry were considered for the UHPC composition. Regarding the fibers to be incorporated for creating a UHPFRC with the proper ductility, a broad spectrum of fibers were considered. The end objective focuses on the experimental validation of the ANN optimized UHPFRC. The threshold values of ductility (i.e., $g \geq 50$ kJ/m^3 and $\varepsilon_{pc} \geq 0.3\%$) to be used as seismic retrofitting material were established in [22, 23]. It is important to note that these ductility requirements are more restrictive than those presented in Table 1. Therefore, it is to be assumed that the material optimized with these requirements can be used as a seismic reinforcement material.

The research herein presented is developed in two phases. First, the UHPC without fibers is optimized, selecting the components under the criteria of minimum cement content and a silica fume content limited to 100 kg/m^3 while reaching resistance at 28 days equal to or greater than 150 MPa. Second, the dosage of fibers is optimized for achieving the threshold ductility parameters at a lower cost. Computational tools such as DoE, ANN, RSM, and multi-objective optimization are used to reduce the number of trials needed.

2 Methodology

For the algorithmic methodologies, R statistical language was utilized.

2.1 Artificial Neural Networks

ANN applications have spread in the last decades due to their remarkable ability to reproduce non-linear and unknown relationships between input and output data pairs [24, 25], thereby demonstrating their ability in complex engineering issues [24–26]. In this research, three ANN models were developed as part of a pre-selecting of components procedure aiming to reduce the experimental campaign. The first ANN model goal is to predict the compressive strength of UHPC, considering supplementary cementitious materials (SCM). The model was trained with 927 data from experiments and literature resources. The regression ANN model and the regression plot are put forward in Fig. 2. For further information about this model, input variables, and the database utilized, references [25, 27] can be consulted. Two more ANN models were created to predict the energy absorption capacity (g) and maximum post-cracking strain (ε_{pc}), using 600 data from experiments and literature resources. Figure 3 presents these models. Interested readers are encouraged to review the following references for a detailed explanation [22, 23, 28].

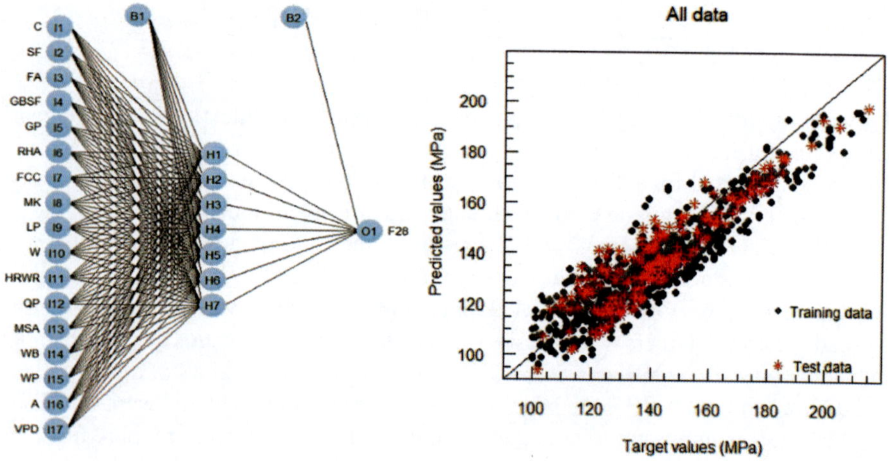

Fig. 2. ANN model for compressive strength of UHPC and its regression plot.

2.2 Multi-objective Optimization

A multi-objective algorithm based on the desirabilities approach presented by Derringer & Suich [29] was employed for the UHPC and UHPFRC optimizations.

2.3 Design of Experiments (DoE)

To "land" the dosages indicated by the ANN models, DoE was utilized. With DoE, the number of tests is kept as low as possible, and the most informative combination of the factors is selected. Therefore, DoE is an effective and economical solution [6, 30, 31].

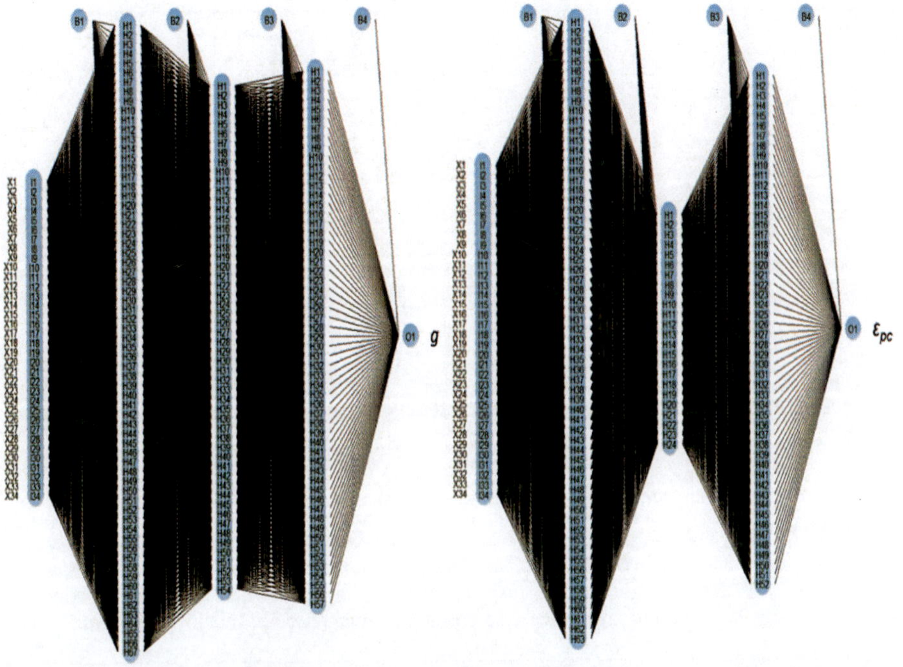

Fig. 3. ANN models for predicting the energy capacity absorption (up) and maximum post-cracking strain (down) of UHPFRC under direct tensile loading.

2.4 Materials and Experimental Procedures

The materials used to develop the UHPC were locally available in Colombia. Cement (C), silica fume (SF), and silica sand (SS) were employed, along with locally available fly ash (FA), ground granulated blast slag furnace (GGBFS), electric arc slag furnace (EASF), recycled glass powder (RGP), recycled glass flour (RGF), rice husk ash (RHA), fluid catalytic cracking residue (FC3R), metakaolin (MK), two sizes of limestone powder (MLP and LP), and quartz powder (QP). Table 2 presents the physical and chemical properties of these components. Reference [23] can be consulted for further information about these components.

For its part, the detailed information of the fibers used in this research is depicted in Table 3.

The blending procedure, casting, and curing conditions are explained in references [32, 33] for plain UHPC, while references [22, 28] can be consulted for UHPFRC.

Compressive strength tests were performed according to ASTM C109 [34], while direct tensile trials followed the recommendation of the JSCE-08 [35].

Table 2. Physical and chemical properties of the components considered for the UHPC matrix.

	C	SF	FA	GGBFS	EASF	GP	GF	RHA	FC3R	MK	LP	MLP	QP	SS
Specific gravity	3.15	2.20	2.32	2.95	3.15	2.55	2.55	2.16	2.76	2.66	2.73	2.73	2.65	2.65
Loss of ignition, %	2.58	0.60	12.54	0.00	4.67	0.80	1.00	4.82	10.61	2.21	42.41	42.21	0.19	0.20
SiO_2, %	19.42	92.29	50.09	36.93	21.70	75.47	72.89	88.59	39.61	52.17	0.90	0.90	95.80	99.80
Al_2O_3, %	4.00	0.59	22.26	16.45	6.20	1.09	1.67	0.31	42.47	39.11	0.10	0.10	0.11	0.14
CaO, %	64.42	3.89	2.19	33.78	33.13	9.02	9.73	0.74	2.85	0.78	55.51	55.71	0.38	0.17
MgO, %	1.52	0.26	0.53	3.91	10.60	1.97	2.08	0.66	0.07	0.19	0.70	0.70	0.20	0.01
SO_3, %	1.93	0.07	0.03	2.51	1.14	0.00	0.01	0.26	0.62	0.22	0.10	0.10	0.52	-
Na_2O, %	0.19	0.31	0.31	0.39	0.15	11.65	12.54	0.26	0.61	0.24	0.03	0.03	0.25	-
K_2O, %	0.39	0.54	0.99	0.40	0.03	0.75	0.76	2.46	0.06	0.20	0.01	0.00	3.49	0.05
TiO_2, %	0.38	0.01	1.05	0.57	0.42	0.04	0.04	0.02	0.67	1.48	0.00	0.00	0.25	-
Mn_3O_4, %	0.05	0.01	0.01	3.41	1.98	0.01	0.01	0.21	0.00	0.00	0.00	0.01	0.01	-
Fe_2O_3, %	3.61	0.24	9.33	1.24	18.92	0.79	0.81	0.29	0.69	1.07	0.05	0.05	0.09	0.04
Cost, COP/kg	340	1,700	80	90	65	160	80	1,200	60	800	100	350	1,600	160

*COP: Colombian pesos

Table 3. Notation, properties, and cost of the fibers used in this investigation.

Notation	Form	d_f (mm)	l_f (mm)	d_f/l_f	Material	Tensile strength (MPa)	Cost (COP/kg)
S_1	Straight	0.20	13	65	Steel	≈ 2600	11,000
S_2	Straight	0.20	6	30	Steel	≈ 2600	10,500
H_1	Hooked	0.50	35	70	Steel	≈ 2000	9,000
H_2	Hooked	0.75	60	80	Steel	≈ 1600	4,855
T	Twisted	0.50	13	26	Steel	≈ 1700	7,500
PP	Straight	0.72	48	75	Polypropylene	≈ 650	30,000
PE	Straight	0.67	50	67	Polyethylene	≈ 550	27,000
PVA	Straight	0.025	6	240	Polyvinyl alcohol	≈ 1600	35,000

3 Results

3.1 UHPC Cementitious Matrix

Before continuing, it is essential to name two limitations of the procedure: (i) the ANN models only work correctly if the components with which it has been trained have similar quality as those whose response we want to predict. This way, FA considered in the database for training the models had an LOI valuer under 5% [36]. However, FA available in the Colombian market has an LOI value of over 12% [37]. Therefore, even though the ANN models pointed out FA-dosage (DoE01) as the most suitable for fulfilling the requirements, DoE03 experimentally demonstrated to be the proper one; (ii) the ANN models could not model the cement-superplasticizer. Therefore, a superplasticizer was experimentally selected for each dosage. Table 4 presents the experimental results of the optimization of the UHPC dosages.

Table 4. Results of the experimental optimization of the UHPC cementitious matrix

DoE	Components	Cement (kg/m^3)	Model R28 (MPa)	Experimental R28 (MPa)	Cost/m^3
DoE01	C, SF, MLP, FA, SS	711	151	148	890
DoE02	C, SF, RGF, RGP, MLP, SS	603	155	152	809
DoE03	C, SF, RGF, MLP, SS	590	157	156	806
DoE04	C, SF, RGP, MLP, RHA, SS	615	155	154	995
DoE05	C, SF, GGBSF, RGF, SS	674	154	157	841
DoE06	C, SF, RGP, MLP, MK, SS	613	153	149	964
DoE07	C, SF, RGP, MLP, FC3R, SS	654	153	150	901
DoE08	C, SF, RGF, MLP, EASF, SS	621	158	156	848
DoE09	C, SF, LP, MLP, EASF, SS	651	155	153	900
DoE10	C, SF, LP, MLP, SS	655	154	150	877

3.2 UHPFRC

After the described procedure, the optimal fiber combination was composed of 0.34% of S_1 fibers and 1.36% of H_2 fibers. Table 5 puts forward the ductility parameters experimentally obtained of the combination of fibers indicated by the algorithmic procedure. Figure 4 presents the tensile loading test set-up and the strain stress response of the optimized UHPFRC under direct tensile loading.

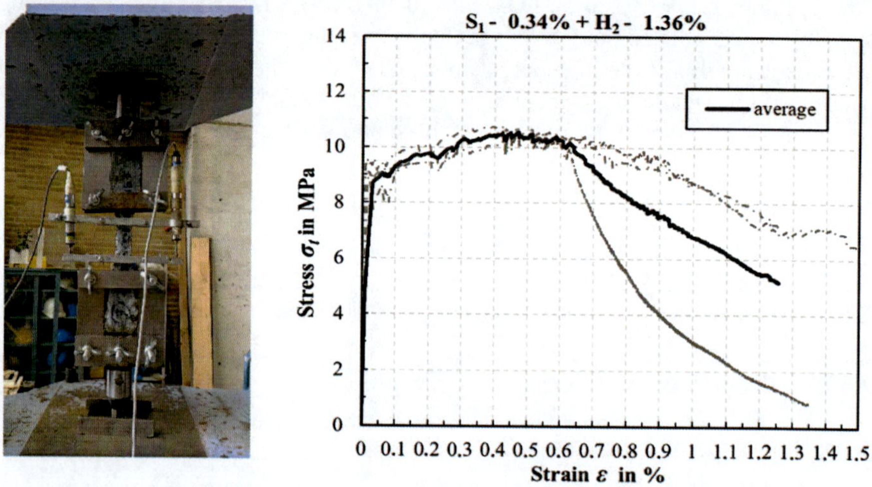

Fig. 4. Direct tensile test set-up and strain-stress behavior of the optimized UHPFRC.

Table 5. Ductility parameters of the optimized UHPFRC

g (kJ/m^3)		ADR (%)	ε_{pc} (%)		**ADR (%)**
Experimental	ANN		Experimental	ANN	
45.177	50.448	11.667	0.458	0.424	7.424

4 Conclusions

The research work developed an innovative study of optimization UHPFRC for seismic retrofitting applications using an experimental and analytical procedure based on the design of experiments, multi-objective simultaneous optimization, and artificial neural networks. This research demonstrated that it is possible to achieve a UHPFRC with the necessary ductility for seismic retrofitting applications when using a low-cost fiber cocktail and an optimized cementitious matrix under sustainability criteria by incorporating sustainable materials in partial replacement of cement and silica fume.

References

1. De Domenico, D., Impollonia, N., Ricciardi, G.: Seismic retrofitting of confined masonry-RC buildings: The case study of the university hall of residence in Messina, Italy. Ing. Sismica **36**, 54–85 (2019)
2. Dogan, E., Krstulovic-Opara, N.: Seismic retrofit with continuous slurry-infiltrated mat concrete jackets. ACI Struct. J. **100**, 713–722 (2003)
3. Abellán, J., Fernández, J., Torres, N., Núñez, A.: Development of cost-efficient UHPC with local materials in Colombia. In: Middendorf, B., Fehling, E., Wetzel, A. (eds.) Proceedings of Hipermat 2020 - 5th International Symposium on UHPC and Nanotechnology for Construction Materials. University of Kassel, Kassel, Germany, pp. 97–98 (2020)
4. Garcia, L.E.: Desarrollo de la normativa sismo resistente colombiana en los 30 años desde su primera expedición. Rev. Ing. **41**, 71–77 (2014)
5. Sísmica A de I: Reglamento Colombiano de construcciónsismo resistente. NSR-10 (2010)
6. Abellán, J., Fernández, J., Torres, N., Núñez, A.: Statistical optimization of ultra-high-performance glass concrete. ACI Mater. J. **117,** 243–254 (2020). https://doi.org/10.14359/51720292
7. Abellán-García, J., Núñez-López, A., Torres-Castellanos, N., Fernández-Gómez, J.: Effect of FC3R on the properties of ultra-high-performance concrete with recycled glass. Dyna **86**, 84–92 (2019). https://doi.org/10.15446/dyna.v86n211.79596
8. Abellán, J., Torres, N., Núñez, A., Fernández, J.: Ultra high preformance fiber reinforced concrete: state of the art, applications and possibilities into the latin american market. In: XXXVIII Jornadas Sudamericanas de Ingeniería Estructural. Lima, Peru (2018)
9. ACI Committe 239R, ACI Committe 239: ACI – 239 Committee in Ultra-High Performance Concrete. ACI, Toronto (2018)
10. Kwon, S., Nishiwaki, T., Kikuta, T., Mihashi, H.: development of ultra-high-performance hybrid fiber- reinforced cement-based composites development of ultra-high-performance hybrid fiber- reinforced cement-based composites (2014). https://doi.org/10.14359/51686890
11. Massicotte, B., Dagenais, M.-A., Lagier, F.: Performance of UHPFRC jackets for the seismic strengthening of bridge piers. In: RILEM-fib-AFGC International Symposium Ultra-High Perform Fibre-Reinforced, pp. 89–98 (2013)
12. Soranakom, C., Mobasher, B.: Correlation of tensile and flexural responses of strain softening and strain hardening cement composites. Cem. Concr. Compos. 465–477 (2008).https://doi.org/10.1016/j.cemconcomp.2008.01.007
13. Wille, K., El-tawil, S., Naaman, A.E.: Properties of strain hardening ultra high performance fiber reinforced concrete (UHP-FRC) under direct tensile loading. Cem. Concr. Compos. **48**, 53–66 (2014). https://doi.org/10.1016/j.cemconcomp.2013.12.015
14. Wille, K., Kim, D.J.D., Naaman, A.E.: Strain hardening UHP-FRC with low fiber contents. Mater. Struct. **44**, 538–598 (2011). https://doi.org/10.1617/s11527-010-9650-4
15. Pyo, S., El-Tawil, S., Naaman, A.E.: Direct tensile behavior of ultra high performance fiber reinforced concrete (UHP-FRC) at high strain rates. Cem. Concr. Res. **88**, 144–156 (2016). https://doi.org/10.1016/j.cemconres.2016.07.003
16. Martin-Sanz, H., Chatzi, E., Brühwiler, E.: The use of ultra high performance fibre reinforced cement-based composites in rehabilitation projects: a review. In: Saouma, V., Bolander, J., Landis, E. (eds.) 9th International Conference on Fracture Mechanics of Concrete and Concrete Structures (2016)
17. Naaman, A.E., Reinhart, H.W.: Proposed classification of HPFRC composites based on their tensile response. Mater. Struct. **39**, 547–555 (2006). https://doi.org/10.1617/s11527-006-9103-2

18. Yoo, D.Y., Kim, M.J.: High energy absorbent ultra-high-performance concrete with hybrid steel and polyethylene fibers. Constr. Build. Mater. **209**, 354–363 (2019). https://doi.org/10.1016/j.conbuildmat.2019.03.096
19. Dagenais, M.A., Massicotte, B., Boucher-Proulx, G.: Seismic retrofitting of rectangular bridge piers with deficient lap splices using ultrahigh-performance fiber-reinforced concrete. J. Bridg. Eng. **23**, 1–13 (2018). https://doi.org/10.1061/(ASCE)BE.1943-5592.0001173
20. Tayeh, B.A., Abu Bakar, B.H., Megat Johari, M.A., Voo, Y.L.: Utilization of ultra-high performance fibre concrete (UHPFC) for rehabilitation - a review. Proc. Eng. **54**, 525–538 (2013). https://doi.org/10.1016/j.proeng.2013.03.048
21. Khan, M.I., Al-Osta, M.A., Ahmad, S., Rahman, M.K.: Seismic behavior of beam-column joints strengthened with ultra-high performance fiber reinforced concrete. Compos. Struct. **200**, 103–119 (2018). https://doi.org/10.1016/j.compstruct.2018.05.080
22. Abellán-García, J., Guzmán-Guzmán, J.S.: Random forest-based optimization of UHPFRC under ductility requirements for seismic retrofitting applications. Constr. Build. Mater. **285** (2021). https://doi.org/10.1016/j.conbuildmat.2021.122869
23. Abellán-García, J.: Dosage optimization and seismic retrofitting applications of ultra-highperformance fiber reinforced concrete (UHPFRC). Polytechnic University of Madrid (2020)
24. Abellán-García, J., Fernández-Gómez, J., Torres-Castellanos, N.: Properties prediction of environmentally friendly ultra-high-performance concrete using artificial neural networks. Eur. J. Environ. Civ. Eng. 1–25 (2020).https://doi.org/10.1080/19648189.2020.1762749
25. Abellán-García, J.: Four-layer perceptron approach for strength prediction of UHPC. Constr. Build. Mater 256 (2020).https://doi.org/10.1016/j.conbuildmat.2020.119465
26. Khashman, A., Akpinar, P.: ScienceDirect non-destructive prediction of concrete compressive strength using neural networks prediction of concrete compressive strength using neural networks. Proc. Comput. Sci. **108**, 2358–2362 (2017). https://doi.org/10.1016/j.procs.2017.05.039
27. Abellán-García, J.: Artificial neural network model for strength prediction of ultra-high-performance concrete. ACI Mater. J. **118**, 3–14 (2021). https://doi.org/10.14359/51732710
28. Abellán-Garcia, J., Sánchez-Díaz, J., Ospina-Becerra, V.: Neural network-based optimization of fibers for seismic retrofitting applications of UHPFRC. Eur. J. Environ. Civ. Eng. (2021). https://doi.org/10.1080/19648189.2021.1938687
29. Derringer, G., Suich, R.: Simultaneous optimization of several response variables. J. Qual. Technol. **21**, 214–219 (1980)
30. Ghafari, E., Costa, H., Nuno, E., Santos, B.: RSM-based model to predict the performance of self-compacting UHPC reinforced with hybrid steel micro-fibers. Constr. Build. Mater. **66**, 375–383 (2014). https://doi.org/10.1016/j.conbuildmat.2014.05.064
31. Upasani, R.S., Banga, A.K.: Response surface methodology to investigate the iontophoretic delivery of tacrine hydrochloride. Pharm. Res. **21**, 2293–2299 (2004)
32. Abellán-García, J.: K -fold validation neural network approach for predicting the one-day compressive strength of UHPC. Adv. Civ. Eng. Mater. **10**, 223–243 (2021). https://doi.org/10.1520/ACEM20200055
33. Abellán-García, J., Núñez-López, A., Torres-Castellanos, N., Fernández-Gómez, J.: Factorial design of reactive powder concrete containing electric arc slag furnace and recycled glass powder. Dyna **87**, 42–51 (2020). https://doi.org/10.15446/dyna.v87n213.82655
34. ASTM: Standard test method for compressive strength of hydraulic cement mortars (Using 2-in . or [50-mm] Cube Specimens). Am. Soc. Test. Mater. C-109/109M 1–9 (2010)
35. Yokota, H., Rokugo, K., Sakata, N.: (JSCE-2008) Recommendations for design and construction of high performance fiber reinforced cement composites with multiple fine cracks (HPFRCC) (2008). https://doi.org/10.1016/j.dci.2010.01.003

36. Alsalman, A., Dang, C.N., Micah Hale, W.: Development of ultra-high performance concrete with locally available materials. Constr Build Mater **133**, 135–145 (2017). https://doi.org/10.1016/j.conbuildmat.2016.12.040
37. Abellán-García, J., Torres-Castellanos, N., Fernández-Gómez, J.A., Núñez-López, A.M.: Ultra-high-performance concrete with local high unburned carbon fly ash. Dyna **88**, 38–47 (2021). https://doi.org/10.15446/dyna.v88n216.89234

Long Term Prediction of the Delayed Behavior of Concrete Structures – The Case of the VERCORS Mock-Up

Abudushalamu Aili[1] and Jean Michel Torrenti[2](✉)

[1] Graduate School of Environmental Studies, Nagoya University, Nagoya, Japan
[2] Univ Gustave Eiffel, CEREMA, UMR MCD, 77454 Marne-la-Vallée, France
jean-michel.torrenti@univ-eiffel.fr

Abstract. The prediction of the long-term behavior of prestressed concrete structures is important in order to assess and/or extend the service life of such structures. Here, a modelling of the delayed strains (creep and shrinkage of concrete and relaxation of steel), based on the relations of the next Eurocode 2 (EC2), is used to predict the behavior of the internal vessel of the VERCORS mock-up of a French NPP. This modelling considers the influence of the temperature and of the relative humidity (varying under service conditions). The predicted delayed strains are compared with the measurements and different hypotheses are tested to improve the prediction of the delayed strains.

Keywords: Concrete · Creep · Shrinkage · Nuclear power plant

1 Introduction

The prediction of the long-term behavior of prestressed concrete structures is important in order to assess and/or extend the service life of such structures. In the case of nuclear power plants (NPPs) where the internal vessel is a biaxially prestressed structure, the prediction of the delayed deformations is also a safety concern. Indeed, in these structures, the prestressing is there to avoid tensile stresses and cracking in case of a severe accident where internal pressure and high temperature could be generated. The prediction of the evolution of delayed strains due to creep and shrinkage of concrete and of relaxation of steel is in this case very important. Here, a modelling of the delayed strains (creep and shrinkage of concrete and relaxation of steel), based on the relations of the next Eurocode 2 (EC2), is used to predict the behavior of the internal vessel of the VERCORS mock-up of a French NPP. This modelling considers the influence of the temperature and of the relative humidity (varying under service conditions). The predicted delayed strains are compared with the measurements and different hypotheses are tested to improve the prediction of the delayed strains.

P. Rossi and J.-L. Tailhan (Eds.): SSCS 2022, RILEM Bookseries 38, pp. 12–21, 2023.
https://doi.org/10.1007/978-3-031-07746-3_2

2 Modelling of Delayed Strains

As proposed in the future version of the Eurocode 2 (EC2) [1], the delayed strains of concrete are split into four components that are adjusted to experimental results available for the VERCORS containment [2]: basic shrinkage, drying shrinkage, basic creep and drying creep. The relaxation of the prestressing is also considered. An elevated temperature during service conditions (around 35 °C) is observed in NPPs. This temperature affects the delayed strains. The effect of temperature is not considered in EC2. For basic creep, a thermo-activation is considered and the magnitude of drying creep is affected by the temperature as proposed in the fib MC2010 [3]. For steel relaxation, the relation proposed by the fib MC2010 is used.

2.1 EC2 Equations for Creep and Shrinkage

The delayed strains of concrete ε^c are decomposed in the future EC2 into basic shrinkage ε^{bs}, drying shrinkage ε^{ds}, basic creep ε^{bc} and drying creep ε^{dc}:

$$\varepsilon^c = \varepsilon^{bs} + \varepsilon^{ds} + \varepsilon^{bc} + \varepsilon^{dc} \tag{1}$$

Shrinkage. Basic shrinkage ε^{bs} and drying shrinkage ε^{ds} are given as a function of mean compressive strength f_{cm} and drying time $t - t_s$ by the following equation:

$$\varepsilon^{bs} = \xi_{cbs1}\alpha_{bs}\left(\frac{0,1f_{cm}}{6+0,1f_{cm}}\right)\left(1 - e^{-0,2\xi_{cbs2}\sqrt{t}}\right) \tag{2}$$

$$\varepsilon^{ds} = \xi_{cds1}\left[(220 + 110\alpha_{ds1})e^{-\alpha_{ds2}f_{cm}}\right]\beta_{RH}\left[\frac{(t-t_s)}{0,035\xi_{cds2}h^2 + (t-t_s)}\right]^{0,5} \tag{3}$$

where α_{bs}, α_{ds1} and α_{ds2} are parameters that depend on the type of cement, h is the notional size, t_s is the age of the concrete at the start of drying; ξ_{cbs1}, ξ_{cds1}, ξ_{cbs2} and ξ_{cds2} are parameters to adjust the predictions to the experimental results (the default values are equal to 1). β_{RH} is a function of the relative humidity RH.

Creep. The basic creep function reads:

$$\varphi_{bc}(t, t_0) = \xi_{bc1}\frac{1}{C}ln\left(1 + \frac{(t-t_0)}{\tau\xi_{bc2}}\right) \tag{4}$$

where τ is a characteristic time depending on the age at loading of the concrete; ξ_{bc1} and ξ_{bc2} are parameters that may be adjusted to the experimental results. C is the creep modulus and is a function of f_{cm}.

The drying creep function is:

$$\varphi_{dc}(t, t_0) = \xi_{dc1}\beta_{dc}(f_{cm}, RH, t_0)\beta_{dc,t-t0} \tag{5}$$

with

$$\beta_{dc,t-t0} = \left[\frac{t-t_0}{\xi_{dc2}\beta_h + t - t_0}\right]^{\gamma(t_0)} \tag{6}$$

Again, ξ_{dc1} and ξ_{dc2} are parameters that may be adjusted according to the experimental results (the default values are equal to 1).

2.2 Influence of Temperature on Creep and Shrinkage

Shrinkage. Shrinkage is certainly affected by temperature. MC2010 proposes relations to consider the effect of temperature on shrinkage. But these relations are given for a constant temperature which is far from the real temperature history in the VERCORS mock-up (see Fig. 5). It is of course possible to evaluate an equivalent time considering the thermo-activation of hydration for the basic shrinkage and of drying for drying shrinkage. But, here, for a sake of simplicity, we neglect the effect of temperature on shrinkage.

Creep. For drying creep, the relations proposed by MC2010 are used (for the amplitude and the kinetics) (see [3] and [4]). For basic creep, we consider 3 cases:

- No thermo-activation of basic creep
- The creep modulus C (see Eq. 4) is affected by thermo-activation (as proposed in [5, 6 and 7]
- The characteristic time τ (see Eq. 4) is affected by thermo-activation as proposed by Frech-Baronet on the basis on micro-indentation creep tests [8]

2.3 Comparison with Laboratory Tests

Creep and shrinkage tests were performed on the concrete of the VERCORS mock-up [9]. Using the possibility to adjust the relations of EC2 to these tests, Figs. 1 and 2 show that it is possible to obtain a very good agreement with the experiments. Table 1 gives the fitted values of the parameters used in Eqs. 2 to 6.

Table 1. Fitting parameters of MC2010 shrinkage and creep models

Basic shrinkage		Drying shrinkage		Basic creep		Drying creep	
ξ_{cbs1}	ξ_{cbs2}	ξ_{cds1}	ξ_{cds2}	ξ_{bc1}	ξ_{bc2}	ξ_{cds1}	ξ_{cds2}
1.5	0.3	1.0	1.4	2.0	1.4	1.2	0.4

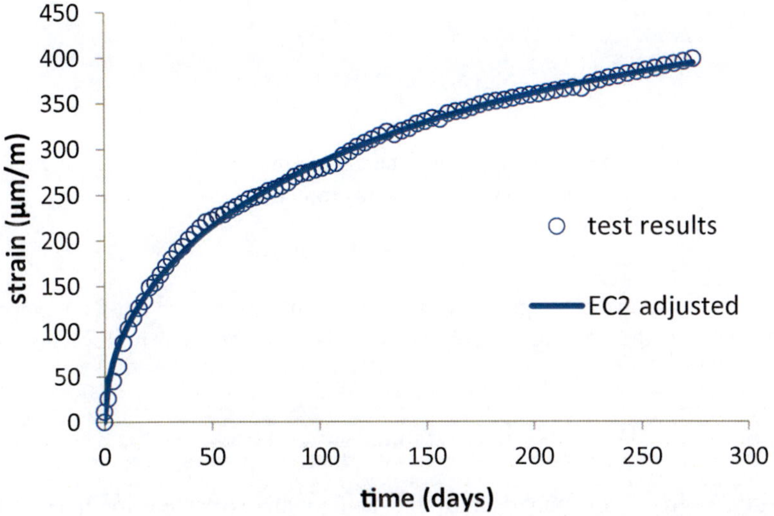

Fig. 1. Comparison between experimental results and modeling of drying shrinkage

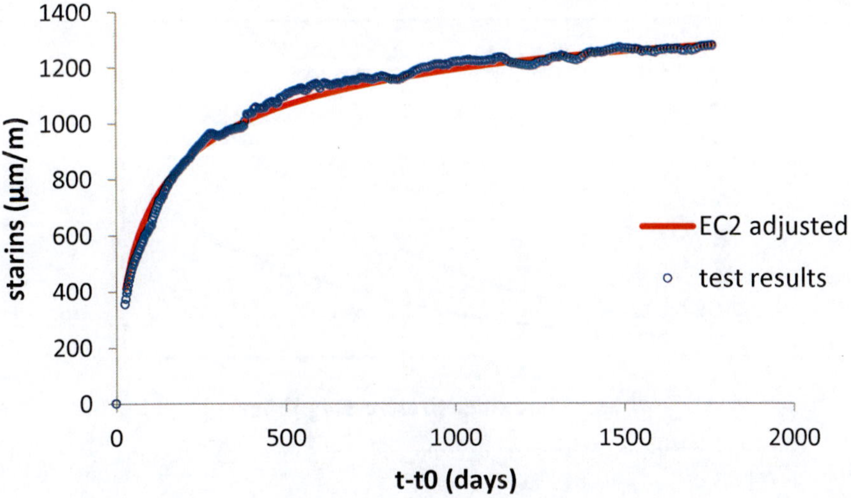

Fig. 2. Comparison between experimental results and modeling of total creep (basic + drying)

2.4 Relaxation

The relaxation kinetics of prestressing in prestressing tendons is expressed by the time evolution of relaxation loss $\rho(t) = (\sigma_0 - \sigma(t))/\sigma_0$, where σ_0 and $\sigma(t)$ are the stress at the time of prestressing and at time t, respectively. If the relaxation loss after 100 h and after 1000 h is denoted as ρ_{100} and ρ_{1000} respectively, the time evolution of relaxation loss is given by:

$$\rho(t) = \rho_{1000}\left(\frac{t}{1000}\right)^k \tag{7}$$

where $k = \log(\rho_{1000}/\rho_{100})$. Supposing that elastic modulus of prestressing tendons is constant, the stress evolution under a constant strain ε_0 reads as:

$$\sigma(t) = E_s\varepsilon_0(1 - \rho(t)) \tag{8}$$

The Eq. 7 is applicable to predict the relaxation under the same constant temperature. For temperature other than 20 °C, the relaxation loss is obtained by multiplying the stress loss at 20 °C by the following temperature-dependent parameter:

$$\alpha_T(T) = \frac{T}{20\,°\mathrm{C}} \tag{9}$$

Figure 3 shows the comparison of this model with experiments on the prestressing tendons used for the VERCORS mock-up [10]. For the cases with varying temperature, the Eq. 9 will be used in the application of the superposition principle.

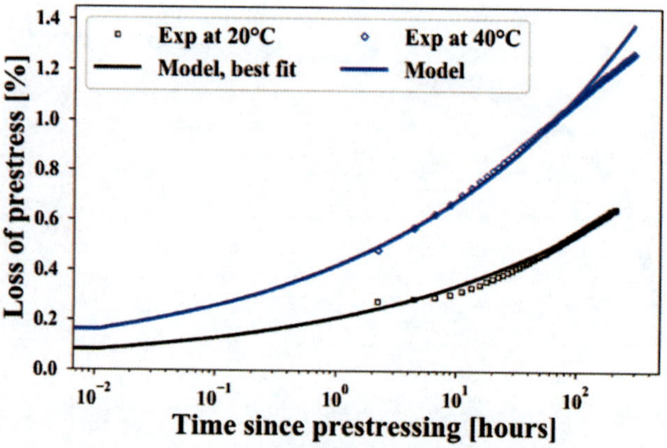

Fig. 3. Comparison between experimental results and modeling of the relaxation of the tendons at 20 °C and 40 °C; $\rho_{100} = 0.551$ and $\rho_{1000} = 0.893$

3 Modelling of the Containment

3.1 General Considerations

Considering the history of temperature and relative humidity (see 3.2), and the biaxial stress state, we compute the total delayed strain of concrete in horizontal (or tangential) and vertical directions using the superposition principle and the relations for creep, shrinkage and relaxation that were presented before. Due to the biaxial stress state,

Poisson's effects are to be considered. The Poisson's ratio for basic creep ν_b is considered to be equal to 0.2 [11, 12] and we assume that the Poisson's ratio for drying creep is also 0.2, though other values such as 0 was also considered elsewhere [13]. The method is fully described in [4]. Note that, except for the fitting of laboratory tests, the prediction of the delayed behavior of the mock-up is done without adjustment on the in-situ measurements.

3.2 In-Situ Measurements

The VeRCoRs Mock-up

To better predict the delayed behavior of the containment of NPPs, the company operating the French nuclear power plants (EDF) has built a mock-up, named VERCORS, of a double containment building, at the 1/3 scale [9]. The ratio 1/3 has been chosen in order to accelerate the drying phenomenon.

Boundary Conditions

Since our modeling concerns only a material point, the relative humidity was approximated from the recorded value of relative humidity in the inner containment and in the annular space. Considering that the fluctuation of relative humidity inside concrete would be much less than that of the air, we took a simplified relative humidity history as shown in Fig. 4.

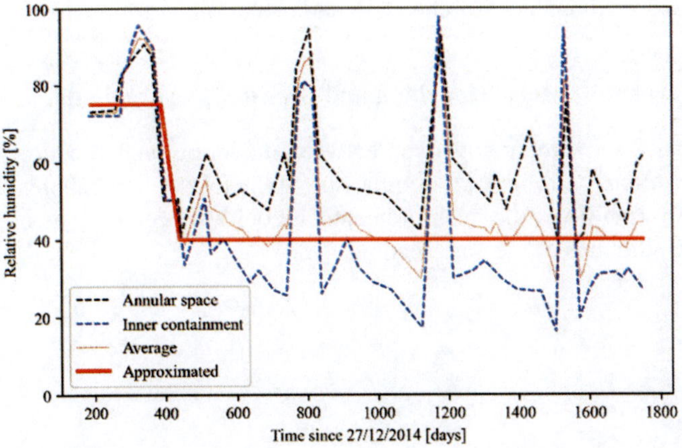

Fig. 4. Measured relative humidity in the annular space and inner space, simplified relative humidity for the modeling

Based on the measurement of the temperature at the annular space and inner space, we took the simplified temperature history shown in Fig. 5 for the modeling unit. The rationale behind this simplification is partly lie in the fact that temperature variation inside the thick concrete shall be slower comparing to that in the air, given the low heat coefficient of concrete. Note that the mean temperature is below 30 °C which is lower than in a real containment (where it is closer to 40 °C [14]).

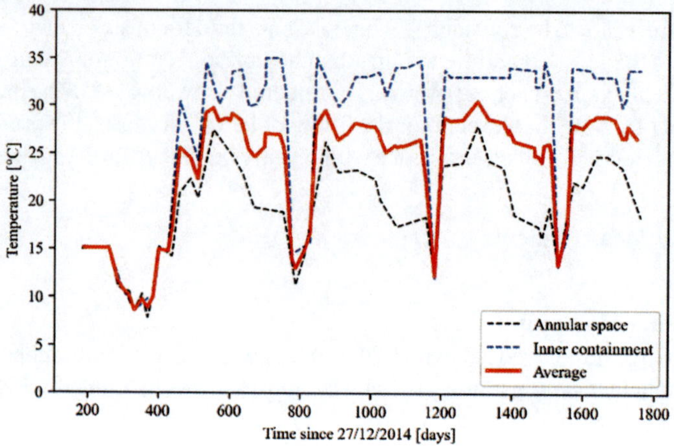

Fig. 5. Temperature history applied in the simulation

Sensors

The VERCORS mock-up is equipped with a comprehensive monitoring system, including a large number of strain sensors (vibrating wires). In our case, we compare our modelling with the measurements of several sensors situated at mid-height and in an area far from singularities. The measurements in the vertical direction (V) and the horizontal direction (T) are considered. Two measures of the prestressing tendons are also available and are compared with the predicted evolution.

3.3 Comparison Between Modelling and In-Situ Measurements

Figures 6 and 7 show the comparison between modelling and experimental measurements of the strains and Fig. 8 the comparison of the loss of vertical prestressing. Compared to the variability of the measurements, it could be seen that the modelling gives acceptable predictions of the deformations.

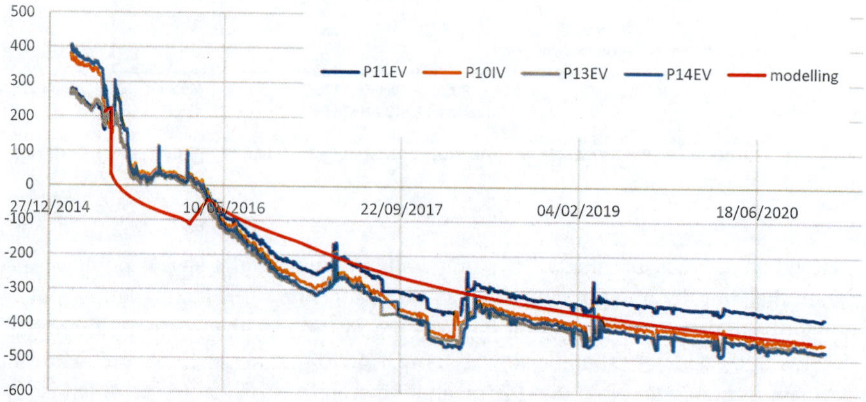

Fig. 6. Modeled vertical strain compared with the measurement of vibrating wires

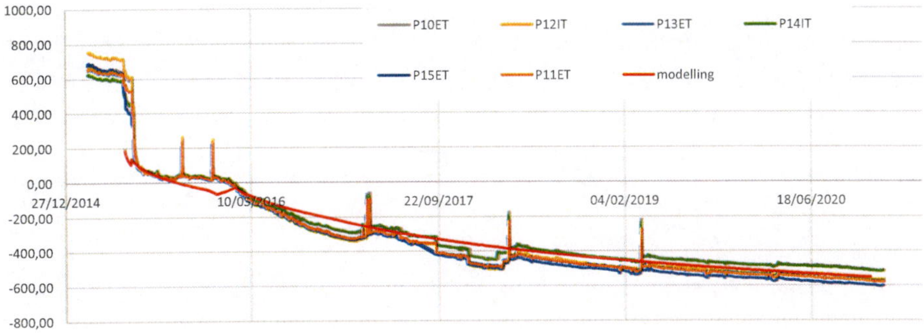

Fig. 7. Modeled tangential strain compared with the measurement of vibrating wires

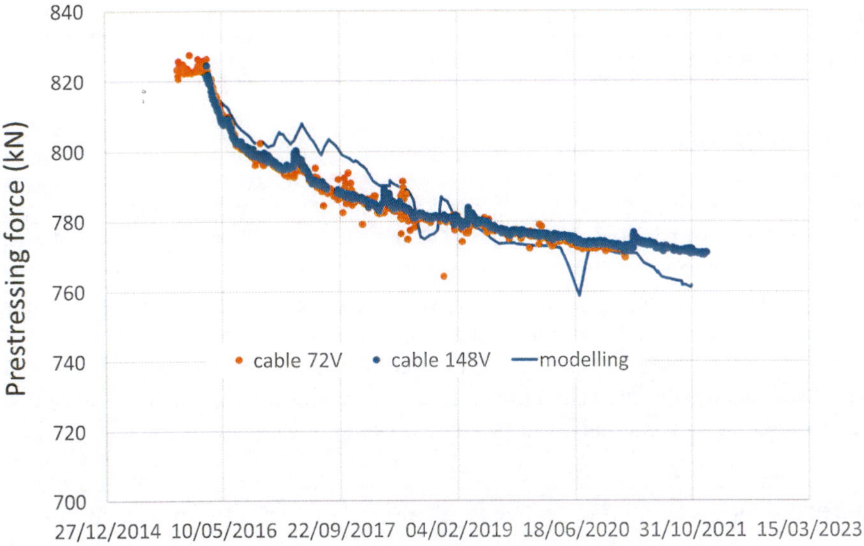

Fig. 8. Evolution of the prestressing force

3.4 Discussion of the Results

This section discusses the impact of thermo-activation on the delayed strain of the mock-up. In addition to the simulation with accounting thermo-activation on C in Eq. 4, two other simulations were performed: one without thermo-activation and the other with thermo-activation on characteristic time τ in Eq. 4. The basic creep in horizontal direction from the three simulations are displayed in Fig. 9, showing the impact of thermo-activation is rather small for the given temperature history in the case of the mock-up. Therefore, thermo-activation on basic creep can be neglected for the given temperature history in the case of the mock-up.

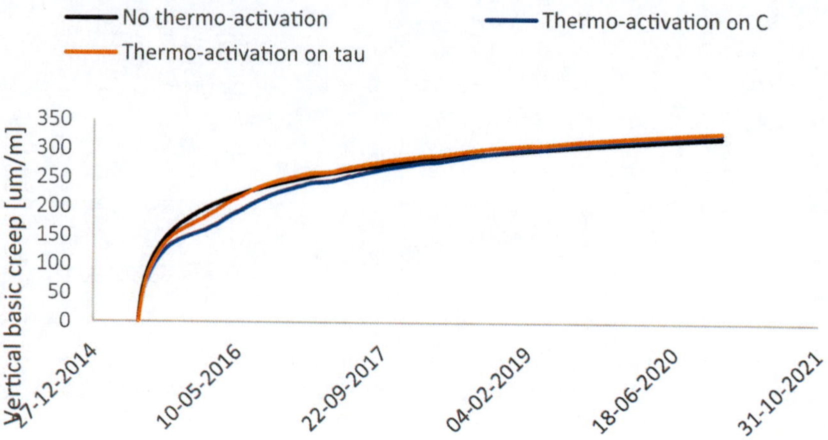

Fig. 9. Comparison of the contribution of basic creep on tangential strains with different assumptions for the thermo-activation of basic creep.

4 Conclusions

Delayed deformations of the VERCORS mock-up are estimated using shrinkage and creep relations of the future Eurocode-2 [1] and MC2010 [3]. Analytical simulation of a single material point considering the coupling between strain of concrete and stress relaxation in prestressing tendons is able to consider impact of varying relative humidity and temperature. Modeling results of the single point are comparable to the strain measurement on the mock-up hence could be used to quick-estimate the margin of safety. It is found that the thermo-activation on basic creep does not play significant role in the case of the mock-up but this result could be different for a real NPP.

References

1. prEN 1992-1-1 ver. 2021-01, Eurocode 2: Design of concrete structures - Part 1-1: General rules - Rules for buildings, bridges and civil engineering structures, CEN, 2021
2. Charpin, L., et al.: Ageing and air leakage assessment of a nuclear reactor containment mock-up: VERCORS 2nd benchmark. Nucl. Eng. Des. **377**, 111136 (2021)
3. fib model code for concrete structures 2010, Ernst und Sohn, 2013
4. Aili, A., Torrenti, J.M.: Modeling long-term delayed strains of prestressed concrete with real temperature and relative humidity history. J. Adv. Concr. Technol. **18**, 396–408 (2020)
5. Bažant, Z.P., Cusatis, G., Cedolin, L.: Temperature effect on concrete creep modeled by microprestress- solidification theory. J. Eng. Mech. **130**(6), 691–699 (2004)
6. Sellier, A., Multon, S., Buffo-Lacarrière, L., Vidal, T., Bourbon, X., Camps, G.: Concrete creep modelling for structural applications: non-linearity, multi-axiality, hydration, temperature and drying effects. Cem. Concr. Res. **79**, 301–315 (2016)
7. Torrenti, J.M.: Basic creep of concrete-coupling between high stresses and elevated temperatures. Eur. J. Environ. Civ. Eng. **22**(12), 1419–1428 (2018)
8. Frech-Baronet, J., Sorelli, L., Chen, Z.: A closer look at the temperature effect on basic creep of cement pastes by microindentation. Constr. Build. Mater. **258**, 119455 (2020)

9. Oukhemanou, E., Desforges, S., Buchoud, E., Michel-Ponnelle, S., Courtois, A.: VeRCoRs Mock-Up: comprehensive monitoring system for reduced scale containment model. In: 3rd Conference on Techonological Innovation in Nuclear Civil Engineering (TINCE-2016) (2016)
10. Toumi Ajimi, W., Chataigner, S., Falaise, Y., Gaillet, L.: Experimental investigations on the influence of temperature on the behavior of steel reinforcement (Strands and Rebars). In: Concrete under Severe Conditions - Environment and Loading. Trans Tech Publications Ltd, vol. 711, pp. 908–915 (2016)
11. Aili, A., Vandamme, M., Torrenti, J.-M., Masson, B.: Theoretical and practical differences between creep and relaxation Poisson's ratios in linear viscoelasticity. Mech. Time-Dependent Mater. **19**(4), 537–555 (2015). https://doi.org/10.1007/s11043-015-9277-5
12. Aili, A., Vandamme, M., Torrenti, J.-M., Masson, B., Sanahuja, J.: Time evolutions of non-aging viscoelastic Poisson's ratio of concrete and implications for creep of C-S-H. Cem. Concr. Res. **90**, 144–161 (2016)
13. Charpin, L., et al.: A 12 year EDF study of concrete creep under uniaxial and biaxial loading. Cem. Concr. Res. **103**, 140–159 (2018)
14. Bouhjiti, D.M., Boucher, M., Briffaut, M., Dufour, F., Baroth, J., Masson, B.: Accounting for realistic Thermo-Hydro-Mechanical boundary conditions whilst modeling the ageing of concrete in nuclear containment buildings: Model validation and sensitivity analysis. Eng. Struct. **166**, 314–338 (2018)

HPC Finite Element Solvers for Phase-Field Models for Fracture in Solids

Mohd Afeef Badri[1] and Giuseppe Rastiello[2](✉)

[1] Université Paris-Saclay, CEA, Service de thermo-hydraulique et de mécanique des fuides, 91191 Gif-sur-Yvette, France
mohd-afeef.badri@cea.fr

[2] Université Paris-Saclay, CEA, Service d'études mécaniques et thermiques, 91191 Gif-sur-Yvette, France
giuseppe.rastiello@cea.fr

Abstract. Finite element approximations of phase-field models for fracture are often used to simulate cracking processes in solids and structures. However, non-linearity, the requirement of extremely fine meshing, and/or large-scale simulations make numerical crack prediction a tedious task with this family of models. Such problems demand enormous computational resources. Under a sequential computing framework, these lead to extremely slow computations with non-feasible computer processing times. As a remedy, domain decomposition approaches that facilitate parallel computing can subdue these issues and significantly decrease computational time and memory. This contribution discusses and compares two ways of setting up domain decomposition for phase-field models under a distributed computing framework. Notably, in the context of parallel computing, a monolithic strategy set up via the vectorial finite elements is compared to a staggered finite element strategy for a hybrid phase-field model. A detailed comparison of performance, scalability, and efficiency on thousands of parallel processors is established for a large-scale fracture mechanics problem with millions of unknowns.

Keywords: Phase-field modeling · Staggered/monolithic · HPC

1 Introduction

Numerous numerical models to predict fracture within solids (mostly based on the Finite Element Method, FEM) evolved over the years, and new methods continue to emerge. These methods can be classified into discrete or diffused/smeared fracture models. Discrete models approximate fracture as a kinematic discontinuity (i.e., a discontinuous displacement field is explicitly introduced in the formulation). Complementary to discrete fracture models are smeared models (e.g., regularized damage and plasticity models). Among them, Phase-Field Models (PFMs) [3,8,16] are widely used due to their strong theoretical background [10]. Based on

P. Rossi and J.-L. Tailhan (Eds.): SSCS 2022, RILEM Bookseries 38, pp. 22–32, 2023.
https://doi.org/10.1007/978-3-031-07746-3_3

a variational framework, the fracture prediction problem to be solved is a global energy functional minimization problem alongside the displacement and damage (phase-) fields. Fracture is considered to exist in a hypothetical continuum space. To elaborate on this, the model involves sorting the state of material via a scalar damage variable (ranging from zero to unity) which is let to evolve in space and time. Naturally, crack boundaries are diffused over a finite length in the domain.

Several PFMs were proposed in the literature and gained popularity. Rightly so, because the model is simple to set up and has been proven to accurately model vast variety of fracture problems [1,2,7,17,21]. Further, several studies (e.g., [18,19]) have provided experimental validations of such formulations. The main downside of PFMs is that they require extremely refined meshes to describe the cracking process accurately.

Mathematically, PFMs are described by non-linearly coupled equations: a modified momentum balance equation and a diffusion-type (Helmholtz-like) equation. To handle non-linearities, the two equations should be solved in a staggered fashion or a monolithic one. This implies solving consecutively two non-linearly coupled systems of equations (until convergence) on top of an extremely fine mesh. As such, performing PFM simulations becomes a tedious task (particularly for large-scale simulations). Often the time to solution for such simulations is not feasible. Naturally, domain decomposition approaches that facilitate parallel computing can be used to subdue these issues and significantly decrease computational time and memory.

Recently, researchers have focused on parallel computing to solve PFMs for fracture. A Graphics Processing Unit-based parallel solver for dynamic brittle fracture was developed in [24] using a staggered FE approach. A parallel staggered approach-based PFM for brittle materials was also presented in [6]. Using geometric multigrid solvers, these authors demonstrated a quasi-linear scaling (strong) of their parallel solver, with scaling being established up to 320 processing units. Built-in parallelization capabilities of the FEM software Abaqus [12] were explored within the framework of staggered and monolithic approaches for the PFM applied to quasi-static and dynamic fracture analyses [15]. Scaling analysis while using 1 to 16 Central Processing Units (CPUs) was presented by the authors; it was observed that the algorithms scale up to 4 CPUs. Strong scaling analysis up to 128 CPUs for the monolithic FEM implementation of a PFM was recently performed in [13]. Using geometric multigrid solvers with matrix-free capabilities, these authors achieved quasi-linear scaling.

The main motives of this article are to discuss and compare two distinct ways of setting up domain decomposition-based parallelization for the hybrid PFM by [2] under the distributed computing framework. Particularly, a monolithic strategy is set up via the vectorial FEM (previously developed in [4]) and is compared to a staggered FE strategy. Based on numerical evidence, this article intends to answer which scheme – monolithic or staggered – is more adapted for High Performance Computing (HPC). Contrary to previous studies, simulations are performed using massively parallel machines involving thousands of CPUs. Overall the aim is also to show how to perform PFM simulations with millions of Degrees of Freedom (DoFs) within minutes by harvesting the power of modern-day supercomputers.

2 Mathematical Formulations

This section summarizes the mathematical formulation of the hybrid PFM by [2]. This model is defined as "hybrid" [2,22,23] since it combines features of the so-called "isotropic" [8] and "anisotropic" [3,16] PFM formulations.

2.1 Model Formulation

Consider an arbitrary open bounded domain Ω with imposed tractions on Neumann boundary $\partial\Omega_{\mathrm{N}} \subset \mathbb{R}^n$, and provided with essential Dirichlet conditions on boundary $\partial\Omega_{\mathrm{D}} \subset \mathbb{R}^n$. The split of the domain boundary $\partial\Omega \subset \mathbb{R}^n$ is such that $\partial\Omega = \overline{\partial\Omega_{\mathrm{D}} \cup \partial\Omega_{\mathrm{N}}}$ and $\emptyset = \partial\Omega_{\mathrm{D}} \cap \partial\Omega_{\mathrm{N}}$, with over-line denoting a closure.

Damaging is described through a scalar-valued field ϕ ranging from zero (sound material) to unity (completely broken material). Accordingly, the problem unknowns are $\mathbf{u}$, the displacement field, and ϕ.

Admissibility Spaces. Let us introduce the following admissibility spaces for the unknown displacement and damage fields:

$$\mathcal{U}(\bar{\mathbf{u}}) = \left\{\mathbf{w} \in [H^1(\Omega)]^n : \ \forall t \in [0,T] \mid \ \mathbf{w} = \bar{\mathbf{u}} \text{ on } \partial\Omega_{\mathrm{D}}\right\}, \tag{1}$$

$$\mathcal{U}(0) = \left\{\mathbf{w} \in [H^1(\Omega)]^n : \ \forall t \in [0,T] \mid \ \mathbf{w} = 0 \text{ on } \partial\Omega_{\mathrm{D}}\right\}, \tag{2}$$

$$\mathcal{D} = \left\{\omega \in H^1(\Omega) : \ \forall t \in [0,T] \mid \ \omega \in [0,1] \ \text{ in } \Omega\right\}, \tag{3}$$

where $H^1(\Omega)$ denotes a square integrable Sobolev functional space and $\bar{\mathbf{u}}$ is the imposed displacement on $\partial\Omega_{\mathrm{D}}$.

Variational Problem. Under quasi-static conditions and in the absence of body forces, one solves at each time $t \in [0,T]$ the following boundary value problem:

Find $\mathbf{u} \in \mathcal{U}(\bar{\mathbf{u}})$ and $\phi \in \mathcal{D}$ such that:

$$\int_{\Omega} (1-\phi)^2 \left[\boldsymbol{\sigma}(\mathbf{u}) : \boldsymbol{\epsilon}(\mathbf{v})\right] \mathrm{d}V = \int_{\partial\Omega_{\mathrm{N}}} \bar{\mathbf{t}} \cdot \mathbf{v} \, \mathrm{d}S \quad \forall \mathbf{v} \in \mathcal{U}(0), \tag{4}$$

$$\int_{\Omega} \left\{ \left[\frac{G_c}{l_0} + 2\mathcal{H}^+(\mathbf{u})\right] \phi\theta + G_c l_0 \nabla\phi \cdot \nabla\theta \right\} \mathrm{d}V = \int_{\Omega} 2\mathcal{H}^+(\mathbf{u})\theta \, \mathrm{d}V \qquad \forall \theta \in \mathcal{D}, \tag{5}$$

where $\boldsymbol{\sigma}(\mathbf{u})$ denotes the elastic Cauchy's stress tensor, $\boldsymbol{\epsilon}(\mathbf{u})$ is the small strain tensor, $\bar{\mathbf{t}}$ is the imposed traction on $\partial\Omega_{\mathrm{N}}$, G_c is the fracture energy, l_0 is a length-scale parameter (i.e., the internal length), and $\mathcal{H}^+(\mathbf{u})$ is the maximum tensile elastic energy $\psi^+(\mathbf{u})$ experienced by the material during loading (i.e., an history function). Finally, to prevent crack interpenetration, the following constraint is applied, $\psi^+(\mathbf{u}) < \psi^-(\mathbf{u}) \Rightarrow \phi = 0$, with $\psi^-(\mathbf{u})$ denoting the compressive elastic energy. For more details concerning the model formulation, the interested reader can refer to [2].

2.2 Finite Element Formulation

To solve the variational problem (4)–(5) according to the FEM, the domain Ω is discretized through a mesh Ω^h containing N_v nodes. In this paper, fields $\mathbf{u}$ and ϕ are approximated by piecewise linear continuous finite elements $\mathbb{P}_1^h$ with DoFs at mesh nodes (denoting by K a mesh element, $\mathbb{P}_1^h = \{v \in H^1(\Omega) | \forall K \in \Omega^h, v_{|K} \in P_1\}$). These are represented by FE fields $\mathbf{u}^h$ and ϕ^h, the discretized problem unknowns. Since problem (4)–(5) is non-linear, an iterative solution strategy based on Picard's iteration is used [4]. Two solving schemes are employed:

- *Staggered scheme.* One first solves Eq. (4) for $\mathbf{u}^h$ while ϕ^h is frozen at previous iteration, then one solves Eq. (5) for ϕ^h with the updated $\mathbf{u}^h$. This process is iteratively repeated till convergence. At each global iteration, one solves two linear systems of the form:

$$\mathbf{A_u x_u} = \mathbf{b_u} \quad \text{and} \quad \mathbf{A}_\phi \mathbf{x}_\phi = \mathbf{b}_\phi, \tag{6}$$

  where $\mathbf{A_u} \in \mathbb{R}^{m \times m}$, $\mathbf{x_u} \in \mathbb{R}^m$, $\mathbf{b_u} \in \mathbb{R}^m$, $\mathbf{A}_\phi \in \mathbb{R}^{N_v \times N_v}$, $\mathbf{x}_\phi \in \mathbb{R}^{N_v}$, and $\mathbf{b}_\phi \in \mathbb{R}^{N_v}$ (with $m = n \times N_v$).
- *Monolithic scheme.* Equations (4) and (5) are solved as a whole for $\mathbf{U}^h = \{\mathbf{u}^h, \phi^h\}$. According to the so-called vectorial FEM, $\mathbf{u}^h$ and ϕ^h belong to the same vectorial FE space:

$$\mathcal{W}^h = \left\{\mathbf{w}^h \in [H^1(\Omega^h)]^{n+1} : \ \{w_i^h\}_{i=1}^n = \overline{\mathbf{u}} \text{ on } \partial\Omega_{\mathrm{D}}^h \ , \ w_{n+1}^h \in [0,1]\right\}, \tag{7}$$

  The linear system assembled is of the form:

$$\mathbf{A_U x_U} = \mathbf{b_U}, \tag{8}$$

  where $\mathbf{A_U} \in \mathbb{R}^{p \times p}$, $\mathbf{x_U} \in \mathbb{R}^p$, and $\mathbf{b_U} \in \mathbb{R}^p$ (with $p = (n+1) \times N_v$).

These two different schemes lead to matrix systems with different sparsity patterns (Fig. 1). Such a difference should be considered while designing an efficient parallel solver and plays a crucial role in the choice of preconditioning for linear systems.

3 Parallelization

Since there is a need to solve large-scale linear systems repeatedly due to pseudo-time dependence, for time efficiency (during the FE assembly and the solution phases), parallelization becomes inevitable. This study employs end-to-end parallelization – from meshing to visualization. Briefly, the following procedure/tools/methodologies are implemented:

- Domain Ω is meshed to obtain a triangular/tetrahedral mesh Ω^h by employing parallel *meshing* algorithms (shared memory-based) provided within the open-source tool `SALOME` [9].

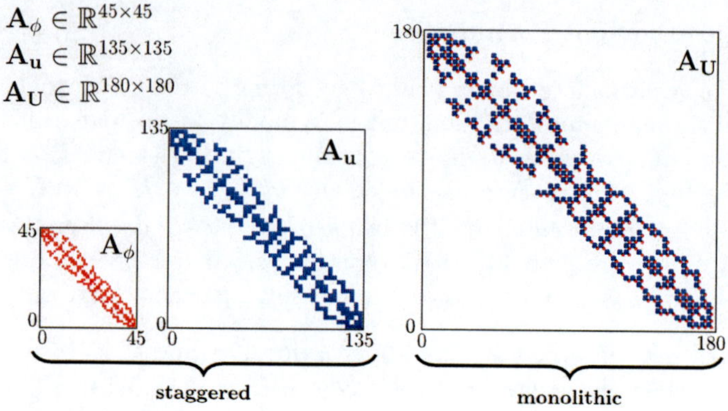

Fig. 1. Sparsity structures of different matrices belonging to staggered or monolithic FE approaches. Spatial mesh used contained $N_v = 45$.

- Mesh Ω^h is split – *mesh partitioning* – into N_p small non-overlapping meshes $\{\Omega_i^h\}_{i=1}^{N_p}$ (mesh partitions) using the open-source parallel mesh partitioning tool `ParMETIS` [14].
- The mesh partitions are then assigned to N_p Message Passing Interface (MPI) processes, which are then used to individually assemble the local FEM matrix systems. The FE kernel `FreeFEM` [11] is used for setting-up the parallel FEM linear system assembly phase using MPI.
- The global problem is then solved iteratively using the Krylov-subspace Conjugate Gradient (CG) algorithm set up in parallel. The local solving is independently performed within each MPI process with a small communication overhead (shared nodes between MPI ranks for a partitioned mesh, see Fig. 2-right) for the CG iterations overall. To obtain good parallel scaling, preconditioning is used for the solution phase. The open-source library `PETSc` [5] is employed for defining suitable preconditioners and carry out the *linear system solution* in parallel.
- Finally, for *visualization*, the partitioned FEM fields are saved on the disk as partitioned visualization files (`pvtu`) native to 3D computer graphics library `VTK` [20]. The open-source tool `SALOME` is again employed for visualizing the results using MPI-based parallelism.

These steps are performed via an in-house wrapper code – `PSD` Parallel Solid/Structural/Seismic Dynamics – written in C++.

4 Numerical Experiments

This section presents detailed results from numerical experiments designed to investigate the implemented parallelized monolithic and staggered strategies. Computations were executed on the French supercomputer Joliot-Curie with

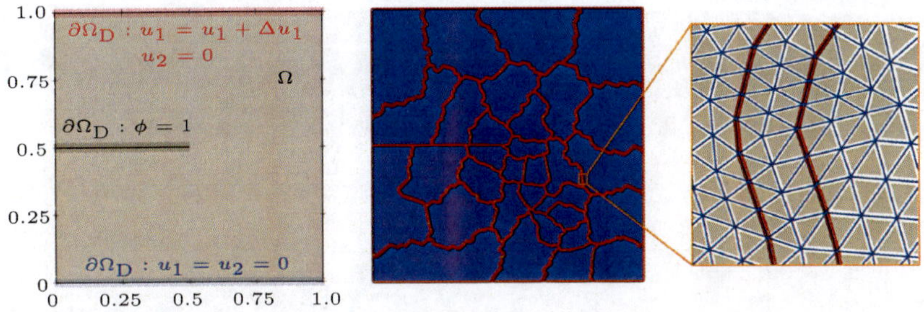

Fig. 2. Simulation domain for the shear cracking test. Left: domain Ω with boundary conditions (coordinates x and y in mm); middle: partitioned mesh $\{\Omega_i\}_{i=1}^{32}$; right: zoomed in view of the mesh.

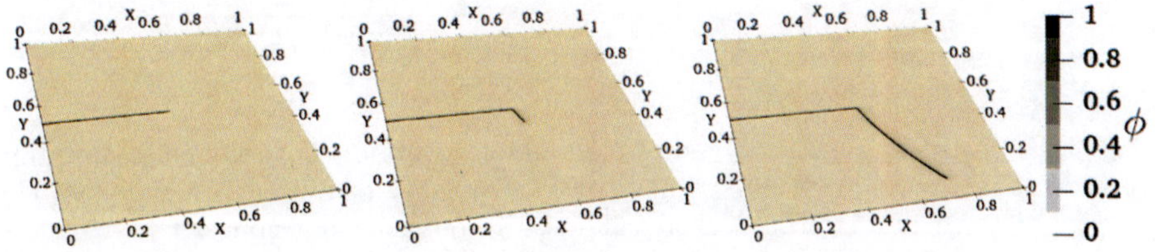

Fig. 3. Crack propagation within the domain at different (pseudo)-time-steps (coordinates x and y in mm).

the Intel Skylake architecture[1]. The preconditioned CG method with block Jacobi preconditioner (setup on left) with incomplete Cholesky factorization in the blocks was employed as the default linear system solver for the two methods[2]. This choice was motivated by the fact that neither the unpreconditioned CG solver nor the Jacobi preconditioned CG solver converged[3] for the scales of problems considered in this work.

Problem Setting. Two-dimensional (2D) numerical experiments are performed for simplicity. A 2D square plate shear cracking problem is simulated (Fig. 2). This test is commonly used in literature (and is considered a challenging one) since it involves mode I and II cracking. Moreover, a strong non-linear coupling exists between the two equations of the PFM.

Under shear loading, one observes damaging as illustrated Fig. 3. It was verified that both monolithic and staggered solvers produce the same results. The monolithic solver used in this study has been parallelized and prevalidated in [4]. The staggered solver was cross-validated against several test cases from the literature. For the sake of conciseness, these tests are not shown here.

---

[1] Intel Xeon 8168@2.7 GHz with 48 cores per node.

[2] Setup in PETSc with `-ksp_type cg -pc_type bjacobi -sub_pc_type icc`.

[3] Within the limit of 10,000 iterations and standard relative residual tolerance of 10^{-8}.

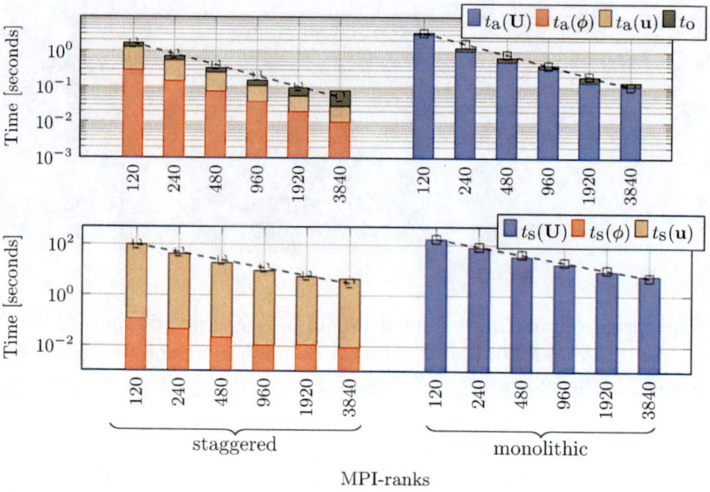

Fig. 4. Strong scaling analysis of staggered and monolithic schemes. The plot on top, provides linear system assembly timings (t_a) and other operation timings (t_o). The bottom plot provides information on the linear system solving timings (t_s).

Solvers Scalability. For numerical solvers designed to harvest the modern-day HPC architectures – as is the case here in this study – scalability is a widely used metric to indicate their efficiency. Scalability is the ability of the solver to deliver higher computational power, and consequently lower solving times, when the amount of processing power (CPUs or MPI processes) is increased.

Here, scalability is used to compare the monolithic and staggered solvers. The square plate shear cracking problem is resolved on a fine mesh with roughly 10 million nodes and 21 million linear triangular elements. This translates into a problem of solving 30 million DOFs per iteration, i.e., 10 million DOFs for $\mathbf{x}_\phi$ and 20 million DOFs for $\mathbf{x}_\mathbf{u}$ with the staggered solver, or a 30 million DOFs for $\mathbf{x}_\mathbf{U}$ with the monolithic one.

In Fig. 4 the scalability (strong scaling) of the two solvers is demonstrated: computational time for linear system assemblies (t_a) and solving time (t_s) decrease linearly when the number of MPI processes is increased. This indicates that both monolithic and staggered approaches lead to embarrassingly-parallel schemes when parallelized with domain decomposition methods.

This result is further attested by Table 1, which indicates the optimal speed-up and the parallel efficiency of the two methods. With 3840 MPI processes, the speed-up and parallel efficiency drop to 87 % (resp. 68 %) for the monolithic (resp. staggered) methods. However, one should note that the total computation time per iteration with these many processors is too low (a few seconds) to expect further scalability.

Table 1. Speed-up and parallel efficiency.

Method/MPI-ranks (N_i)	120	240	480	960	1920	3840
Speed-up, $S_i = t_{120}/t_{N_i}$						
Monolithic	1	2.2	4.8	10.1	18.7	27.9
Staggered	1	2.3	5.0	10.7	19.7	21.9
Efficiency, $E_i = S_i/\overline{S}_i$						
Monolithic	100%	108%	120%	126%	117%	87%
Staggered	100%	116%	125%	133%	123%	68%

Analyzing in-depth Fig. 4, as expected at such scales, we note that FEM linear system assembly timings remain much lower (1–2% of overall solving time) than the linear system solving timing (98–99% of overall solving time). Additional CPU wall time per iteration (t_o), which involves several tasks (field updates, non-linear criteria checking (MPI reduction), saving the solution fields for I/O, etc.), has also been plotted in Fig. 4. Solver operations that concern t_o are cheap to perform when compared with assembly time or solving time, i.e., $t_\mathrm{o} \ll t_\mathrm{a} \ll t_\mathrm{s}$. As a consequence, t_o **has no effect on scaling.**

Comparing the timings for the assembly and solving phases, the staggered solver outperforms the monolithic one. The FEM assembly routines for the two methods require approximately similar CPU resources and time. The only difference is that the staggered solver splits the FEM assembly time between the linear system assemblies for $\mathbf{x_u}$ and $\mathbf{x}_\phi$. Conversely, the monolithic solver assembles a larger linear system for $\mathbf{x_U}$. The larger dimension of the linear system (consequently leading to costlier matrix-vector products during the CG solving iterations) that is solved in a monolithic approach, in comparison to two smaller dimensional matrices solved in the staggered approach, plays the key role in high linear system solving time observed for monolithic solver in comparison to the staggered one. This is demonstrated in Fig. 5, which plots the number of CG iterations required for convergence during the simulation. The monolithic solver requires fewer CG iterations to converge but with much costlier matrix-vector products. On the other hand, the staggered solver requires a higher number of CG iterations for solving $\mathbf{x_u}$ with a cheaper matrix-vector product and relatively few CG iterations for $\mathbf{x}_\phi$. In other words, it is advantageous to solve for $\mathbf{x}_\phi$ as a separate entity – as is done in a staggered approach. These results motivate using a staggered scheme over the monolithic one when designing HPC FEM solvers for the PFM for fracture.

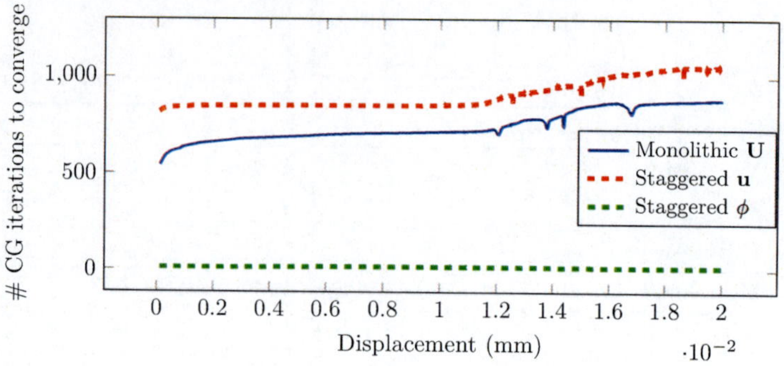

Fig. 5. Evolution of Krylov-subspace CG iterations to convergence.

5 Conclusion

Monolithic and staggered approaches for solving a hybrid PFM [2] for fracture from an HPC perspective were analyzed. Numerical experiments were performed to compare the two methods based on computational time, parallel scaling, speed-up, Krylov iterations to converge, and matrix structures. The following conclusions are made: *scaling-wise*, both monolithic and staggered solvers exhibit linear scaling capabilities when domain decomposition-based parallelism is used with distributed memory architectures. *Timing-wise*, the staggered solver is faster than the monolithic one, both for linear system assemblies or solving. The monolithic solver produces comparatively better-conditioned matrices; it requires fewer Krylov iterations to converge compared to the staggered solver. However, since the per-iteration cost of a monolithic solver is higher (larger matrix dimensions) than a staggered solver, this leads to a higher computational time. Overall, all these factors indicate that a staggered scheme is advantageous compared to a monolithic one for setting up parallel FEM solvers for hybrid PFM simulations.

Acknowledgments. The authors acknowledge the financial support of the Cross-Disciplinary Program on Numerical Simulation of CEA (France), the French Alternative Energies and Atomic Energy Commission. We thank TGCC-CEA, Bruyères-Le-Châtel, France, for providing us with compute time on the Irene Skylake partition on the French supercomputer Joliot-Curie. G. Rastiello was also supported by the SEISM Institute (http://www.institut-seism.fr).

References

1. Ambati, M., Gerasimov, T., De Lorenzis, L.: Phase-field modeling of ductile fracture. Comput. Mech. **55**(5), 1017–1040 (2015). https://doi.org/10.1007/s00466-015-1151-4

2. Ambati, M., Gerasimov, T., De Lorenzis, L.: A review on phase-field models of brittle fracture and a new fast hybrid formulation. Comput. Mech. **55**(2), 383–405 (2014). https://doi.org/10.1007/s00466-014-1109-y
3. Amor, H., Marigo, J.J., Maurini, C.: Regularized formulation of the variational brittle fracture with unilateral contact: numerical experiments. J. Mech. Phys. Solids **57**(8), 1209–1229 (2009)
4. Badri, M., Rastiello, G., Foerster, E.: Preconditioning strategies for vectorial finite element linear systems arising from phase-field models for fracture mechanics. Comput. Methods Appl. Mech. Eng. **373**, 113472 (2021)
5. Balay, S., et al.: PETSc web page (2017). http://www.mcs.anl.gov/petsc
6. Bilgen, C., Kopaničáková, A., Krause, R., Weinberg, K.: A phase-field approach to conchoidal fracture. Meccanica **53**(6), 1203–1219 (2017). https://doi.org/10.1007/s11012-017-0740-z
7. Borden, M.J., Verhoosel, C.V., Scott, M.A., Hughes, T.J., Landis, C.M.: A phase-field description of dynamic brittle fracture. Comput. Methods Appl. Mech. Eng. **217**, 77–95 (2012)
8. Bourdin, B., Francfort, G., Marigo, J.J.: Numerical experiments in revisited brittle fracture. J. Mech. Phys. Solids **48**(4), 797–826 (2000)
9. CEA-EDF: SALOME Platform - the open-source platform for numerical simulation (2022). https://www.salome-platform.org
10. Francfort, G.A., Marigo, J.J.: Revisiting brittle fracture as an energy minimization problem. J. Mech. Phys. Solids **46**(8), 1319–1342 (1998)
11. Hecht, F.: New development in FreeFem++. J. Numer. Math. **20**(3–4), 251–266 (2012)
12. Hibbitt, H., Karlsson, B., Sorensen, E.: ABAQUS-EPGEN: a general-purpose finite element code. Volume 1 (Revision 2). User's manual. Technical report, Hibbitt, Karlsson and Sorensen, Providence, RI, USA (1984)
13. Jodlbauer, D., Langer, U., Wick, T.: Matrix-free multigrid solvers for phase-field fracture problems. arXiv e-prints **54**, February 2019
14. Karypis, G., Schloegel, K., Kumar, V.: ParMETIS: parallel graph partitioning and sparse matrix ordering library. Version 1.0, Department of Computer Science, University of Minnesota, p. 22 (1997)
15. Liu, G., Li, Q., Msekh, M.A., Zuo, Z.: Abaqus implementation of monolithic and staggered schemes for quasi-static and dynamic fracture phase-field model. Comput. Mater. Sci. **121**, 35–47 (2016)
16. Miehe, C., Welschinger, F., Hofacker, M.: Thermodynamically consistent phase-field models of fracture: variational principles and multi-field FE implementations. Int. J. Numer. Meth. Eng. **83**(10), 1273–1311 (2010)
17. Miehe, C., Schänzel, L.M.: Phase field modeling of fracture in rubbery polymers. Part I: Finite elasticity coupled with brittle failure. J. Mech. Phys. Solids **65**, 93–113 (2014)
18. Nguyen, T.T., et al.: On the choice of parameters in the phase field method for simulating crack initiation with experimental validation. Int. J. Fract. **197**(2), 213–226 (2016). https://doi.org/10.1007/s10704-016-0082-1
19. Pham, K.H., Ravi-Chandar, K., Landis, C.M.: Experimental validation of a phase-field model for fracture. Int. J. Fract. **205**(1), 83–101 (2017). https://doi.org/10.1007/s10704-017-0185-3
20. Schroeder, W., Martin, K.M., Lorensen, W.E.: The Visualization Toolkit an Object-Oriented Approach to 3D Graphics. Prentice-Hall, Inc., Upper Saddle River (1998)

21. Verhoosel, C.V., de Borst, R.: A phase-field model for cohesive fracture. Int. J. Numer. Meth. Eng. **96**(1), 43–62 (2013)
22. Wu, J.Y.: A unified phase-field theory for the mechanics of damage and quasi-brittle failure. J. Mech. Phys. Solids **103**, 72–99 (2017)
23. Wu, J.Y., Nguyen, V.P.: A length scale insensitive phase-field damage model for brittle fracture. J. Mech. Phys. Solids **119**, 20–42 (2018)
24. Ziaei-Rad, V., Shen, Y.: Massive parallelization of the phase field formulation for crack propagation with time adaptivity. Comput. Methods Appl. Mech. Eng. **312**, 224–253 (2016)

Integrating Hybrid Reinforced Concrete Technology and Advanced FEM-Based Numerical Modelling for Crack Control in Long Concrete Foundations Without Joints

Joaquim A. O. Barros[1](✉), Tiago D. S. Valente[2], Inês G. Costa[2], and Felipe J. S. A. Melo[2]

[1] ISISE, Department of Civil Engineering, University of Minho, Azurém, 4800-058 Guimarães, Portugal
barros@civil.uminho.pt

[2] CiviTest – Pesquisa de Novos Materiais para a Engenharia Civil, Lda., 4770-160 Vila Nova de Famalicão, Portugal

Abstract. The present work describes the analysis of a long and jointless concrete foundation reinforced with conventional steel meshes and discrete polypropylene fibers. A thermo-mechanical nonlinear transient simulation is performed to assess the cracking risk and magnitude of the fiber reinforced concrete (FRC) due to the heat development generated from the cement hydration in the early stages of the concrete hardening phase. The thermal and cracking material data considered in the constitutive model are calibrated from the experimental program conducted when casting the concrete foundation. The concrete shrinkage, viscoelasticity and maturity concepts are also considered in the analysis.

The results of the numerical simulations revealed an adequate performance of the hybrid reinforcement to limit the crack opening of the concrete foundation since early ages, while significantly reducing the conventional steel reinforcement ratio.

Keywords: Hybrid reinforcement · Polypropylene fibers · Finite Element Method · Thermo-mechanical analysis · Early ages

1 Introduction

The use of hybrid reinforcements in concrete elements, combining discrete fibers and conventional steel bars, is seen as a promising solution to maximize the potentialities of both types of reinforcements. The use of this hybrid reinforcement was reported in foundations, beams, slabs, shells, among other structures [1–3].

It is well known that the addition of short and randomly distributed fibers in concrete increases its post-cracking resistance, ductility and energy absorption [4, 5]. Furthermore, the restrain to crack opening, provided by the different fiber reinforcement mechanisms at fracture, enhances the durability and integrity of concrete members. Based on

P. Rossi and J.-L. Tailhan (Eds.): SSCS 2022, RILEM Bookseries 38, pp. 33–46, 2023.
https://doi.org/10.1007/978-3-031-07746-3_4

these benefits, fiber reinforcements demonstrate a great potential to replace conventional steel reinforcements, particularly in structural members of high support redundancy.

The present work focusses in the study of a long and jointless concrete foundation member of an industrial equipment, submitted to relatively low level of external actions.

The geometry of the foundation is presented in Fig. 1. The cross-section has a hollow concrete shape, with concrete walls with 200 mm of thickness. The total length of the foundation is 22.4 m. Reinforced concrete diaphragms of 100 mm thickness are disposed at the extremities, ¼ and ½ of the foundation length for increasing transversal and torsional stiffness. The hollow core is formed by expanded polystyrene blocks (EPS), with 10 kg/m^3 of density. In order to the reduce the frictional restriction between the foundation and surrounding soil, a double layer of polyethylene sheets lubricated with mineral oil was placed in the bottom and lateral faces of the foundation.

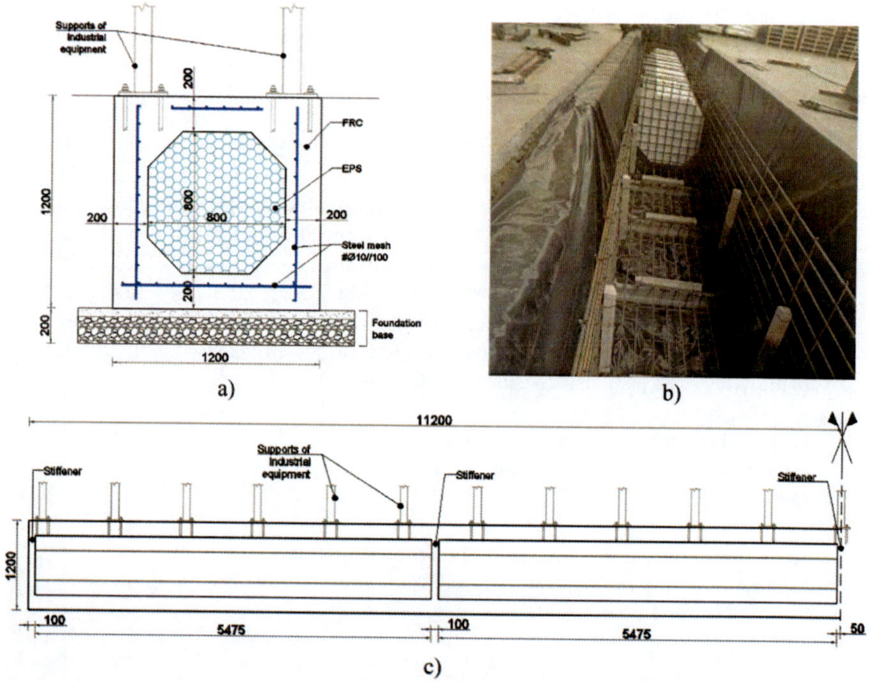

Fig. 1. Geometry of the foundation: a) Cross-section view; b) View of the positioning of the conventional reinforcements and EPS core; c) Longitudinal view of half of the foundation (dimensions in mm).

A hybrid reinforcement for the concrete parts of the foundation was defined at the design stage, formed by conventional steel meshes and discrete polypropylene fibers. The use of fibers aims to reduce the conventional reinforcement ratio, increasing the concrete post-cracking resistance and restraining crack opening, while increasing the durability of the foundation member by only one layer of conventional reinforcement with relatively high concrete cover thickness.

The concrete strength class C30/37 was chosen for the execution of the foundation, while at the design stage the toughness class 3c (classification of the a post-cracking behavior of a fiber reinforced concrete according to *fib* Model Code 2010 [6]) was prescribed, corresponding to a characteristic residual flexural strength for a crack mouth opening displacement (CMOD) equal to 0.5 mm of $f_{R1k} = 3$ MPa and to CMOD = 2.5 mm a value of $f_{R3k} = 2.7$ MPa.

The adopted polypropylene fibers were fabricated by Exporplás – Indústria de Exportação de Plásticos, S.A, referenced as 1EST54, with an equivalent diameter of 0.7 mm, length equal to 54 mm, modulus of elasticity of 6 GPa and ultimate tensile strength of 500 MPa. In order to achieve the toughness class 3c, 6 kg/m^3 of this type of fibers were used for producing the fiber reinforced concrete (FRC).

The adopted conventional reinforcement was formed by Ø10/100 mm squared steel mesh of the strength class A500NR. A minimum cover of 50 mm was adopted in order to minimize the susceptibility of steel bars to corrosion.

Due to the geometry of the structural member, particularly its length, the main cracking risk arises from the tensile strain gradient caused by the heat development generated by the cement hydration in the early stages of the concrete hardening phase and subsequent heat dissipation. The present work provides an integrated analysis, based on experimental and numerical studies, to assess the cracking risk of the foundation during the early ages of the concrete hardening phase.

2 Experimental Study

The construction of the hybrid reinforced concrete (HRC) member included the material characterization of the evolution of the FRC mechanical properties since its early ages. In addition, a monitoring program was conducted to register the evolution of the temperature in some positions of the HRC foundation, during the early stages of the concrete hardening phase.

2.1 Material Characterization

During the execution of the foundation, FRC samples were collected to determine the evolution with time of the compressive strength and post-cracking residual flexural strength.

The compressive strength of the FRC was evaluated on cubic samples of 100 mm edge, according to the standard NP EN 12390-3 [7], at the ages of 1, 4, 21 and 28 days after casting. The post-cracking residual flexural strength was determined from notched prismatic samples, submitted to 3-point bending, according to the standard EN 14651 [8], at the ages of 1, 4 and 21 days after casting.

The evolution of the average values of compressive strength is presented in Fig. 2. The evolution of the compressive strength according to the maturity model presented in Eurocode 2 – Part 1-1 (EC2-1-1) [9] is also presented, which is determined from the following equation:

$$\begin{cases} f_{cm}(t) = \beta_{cc}(t) f_{cm} \\ \beta_{cc}(t) = \exp\left[s\left[1 - \left(\frac{28}{t}\right)^{0.5} \right] \right] \end{cases} \quad (1)$$

where t is time in days, and s is a coefficient that depends on the type of cement (for CEM II 42.5R, $s = 0.2$).

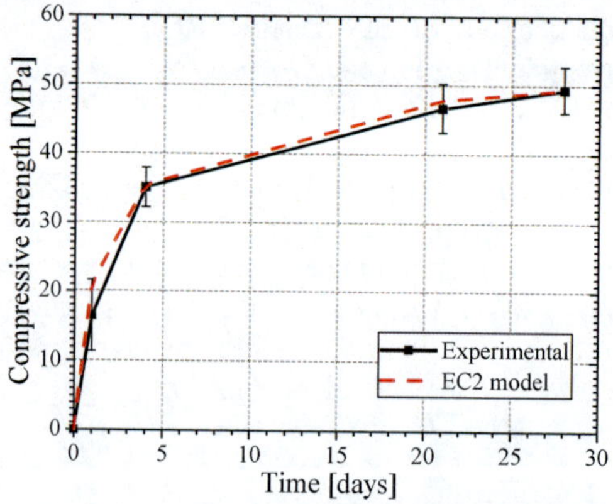

Fig. 2. Evolution of the average compressive strength, including standard deviation of tested samples for each age after casting.

Considering that the ratio between the compressive strength on cubic and cylindric samples is equal to $f_{cm,cubic}/f_{cm,cyl} = 0.8$ and that the characteristic compressive strength is equal to $f_{ck} = f_{cm} - 8MPa$, it is possible to verify that the developed FRC has characteristic compressive strength of 31.5 MPa, therefore it is of strength class C30/37.

The evolution of the average post-cracking residual flexural strength is presented in Fig. 3, namely the average value of the residual flexural strength corresponding to a CMOD equal to 0.5 mm (f_{R1m}), 1.5 mm (f_{R2m}), 2.5 mm (f_{R3m}) and 3.5 mm (f_{R4m}).

The flexural residual strength of the FRC was also estimated for $t = 28$ days, assuming an analogous evolution of FRC flexural residual strength as the one presented by Eq. (1) for the concrete compressive strength. The estimated results for $t = 28$ days are also displayed in Fig. 3.

Considering the ratio between the characteristic and average residual flexural strength equal to $f_{Rik}/f_{Rim} = 0.7$, as proposed in fib Model Code 2010 [6], it is possible to estimate that $f_{R1k} = 2.00$MPa and $f_{R3k} = 2.71$MPa, which revels that the adopted FRC is of toughness class 2e, instead of the 3c, as prescribed in the design stage.

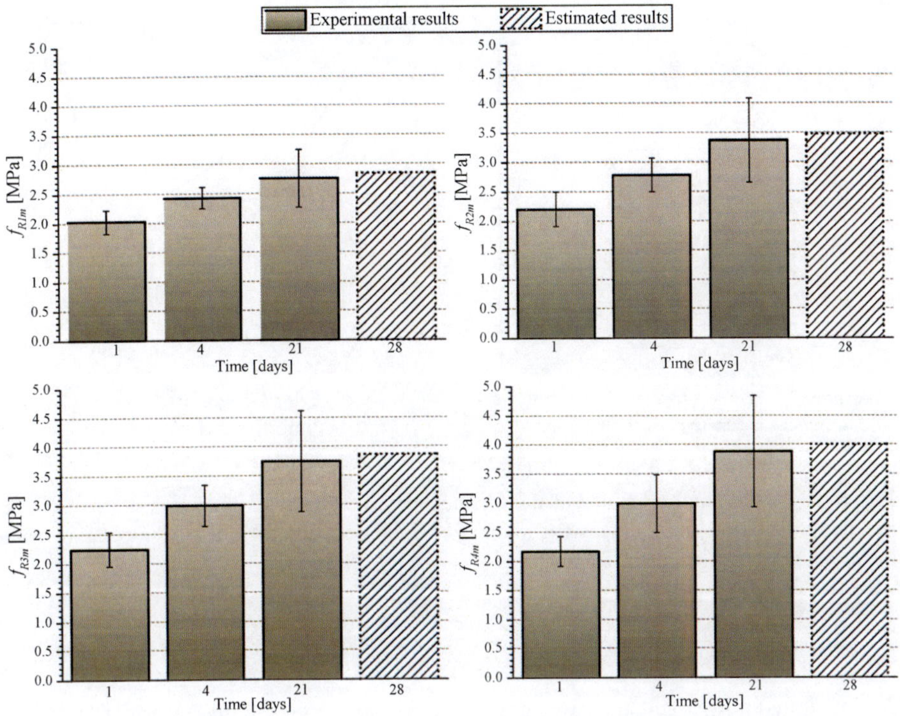

Fig. 3. Evolution of average values of the FRC residual flexural strength, including the standard deviation for each age after casting.

2.2 Measurement of Temperature Evolution in the HRC Foundation

Previously to the FRC casting, thermocouple sensors were applied to capture the temperature evolution inside the HRC foundation, particularly during its initial hardening phase. The thermocouples were placed at 1.5 m from the middle cross-section of the foundation. The position and reference of the thermocouples in the cross-section is presented in Fig. 4a.

Figure 4b presents the evolution of the temperature captured by the thermocouples placed inside the HRC foundation (TC_1 to TC_5) and also the external environment temperature captured by a thermocouple placed in the vicinity of the structural member (TC_6). The temperature was recorded up to 105 h (approximately 4 days) after FRC casting. The peak of temperature registered by the thermocouples was recorded at about 25 h after casting, with a maximum value of 44.1 ºC (TC_1). In this period, the environment temperature ranged between 13.4 ºC and 19.9 ºC.

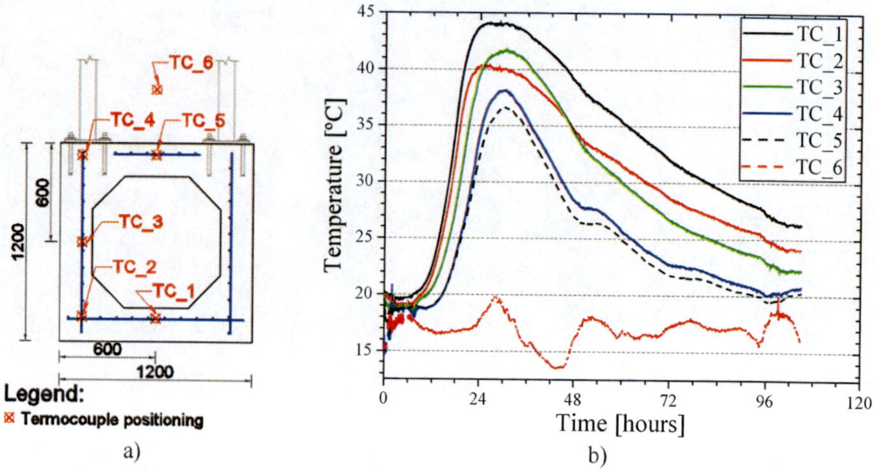

Fig. 4. a) Position and reference of the thermocouples in the cross-section at 1.5 m from the middle of the foundation (dimensions in mm); b) Evolution with time of the temperature inside the prototype and of the environment.

3 Numerical Study

An integrated numerical study was developed for obtaining reliable information to capture the behavior of the HCR element presented in Fig. 1, particularly to assess the cracking risk of FRC element due to the early age heat development.

In this scope, the numerical model parameters that describe the heat development due to the cement hydration of the FRC composition were calibrated based on the data collected from the experimental program.

Based on the results of the 3-point bending tests of the FRC, the stress crack width constitutive law of this FRC was derived by performing an inverse analysis.

The numerical simulation of the HRC foundation is presented considering the numerical models that describe the fracture behavior and heat development due to cement hydration, in order to investigate the behavior of the concrete element under analysis.

3.1 Calibration of Thermal Model Parameters

The numerical model that describes the heat development of concrete due to cement hydration is described in [10], which is based on the mathematical formulation proposed by Reinhardt et al. [11] and on the Arrhenius law.

The unknown main model parameters correspond to the total accumulated heat of the cement hydration, Q_{total}, and to the normalized heat generation rate, $f(\alpha_T)$, that is a function of the degree of heat development. The cement content (kg/m^3), the apparent activation energy, E_a, that is dependent on the cement type, and the Arrhenius rate constant, A_T, are known model parameters: i) the FRC composition adopted 350 kg/m^3 of cement type CEM II/A-L 42.5R; ii) based on the data collected from [12], for this type of cement it can be assumed that $A_T = 52.05 \times 10^6$ and $E_a = 40.31\ kJ/mol$.

In order to derive the unknown model parameters (Q_{total} and $f(\alpha_T)$), a numerical model of the foundation was developed for obtaining the temperature field in the structure and compare the numerical results with the experimental data recorded in the thermocouples installed in the element (Sect. 2.2). A series of nonlinear thermal transient analysis were conducted with different sets of the model parameters to simulate the heat development of the initial 105 h of the cement hydration phase.

The numerical simulations were carried out in the Finite Element Model (FEM) software, FEMIX [13]. Only ¼ of the model was simulated, taking advantage from the double symmetry of the HRC foundation and surrounding soil. The model was formed with by solid finite elements of 20 nodes, namely 6376 to simulate the HRC foundation, 1024 to simulate the EPS core of the foundation, and 3168 to simulate the surrounding soil. It was adopted a Gauss Legendre integration scheme, with 2 × 2 × 2 integration points (IP's). The finite element mesh is presented in Fig. 5.

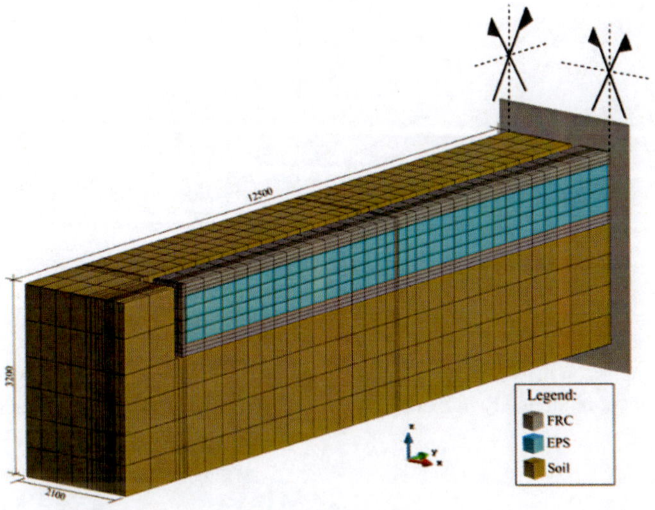

Fig. 5. Finite element mesh of ¼ of the foundation and surrounding soil (dimensions in mm).

The following values for the FRC properties in the thermal analysis were considered: thermal conductivity equal to 2.6 W/m.K; volumetric heat capacity equal to 2400 kJ/m^{3}K. For the environmental temperature it was assumed the hourly average value of the environment temperature registered during the experimental tests.

The value of the heat transfer coefficient assigned to the face in contact with the environment is 10 W/m^{2}K, while for the faces in contact with the plywood formwork (lateral faces of the concrete element) was adopted an heat transfer coefficient equal to 5.0 W/m^{2}K (considering that the plywood thickness is equal to 20 mm and its thermal conductivity is equal to 0.2 W/m.K).

During the transient thermal analysis, it was considered that the concrete element is demolded at 48 h after casting. Afterwards, it was assumed that all faces are exposed to environment and have a heat transfer coefficient equal to 10 W/m^{2}K. During the entire duration of the analysis it is considered a 200 mm gap between the lateral faces of the foundation and the surrounding soil.

For the faces of the elements in the planes of symmetry and on the extremities boundaries of the model it was assumed an adiabatic boundary (heat transfer coefficient equal to 0.0 W/m^2K).

The evolution of the temperature in the nodes of the mesh correspondents to the localization of the thermocouples was compared with the results collected during the experimental tests. As presented in Fig. 6a, a good agreement between the numerical and the experimental results was obtained for $Q_{total} = 400\,kJ/kg$ and for the normalized heat generation rate presented in Fig. 6b.

In Fig. 7 is presented the temperature field in the HRC foundation at $t = 25$ h after casting, which corresponds to the instant that the maximum temperature was registered by the thermocouples in the experimental tests.

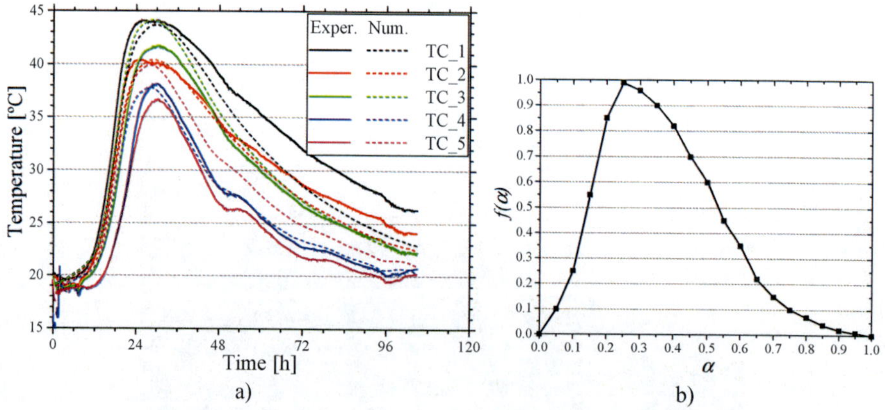

Fig. 6. a) Comparison between temperature registered by the thermocouples in the experimental tests and predicted numerically in their locations; b) Derived normalized heat generation rate.

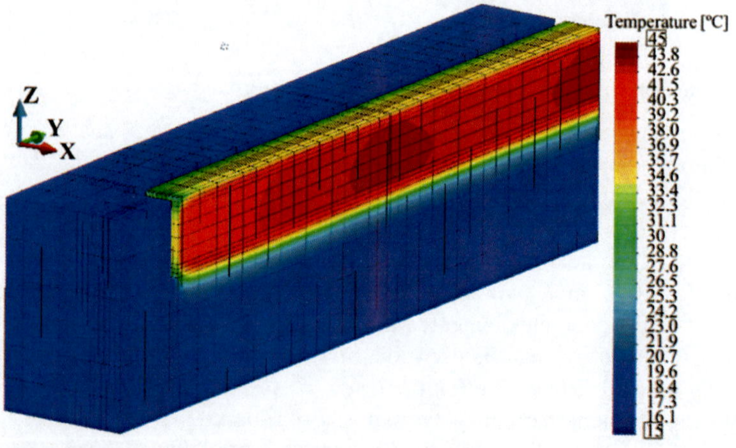

Fig. 7. Temperature field in the HRC foundation at the $t = 25$ h after casting.

3.2 Derivation of FRC Stress-Crack Width Relationship

To simulate the fracture mode I process of the developed FRC, including its evolution since its casting stage, the stress-crack width relationship of the FRC was derived by performing an inverse analysis procedure based on the results obtained in the 3-point bending tests performed experimentally.

The inverse analysis was performed by conducting numerical simulations of the bending tests considering different sets of parameters that define the stress-crack width relationship of the FRC. The numerical simulations were conducted with the inverse analysis software COFIT, whose description can be found elsewhere [14].

The average response of the 3-point bending tests executed at the age of 21 days after casting were used to derive the parameters of a multilinear stress vs. crack width relationship that minimizes the deviation between the experimental and numerical force vs. midspan deflection response in these experimental tests.

The best fit of the stress-crack width relationship of the FRC at the age of 21 days after casting, derived by the inverse analysis procedure, considering a multilinear stress-crack width relationship with 6 branches and a quadrilinear stress-crack width relationship are presented in Fig. 8a, including the FRC fracture energy (G_f). In Fig. 8b is presented the comparison between the average experimental and numerical force vs. midspan deflection of the FRC prisms submitted to 3-point bending at the age of 21 days after casting.

From the inverse analysis procedure, it was also possible to obtain the modulus of elasticity of concrete that promotes the best fit between the numerical and experimental results, namely $E = 35GPa$.

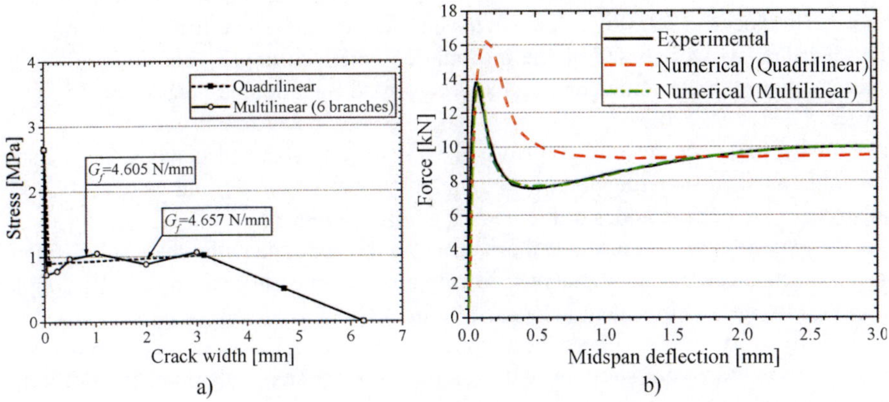

Fig. 8. a) Derived stress vs. crack width relationship of the FRC at the age of 21 days after casting; b) Comparison between the experimental and numerical force vs. midspan deflection of the FRC prisms submitted to 3-point bending at the age of 21 days after casting.

3.3 Assessment of Cracking Risk of the Structure Since Early Ages

To assess the cracking risk and magnitude in the early ages of the curing phase of the HRC foundation, the temperature field obtained from the thermal model is coupled with a mechanical model that can simulate the crack initiation and propagation of concrete members. A multidirectional fixed smeared crack model [15] was adopted to simulate the fracture properties of the FRC, considering the quadrilinear stress-crack width relationship derived from the inverse analysis of the 3-point bending tests of FRC prisms. To ensure that the results are independent of the finite element mesh refinement, a crack bandwidth equal to the cubic root of the integration point's volume was considered.

For simulating the crack shear stress transfer, the concept of shear retention [10] was adopted, which decreases from 1 to zero with the increase of the normal crack tensile strain according to a quadratic degradation parameter.

Considering the relatively small load level that the foundation is submitted, a linear-elastic regime was assumed for the compressive behavior of the FRC.

Surface interface finite elements were disposed between the bottom face of the foundation and the supporting soil in order to simulate their contact conditions. A total of 407 zero-thickness interface finite elements formed by 16-node quadratic elements were added to the finite element mesh presented in Fig. 5 (only in the mechanical model). A Gauss-Legendre integration-scheme with 2×2 IP's was adopted for the interface elements.

The constitutive model adopted for the interface finite elements is based on the work of [16]. The model considers a linear relationship between the normal compressive stresses and vertical displacement, with a stiffness of $k_n = 50MPa/m$, while assuming that normal tensile stresses cannot be transferred between the faces in contact simulated by the interface finite elements. In addition, the model admits a nonlinear response between the shear stresses and sliding of the interface ($\tau - s$). The values of the parameters defining this law were based on the information collected in [17], considering that the interface between the foundation and soil is formed by a double layer of polyethylene sheets.

In order to reduce the time to perform the transient mechanical analysis, the elements to simulate the surrounding soil of the foundation were removed and were replaced by supports in the bottom nodes of the interface finite elements.

The evolution of the mechanical properties of the concrete with time was considered in the mechanical transient analysis, by adopting a concrete maturity model coupled with the concept of equivalent age proposed in [18].

The maturity model adopts the recommendations of Eurocode 2 to simulate the evolution of the tensile strength and modulus of elasticity with time. In addition, it adopts the relationship proposed by [10] that expresses the evolution with time of the fracture energy of the concrete matrix. The maturity model evaluates the tensile strength, modulus elasticity and fracture energy at a specified time after casting as a function of the value of these properties at 28 days of age, and according to the employed type of cement.

The values of the mechanical properties at 28 days were estimated from the inverse analysis procedure of the results of the tests performed at 21 days after casting, and are presented in Table 1.

Table 1. Estimated values of FRC mechanical properties for t = 28 days.

f_{ct}[MPa]	G_f [N/mm]	E[GPa]
2.735	3.640	35.326

The model described in [18] was adopted for the simulation of the concrete aging creep behavior. Due to the inexistence of creep compliance curves for the concrete mixture under study, the B4 model is adopted to predict the creep behavior of the concrete for three loading ages $t_0 = \{1, 3, 5\}$ days, based on concrete strength performance. The creep compliance curves are then used to estimate the creep response of concrete for any loading age, considering 7 Kelvin chains with the retardation times $\tau = \{0.001; 0.01; 0.1; 1; 10; 100; 1000\}$ days and Young's modulus $E_0(t_0 = 1\ \text{day}) =$ 27 GPa, $E_0(t_0 = 3\ \text{days}) = 31$ GPa and $E_0(t_0 = 5\ \text{days}) = 32$ GPa.

In addition, the autogenous and drying shrinkage deformation of the FRC foundation were simulated based on the model proposed by EC2–1-1 that is available in FEMIX. It was considered that the drying shrinkage starts at $t_d = 1$ day.

In addition to the temperature field obtained from the nonlinear transient thermal analysis, the self-weight of the foundation materials was also considered in the mechanical analysis.

During the transient mechanical analysis, it was possible to detect the crack initiation and propagation in the concrete foundation. The maximum computed crack width (evaluated as the product of the maximum normal crack strain with the crack bandwidth) was about 12.9 μm, near the middle section of the foundation, for the time instant $t = 23$ h after casting. The maximum crack width is well below the maximum allowable value (0.3 mm) indicated in EC2–1-1 for the serviceability limit state safety verifications for reinforced concrete structures, and do not constitute a threat to the durability of the HRC foundation.

The crack pattern of the foundation at $t = 23$ h after casting is presented in Fig. 9, where it is noticeable that the cracking arises in the outer elements of the HRC foundation, as result of the stress gradient created due to thermal exchanges of heat from the FRC elements to the surrounding air.

In Fig. 10 is presented the evolution with time of the temperature and crack width in the element of the mesh that registered the maximum crack width during the analysis. Indeed, a peak of crack width opening was observed at $t = 23$ h, which follows the evolution with time of temperature due to cement hydration. Afterwards, with the heat dissipation and concrete volume contraction, the crack width reduces and becomes completely closed at the end of the analysis.

At the end of the mechanical transient analysis ($t = 105$ h after casting) all the micro-cracks that were formed in the early stage of the concrete hardening phase end up to close, which agrees with the observation that no cracks were visible in the HRC foundation.

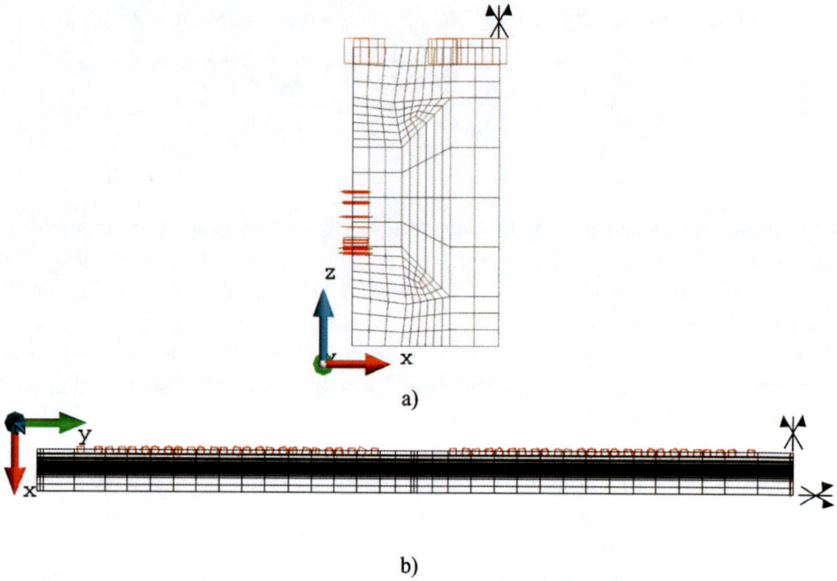

Fig. 9. Crack pattern at t = 23 h after casting: a) XZ view of the HRC foundation (from the left extremity, y = 0); b) XY (plant) view of the HRC foundation. Legend: red – opening (only cracks of width higher than 0.05mm are displayed).

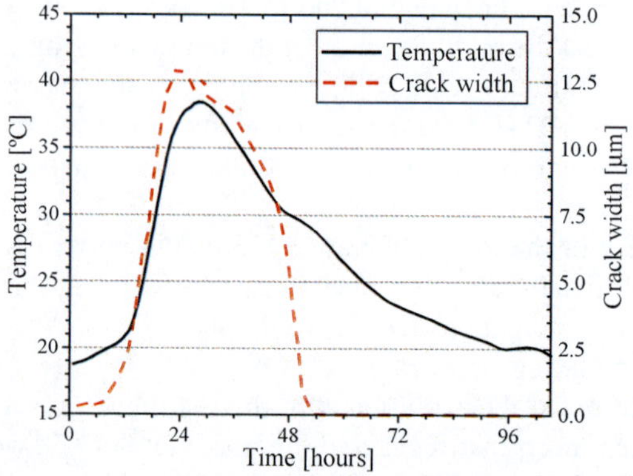

Fig. 10. Evolution of crack width time and temperature at element with maximum crack width during mechanical analysis.

4 Conclusions

In this work is presented the analysis of the cracking risk and magnitude at the early stages of the hardening phase a long jointless concrete foundation hybrid reinforced with conventional steel meshes and discrete polypropylene fibers.

A thermo-mechanical nonlinear transient analysis of the foundation was performed adopting constitutive models to simulate the cracking, viscoelasticity, shrinkage and maturity behavior of the FRC, and a model to simulate the heat development of concrete due to the cement hydration.

The strength parameters adopted in the mechanical models were derived from the results of the experimental tests. Emphasis was made on the derivation of the stress-crack width relationship of the FRC, based on an inverse analysis technique considering the results of 3-point notched beam bending tests.

In order to supply an accurate prediction of the temperature field generated by the cement hydration reactions during the early stages of concrete hardening phase, the parameters of the thermal model were calibrated based on the results collected by thermocouples installed in the concrete foundation.

The numerical simulation revealed that the hybrid reinforcement solution considered for the long jointless concrete foundation exhibited an adequate performance to limit the crack width magnitude to very small openings (less than 15 μm), while significatively reducing the conventional steel reinforcements ratio. Indeed, after the heat dissipation, these cracks have closed due to the contraction of the concrete volume. The obtained numerical results are in agreement with the visual inspection of the HRC foundation where no cracks were detected.

Acknowledgements. The authors acknowledge the support provided by the NG_TPfib –New generation of fibers for the reinforcement of cement-based materials, supported by ANI (FEDER through the Operational Program for competitiveness and internationalization (POCI)), as well as FemWebAI project (PTDC/ECI-EST/6300/2020). The collaboration of Exporplas company is also acknowledge.

References

1. Domingo, A., Lázaro, C., Serna, P.: Construction of JCHYPAR, a steel fiber reinforced concrete thin shell structure. Varenna, Italy, p. 10, September 2004
2. Dinh, H.H.: Shear behavior of steel fiber reinforced concrete beams without stirrup reinforcement. PhD Thesis, University of Michigan, Michigan, USA (2009)
3. Conforti, A., Minelli, F., Plizzari, G.A.: Wide-shallow beams with and without steel fibres: a peculiar behaviour in shear and flexure. Compos. Part B Eng. **51**, 282–290 (2013). https://doi.org/10.1016/j.compositesb.2013.03.033
4. Barros, J.A.O.: Comportamento do betão reforçado com fibras - análise experimental e simulação numérica/Behavior of fiber reinforced concrete - experimental and numerical analysis. PhD Thesis, Department of Civil Engineering, FEUP, Portugal (1995)
5. Barros, J.A.O., Sena-Cruz, J.: Fracture energy of steel fiber-reinforced concrete. Mech. Compos. Mater. Struct. **8**(1), 29–45 (2001)
6. fib-federation internationale du beton: fib Model Code for Concrete Structures 2010. Wiley, Hoboken (2013)
7. European Committee for Standardization: Testing hardened concrete - Parte 3: Compressive strength of test specimens, vol. EN 12390-3 (2009)
8. European Committee for Standardization: Test method for metallic fibered concrete - Measuring the flexural tensile strength (limit of proportionality (LOP), residual), vol. EN 14651 (2005)
9. European Committee for Standardization: Eurocode 2: Design of concrete structures Part 1-1: General rules and rules for buildings, vol. NP EN 1992-1-1 (2010)

10. Ventura-Gouveia, A.: Constitutive models for the material nonlinear analysis of concrete structures including time dependent effects. PhD Thesis, Department of Civil Engineering, University of Minho (2011)
11. Reinhardt, H.W., Blaauwendraad, J., Jongedijk, J.: Temperature development in concrete structures taking account of state dependent properties (1982)
12. Azenha, M.Â.D.: Numerical simulation of the structural behaviour of concrete since its early ages, PhD Thesis, aculdade de Engenharia. Universidade do Porto, School of Enegineering. University of Tokyo (2012)
13. Azevedo, A.F.M., Barros, J.A.O., Sena-Cruz, J., Gouveia, A.V.: Femix 4.0. Porto: Consoft (2013)
14. Matos, L.M., Barros J.A.O., Ventura-Gouveia, A., Calçada, R.A.B.: A new inverse analysis approach for predicting the fracture mode I parameters of FRC from three point notched beam bending tests and round panel tests. Technical report 20-DEC/E-40 (2020)
15. Gouveia, A.V., Barros, J.A., Azevedo, Á.F., Sena-Cruz, J.: Multi-fixed smeared 3D crack model to simulate the behavior of fiber reinforced concrete structures, presented at the CCC 2008 - Challenges for Civil Construction, Porto (2008). http://hdl.handle.net/1822/12802
16. Valente, T.: Advanced tools for design and analysis of fiber reinforced concrete structures. PhD Thesis, University of Minho, Guimarães, Portugal (2019)
17. Chia, W.S., McCullough, B.F., Burns, N.H.: Field evaluation of subbase friction characteristics. Center of Transportation Research Bureau of Engineering Research the University of Texas at Austin, Research Report 401-5 (1986). Accessed 24 May 2016. http://library.ctr.utexas.edu/digitized/texasarchive/phase2/401-5-CTR.pdf
18. Ventura-Gouveia, A., Barros, J.A.O., Azevedo, A.F.M.: Thermo-mechanical model for the material nonlinear analysis of cement based materials, presented at the 9th international conference on fracture mechanics of concrete and concrete structures, Berkeley, California USA (2016)

Multi-physical Simulation of Concrete Hydraulic Facilities Affected by Alkali-Aggregate Reaction: From Material to Structure

Mahdi Ben Ftima[1](✉), Anthony Chéruel[1], and Matthieu Argouges[2]

[1] École Polytechnique de Montréal, Montréal (Qc) H3C 3A7, Canada
mahdi.ben-ftima@polymtl.ca
[2] Hydro-Québec, Montréal (Qc) H2Z 1A4, Canada

Abstract. This work aims to fill the gap between experimentation on laboratory specimens and structural diagnosis of mass concrete hydraulic structures affected by alkali-aggregate reaction (AAR) by conducting numerical and experimental investigations on an existing hydroelectric facility affected by AAR.

A large experimental program was performed to characterize the mechanical properties and the kinetics of the AAR chemical reaction for the original mass concrete used in the construction of the facility (maximum aggregate size of 76 mm). The importance of considering mechanical and chemical size effects is discussed, on the basis of the asymptotic fracture energy and the free AAR expansion curve.

The developed phenomenological hygro-chemo-mechanical approach for AAR modelling in mass concrete hydraulic structures is presented. The efficient and yet simple approach is based on three analyzes: transient thermal analysis, transient hygral analysis, and final multi-physical analysis that includes mechanical loading. The modelling approach was validated with existing benchmarks from the literature and provided very promising results, despite the simplifications made and the assumptions for some uncertain input parameters.

Application of the numerical modelling approach to the existing hydraulic facility demonstrated its feasibility in an industrial context. It also provided fairly similar damage pattern if compared to the existing cracking pattern and improved the understanding of the complex structural behaviour of the facility. Comparison of displacement model predictions and available monitoring data allowed to assess an important size effect between laboratory and in-situ expansions.

Keywords: Alkali aggregate reaction · Concrete hydroelectric facility · Assessment · Size effects · Multi-physical simulation

1 Introduction

Hydroelectric facilities have been essential structures for society for centuries. In the case of Quebec, these major structures are crucial to produce 97% of the province total electricity. Generally designed to last 50 to 60 years, most of these structures are

P. Rossi and J.-L. Tailhan (Eds.): SSCS 2022, RILEM Bookseries 38, pp. 47–60, 2023.
https://doi.org/10.1007/978-3-031-07746-3_5

approaching the end of their design lives. In Canada, for example, 51% of large dams are over 50 years old [1]. Concrete hydroelectric facilities (Fig. 1a) are categorized as strategic infrastructures with very high replacement cost. Due to the high volume of concrete used in construction, extension of their lifespan through planned and well-targeted repair technologies would contribute to reduce carbon dioxide emission for the next decades.

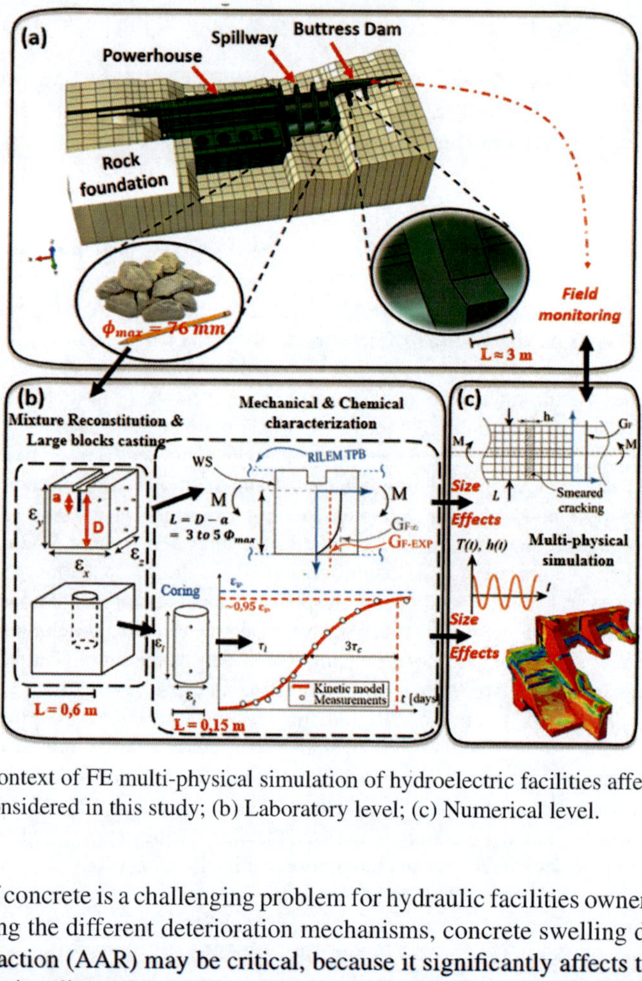

Fig. 1. The context of FE multi-physical simulation of hydroelectric facilities affected by AAR: (a) Facility considered in this study; (b) Laboratory level; (c) Numerical level.

Aging of concrete is a challenging problem for hydraulic facilities owners around the world. Among the different deterioration mechanisms, concrete swelling due to alkali-aggregate reaction (AAR) may be critical, because it significantly affects the structural stability, functionality and durability of the structure. An effective management of hydro-electric facilities affected by AAR requires the use of finite element (FE) technology with dedicated AAR concrete constitutive models. Those are used in combination with comprehensive in-situ monitoring and material testing programs to obtain data for their calibration.

The key objective of AAR FE models is to assist in predicting the future evolution of deterioration related to AAR, that can lead to unacceptable states of functionality, strength and durability. However, a number of key challenges must be addressed before achieving this objective: (1) complexity of the AAR phenomenon, mainly due to its multi-physical nature; (2) difficulty in conducting characterization tests for existing structures. For the specific case of hydraulic structures (Fig. 1a), mass concrete mixtures with coarse aggregate sizes (typically, maximum aggregate size Φ_{max} ranging from 40 to 150 mm) are used, and very large concrete members (typically, nominal member size L is more than 3 m), lightly reinforced or non-reinforced are common, because of stability considerations. Important size effects arise, and results of experimental characterization tests can not be directly used in the FE simulations models (from Fig. 1b – laboratory level to Fig. 1c – full-scale multi-physical simulation).

This work aims to fill the gap between experimentation on laboratory specimens and structural diagnosis of mass concrete hydraulic structures affected by alkali-aggregate reaction (AAR) by conducting numerical and experimental investigations on an existing hydroelectric facility affected by AAR. The existing facility of Fig. 1a is considered and the overall methodology from experimental characterization campaign of the material (Sect. 2), development of phenomenological numerical approach (Sect. 3), to application at the hydroelectric facility level (Sect. 4) is presented.

2 Experimental Characterization Program

A large experimental campaign was considered to assess the main input parameters required for the multi-physical simulation of the existing facility, mainly, the fracture energy of concrete altered and un-altered by AAR effects and the kinetic parameters of the AAR free expansion (Fig. 1b). A representative concrete mixture with large aggregates was reconstructed (ϕ_{max} of 76 mm) for the spillway and buttress dam as shown in Fig. 1a to reproduce as much as possible the mechanical and chemical characteristics of the original concrete used in construction. Since the large aggregates used for the original concrete came from on-site crushed excavated rock, it was possible to recover from the site of the facility pieces of rock to produce aggregates of the mixtures. The mixture proportions and assessed compressive strength are provided in the following table.

2.1 Size Effects

Due to the multi-physical nature of AAR, different types of size effects make it difficult to directly use the results of experimental characterization tests as material inputs into the multi-physical simulation (Figs. 1b and 1c). They are categorized in this study as *mechanical* and *chemical* size effects. Mechanical size effects affecting the mechanical concrete properties such as compressive strength f_c', tensile strength f_t' of concrete are the best known in the literature. For the specific field of mass concrete hydraulic structures, the fracture energy of concrete, noted G_F, is perhaps the most important input parameter due to large dimensions of the members and to the use of little or no reinforcement [2]. Unlike f_c' and f_t', the size-dependency of G_F is increasing with respect to the size L of the tested concrete ligament as clearly shown by the experimental evidences of wedge splitting tests or WS ([3–5]), carried on mass concrete mixtures. This difference in size effect behavior between G_F and f_c'/f_t' is important according to authors opinion.

It intuitively predicates that the source of the size effect for fracture energy is different. According to authors opinion, the size effect in G_F is rather a border effect that shows up in experimental setups involving small ligaments with low values of ligament length to maximum aggregate ratio ($\frac{D-a}{\Phi_{max}}$), where D is the depth of the WS specimen and a is the initial notch depth (Fig. 1b). Asymptotic fracture energy noted $G_{F\infty}$ also known as size independent fracture energy is presented in this study as an intrinsic concrete material parameter that is underestimated by current experimental investigation methods due to boundary effects [6]. The asymptotic fracture energy is reached for ligament ratios $\frac{D-a}{\Phi_{max}} \geq 6\,to\,10$ which are much larger than minimum ligament ratios $\frac{D-a}{\Phi_{max}} \geq 3\,to\,5$ suggested in literature and actually related to the homogeneity of the uncracked ligament (e.g. [7]). In fact, the experimental fracture energy assessed using the work of fracture method (noted G_{F-EXP} in Fig. 1b) can be viewed as the average energy obtained by dividing the total fracture work by the projected cracked area and $G_{F\infty}$ can be viewed as the value of the local fracture energy far from the back boundary [2]. It is demonstrated in [2], that while the underestimation of fracture energy is acceptable for nonlinear FE using smeared crack models for normal reinforced concrete structures, it can be too conservative in the case of mass concrete structures. $G_{F\infty}$ is therefore the input parameter that shall be used for nonlinear FE simulation of the hydroelectric facility.

Chemical size effects refer in this study to the effects of specimen size (L in Fig. 1b) on the parameters of the kinetic model that characterizes the AAR free expansion, namely: the asymptotic volumetric expansion ε_v, characteristic and latency times: τ_c and τ_l. Two main interpretations have been mentioned in the literature: alkali leaching and ASR-gel permeation [8–10]. Alkali leaching and ASR-gel permeation from specimens during the free expansion test may lead to reduced expansion and sometimes to misleading conclusions about the reactivity of the aggregates. It is believed that permeation and leaching mechanisms are negligible at the level of hydroelectric facilities due to large member sizes and therefore, kinetic parameters without size effects noted:$\varepsilon_{v\infty}$, $\tau_{c\infty}$ and $\tau_{l\infty}$ are required for the FE numerical analysis. These properties can be viewed as free expansion AAR parameters assessed from an infinitely large specimen.

2.2 Asymptotic Fracture Energy

The simplified boundary effect method (SBEM) [11] and the recently developed disturbed fracture process zone theory (DFPZ) [12], were used to assess the asymptotic fracture energy. Both experimental methods are based on the Boundary effect theory, assuming a bi-linear distribution of the local fracture energy [6]. The SBEM method uses two configurations of WS tests with two different notch over depth *a/D* ratios, whereas the DFPZ uses a single WS with the lowest *a/D* ratio and digital image correlation monitoring during the test. As shown in Figs. 2d, e and f, large concrete blocks were cast from the reconstructed concrete mixture. The blocks (18 in total) were stored in environmental chamber with controlled temperature and humidity conditions, to accelerate the AAR and were consequently tested using WS test. The notches were performed with saw cutting before the test, considering two notch over depth ratios: $a/D = 0.1$ and $a/D = 0.5$. Some of the specimens were equipped with inserts to monitor the free expansion along the 3 directions as shown in Fig. 1b.

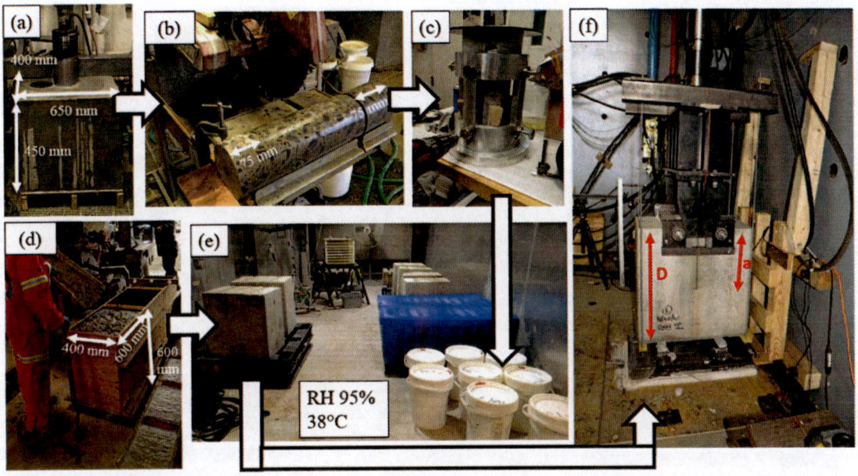

Fig. 2. Experimental characterization campaign: (a) coring of large blocks; (b) core end cutting; (c) drilling for contact studs installation; (d) large WS specimens casting; (e) storage in environmental chamber; (f) WS test of large block.

2.3 AAR Free Expansion

Simultaneously to WS specimens, large concrete blocks from the same mixture (400 mm × 650 mm × 400 mm) were cast as shown in Fig. 2a. Cylindrical cores (Ø150 mm) were drilled from these blocks as shown in Fig. 2a (24 in total). A 75 mm long piece was cut from each end of the cylinder to obtain final specimens of ⌀150mm × 300 mm length. Each cylinder was equipped with 8 studs (Fig. 2c) to monitor the free AAR expansion, and was kept in an environmental chamber at 38 °C, in watertight plastic containers as shown in Fig. 2e. It was therefore possible to compare the volumetric free expansions between the WS large blocks ($\varepsilon_v = \varepsilon_x + \varepsilon_y + \varepsilon_z$) and the relatively smaller cylinder cores ($\varepsilon_v = \varepsilon_l + 2\varepsilon_t$) (see Fig. 1b), to assess the chemical size effects.

2.4 Results

Figure 3 presents some of the experimental results highlighting both mechanical and chemical size effects. Free AAR volumetric expansions results are presented for the three-year monitoring period. Fracture energy results are presented for 6 of the 18 WS blocks (Sect. 2.2) that were tested at the beginning of the campaign, to focus on the purely mechanical size effects (so not altered by AAR). More details about this initial campaign can be found in [2]. Ongoing work on WS tests of specimens altered by AAR demonstrate that fracture energy tends to slightly increase and then remain constant with the progress of AAR.

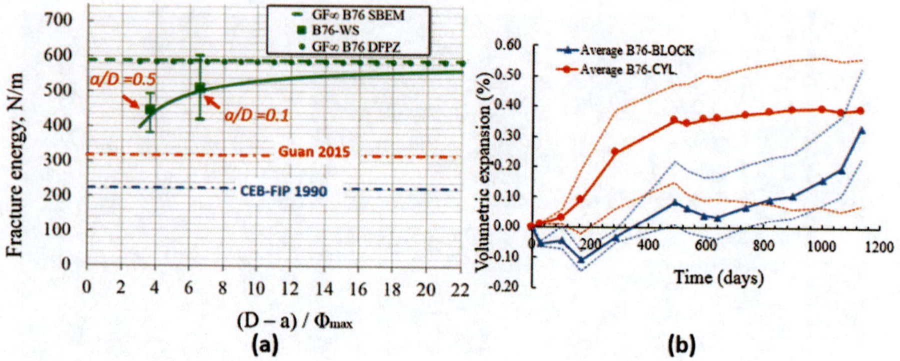

Fig. 3. Results of experimental campaign: (a) mechanical size effects on the assessed fracture energy; (b) chemical size effects on the free volumetric expansion.

Figure 3a highlights the mechanical size effects: an increase of 15% in the assessed experimental fracture energy with an 80% increase in the tested ligament length. It also demonstrates that asymptotic fracture energies assessed by DFPZ and SBEM methods were relatively close, around 590 N/m. It is important to note that this value is 2.6 times higher than the value computed from CEB-FIP 90 empirical equation for structural concrete (using linear extrapolation to $\Phi_{max} = 76$ mm), and 1.8 times higher than the value assessed using the recently developed empirical equation for mass concrete in [4].

Figure 3b highlights the chemical size effects. Slower kinetics are observed for the large blocks, compared to the cylinders. It was not possible to attain an asymptotic volumetric expansion for the blocks and AAR is still going on. This result is certainly dependent on the reactivity of the aggregates. It suggests that for the sake of feasibility within an industrial context, chemical size effects would better be considered at the numerical level whether by using a multi-scale analysis (e.g. Multon and Sellier 2015) or through calibration using field monitoring, as it is suggested in this study (Fig. 1).

Table 2 provides the assessed mechanical and AAR chemical parameters for the reconstructed mass concrete mixture. The AAR parameters were identified based on the cylinder average free expansion curve of Fig. 3b.

Table 1. Mixture proportions and characterization.

	W/C	Cement (kg/m^3)	Type of cement	Water (kg/m^3)	Sand (kg/m^3)	6-20 mm (kg/m^3)	20-40 mm (kg/m^3)	40-76 mm (kg/m^3)	Na2Oeq (kg/m^3)	f'c (MPa)
B76	0.57	224	Low Heat	128	691	440	293	612	3.0	24.0

Table 2. Assessed parameters for the mass concrete mixture (B76).

	Parameter	Value	Unit
AAR parameters	ε_v	0.39	%
	τ_l	250	days
	τ_c	140	days
Mechanical parameters	f_c'	24	MPa
	f_t'	2.0	MPa
	ν	0.18	-
	E_0	25 000	MPa
	G_F	0.5	N/m
Hygral parameters	k	10^{-12}	m/s
	Sr_0	0.85	-
	ϱ_c	0.12	-
	Saturation dependency	Correy's law	-

3 Constitutive Modelling

3.1 Computational Framework

A computational framework was developed for multi-physical simulation of concrete structures involving three different finite element analyses using sequential coupling [13, 14]. For more details about the constitutive modelling, the reader is referred to [14].

In a first step, implicit transient hygral and thermal analyses are performed to compute the variations over time of the degree of saturation field $Sr(t)$ and the temperature field $T(t)$. These fields are then imported into a non-linear quasi-static explicit finite element analysis involving a constitutive concrete model, in addition to the mechanical loads (e.g. hydrostatic and gravity loads) (Fig. 4a). The developed framework showed important advantages over the conventionally developed multi-physical frameworks such as: (i) the ability to handle large industrial models with millions degrees of freedom; (ii) highly non-linear behavior such as mass concrete cracking (without reinforcement) and frictional contact between structural components of the facility; (iii) the ability to consider early age effects due to the very small increment time required by the stability of the explicit algorithm, and (iv) because strain decomposition was used, the framework could be implemented with any existing constitutive concrete model that has undergone a rigorous verification and validation (V&V) process for the mechanical part.

As schematically shown in Fig. 4b, all the three-transport phenomenon: diffusion, adsorption and permeability can simultaneously occur in normal concrete structures. It is believed however, that for mass concrete structures, the adsorption and permeability are the governing mechanisms, since diffusion process occurs within a negligible thickness of the mass structure. The core of the dam can be either partially or fully saturated. For the fully saturated portion in the upstream face, the pressure gradient is the driving transport mechanism using Darcy's law. Darcy's law is extended to the case of partially saturated medium using the concepts of permeability's dependency on the degree of saturation and capillary curve as shown in Fig. 4c. The permeability depends on the saturation of the liquid water, assuming the following relation:

$$\hat{k}(S_r) = k_s(S_r).k \tag{1}$$

where k_s gives the saturation dependency $k_s(S_r = 1) = 1$ and k is the fully saturated permeability of concrete to liquid water. Several relations have been suggested in the literature for partially saturated mediums (mainly developed for soils). Correy's law is used in this study [14]. Capillary curve is obtained by combining desorption isotherms assessed experimentally (relating concrete water content to relative humidity h), to Kelvin equation (relating capillary pressure to the relative humidity h). Semi-empirical equations suggested by Baroghel et al. [22] are used for the desorption isotherm:

$$p_c(S_r) = a\left(S_r^{-b} - 1\right)^{1-\frac{1}{b}} \tag{2}$$

The multi-physical framework is based on the hypothesis of decomposition of the total strain increment $\boldsymbol{\Delta\epsilon}$ into mechanical $\boldsymbol{\Delta\epsilon_{mec}}$, thermal $\boldsymbol{\Delta\epsilon_{th}}$, creep $\boldsymbol{\Delta\epsilon_{cr}}$, shrinkage $\boldsymbol{\Delta\epsilon_{sh}}$ and AAR $\boldsymbol{\Delta\epsilon_{aar}}$ strains, according to the following equation in the incremental form:

$$\boldsymbol{\Delta\epsilon} = \boldsymbol{\Delta\epsilon_{mec}}\left(t, \xi, T, Sr, \boldsymbol{\sigma}_{0,}\right) + \boldsymbol{\Delta\epsilon_{th}}(t, T) + \boldsymbol{\Delta\epsilon_{cr}}\left(t, \boldsymbol{\sigma}_{0,}T\right) + \boldsymbol{\Delta\epsilon_{sh}}(t, Sr) + \boldsymbol{\Delta\epsilon_{aar}}(t, \xi, T, Sr, \boldsymbol{\sigma}_0) \tag{3}$$

This equation shows the coupling between the four fields considered in this framework: the stress field tensor $\boldsymbol{\sigma}_0$, the temperature field T, the degree of saturation field Sr and the AAR advancement field ξ. The subscript 0 in $\boldsymbol{\sigma}_0$ refers to the previous equilibrium configuration in the previous increment.

Larive model [20] is used for the kinetics of AAR advancement field $\xi(t)$. According to this model:

$$\xi(t) = \frac{1 - e^{\frac{-t}{\tau_c}}}{1 + e^{-\frac{t-\tau_L}{\tau_c}}} \tag{4}$$

where τ_l and τ_c are respectively the latency and characteristic times that depend on local temperature, stress and moisture conditions. Using the same notations as Saouma and Perotti [15], a *modified expression* of the incremental AAR strain model of [15] is presented in this study:

$$\Delta\varepsilon_{aar_i}(t, \xi, T, Sr, \sigma_0) = \Gamma_t(COD_{max}) * \Gamma_c(\sigma_0) * W_i(\sigma_0) *$$

$$\Delta\xi(t, T, Sr, \sigma_0) * \varepsilon_v(Sr) \tag{5}$$

where i is a given stress principal direction, Γ_t and Γ_c are retardation factors that depend on the tensile, compressive damage and stress field of the material point in the previous increment. $W_i(\sigma_0)$ is the weighting for redistribution of volumetric AAR expansion in the principal direction i, depending on the stress field σ_0 as described in [15], and thus, allows to consider anisotropic expansion of AAR. ε_v is the asymptotic volumetric AAR strain.

Once all non-mechanical strain increments are computed, the mechanical strain increment $\boldsymbol{\Delta\epsilon_{mec}}$ can be deduced from Eq. 3, because the total strain increment $\boldsymbol{\Delta\epsilon}$ is provided by FE software at the given integration point [13]. Any existing mechanical constitutive law (even linear elastic) can then be used to compute the stress increment $\boldsymbol{\Delta\sigma}$

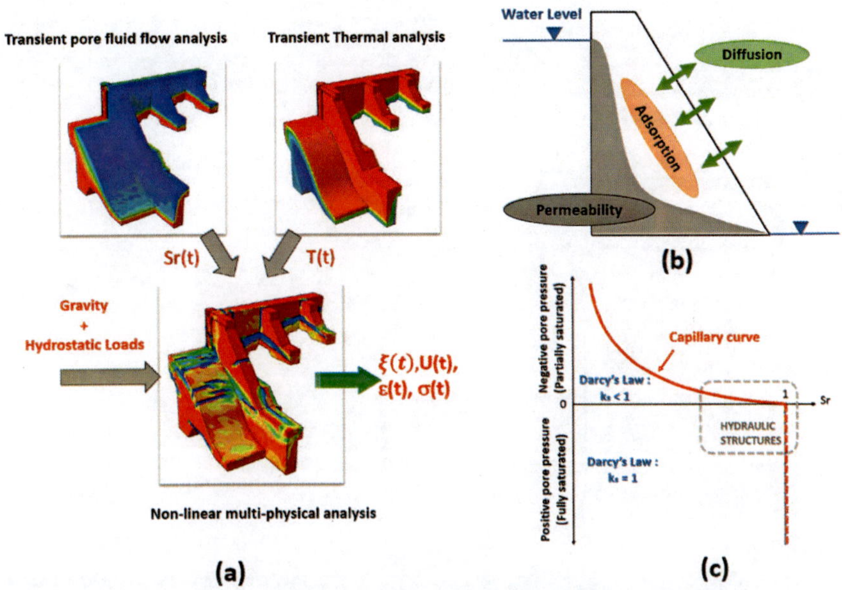

Fig. 4. Computational framework: (a) multi-physical analyses; (b) water transport mechanisms in hydraulic structures (c) pore pressure-degree of saturation relation;

and update the stress ($\boldsymbol{\sigma}_0 \leftarrow \boldsymbol{\sigma}_0 + \Delta\boldsymbol{\sigma}$). In this work, the non-linear concrete constitutive model EPM3D [16] programmed as a user-subroutine VUMAT in ABAQUS, is used.

Based on Fig. 3b and previous literature results (e.g. [8]), it can be said that free expansion tests carried on small laboratory specimens can lead to an under-estimation of the asymptotic swelling and an under-estimation of the characteristic time. In other words, AAR expansion in the mass structure shall be slower and reach a final higher value. In the numerical modelling of this work, scale factors (α_{τ_c} and α_{ε_v}) are suggested for the characteristic time and AAR asymptotic expansion using the following relations:

$$\tau_{c\infty}(S_r, T) = \alpha_{\tau_c}.\tau_c(S_r, T) \tag{6.a}$$

$$\varepsilon_{v\infty}(S_r, T) = \alpha_{\varepsilon_v}.\varepsilon_v(S_r, T) \tag{6.b}$$

where $\tau_{c\infty}$ and $\varepsilon_{v\infty}$ are respectively latency and AAR asymptotic expansion for an infinitely large specimen (or with negligible alkali leaching/gel permeation). In this work, scale factors (α_{τ_c} and α_{ε_v}) are calibrated using available displacement monitoring data at the facility level, provided that similar concrete alkali content is maintained between laboratory specimen and existing facility [14]. As explained in [14], latency time (τ_l) does not seem to be affected by size effects. Latency is related to the time for the gel to fill the available porous volume around reactive aggregates, at the beginning of the reaction. It is therefore indirectly related to gel expansion and is assumed unaffected by size effects in this study, therefore: $\tau_{L\infty} = \tau_L$.

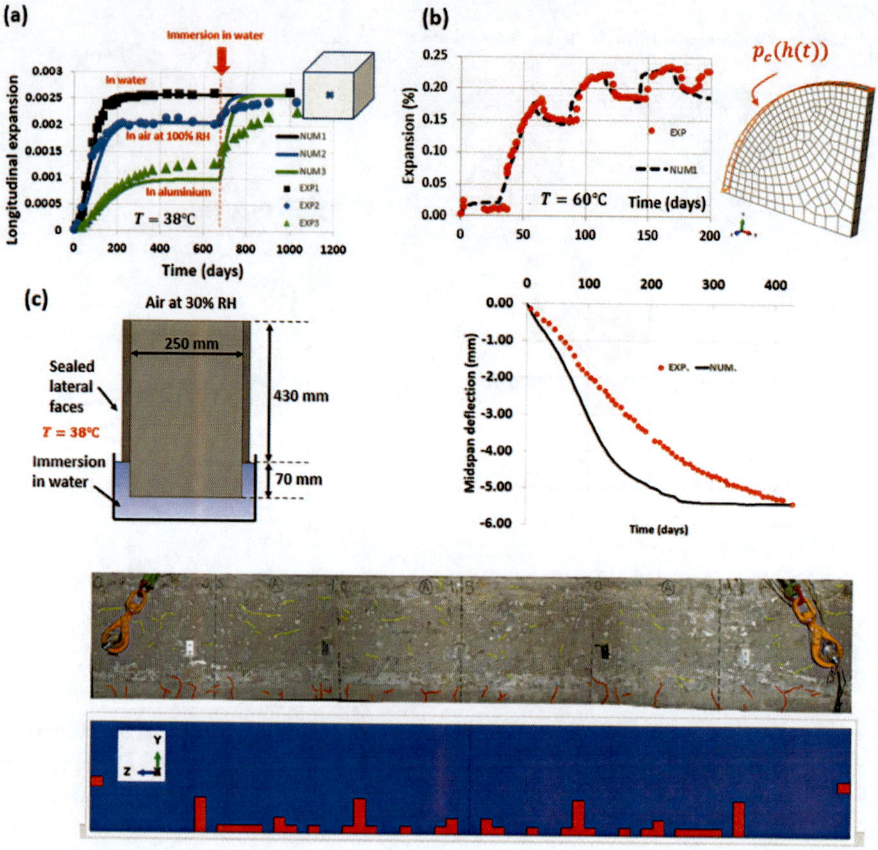

Fig. 5. V&V examples: (a) verification at the material level using experiments of [17]; (b) verification at the small specimen level [18]; (c) validation at the structural level using beam experiment [19].

3.2 V&V

Figure 5 presents some of the V&V results performed for the developed constitutive modelling (for more details, refer to [14]). Figure 5a presents verifications at the material point level with experimental results of [17] involving three different hygral scenarios of expansion: in water (EXP1), a late water supply at 676 days was applied for experiments EXP2 and EXP3, initially kept in air at 100% RH and under watertight aluminum. Figure 5b shows comparison between experimental and numerical results for experiments conducted by [18] in which specimens were subjected to continuous cycling external RH conditions (between 59% and 96% within a period of 28 days). The experimental benchmark of [19] is considered for the validation of the model, where a plain concrete beam made of reactive aggregates is subjected to vertical moisture gradient as shown in Fig. 5c. Figure 5c compares the experimental cracking pattern and the numerical damage pattern (damaged elements are in red). The numerical results for midspan deflection compare very well with the experimental results, as shown in the same figure.

Chemical size effects were neglected for all the V&V examples in this section and therefore, $\alpha_{\tau_c} = \alpha_{\varepsilon_v} = 1.0$ were used in Eq. 6. V&V results are quite encouraging, knowing the simplicity of the computational scheme, the assumptions that have been made and that general semi-empirical laws were used. A trend can however be detected when comparing the numerical and experimental deflection versus time curves (Fig. 5c). The AAR kinetics in the model (calibrated at the material level) appear to be faster than the actual kinetics of the experiment (at the element level of a beam). This same trend was found in other studies and using other numerical models. Chemical size effects on the free AAR accelerated expansion test could be the reason for this difference. The assessment of chemical size effects at the structural level of the existing hydraulic structure is considered in the next section.

4 Application at the Facility Level

A numerical simulation of the left bank portion of the facility is performed using the B76 identified free AAR expansion kinetic parameters. The finite element model including the concrete and rock foundation is shown in Fig. 6a (a total of 56 000 nodes). The points A, B, C and D represent reference monitoring points. The main parameters required by the analyses are provided in Table 2.

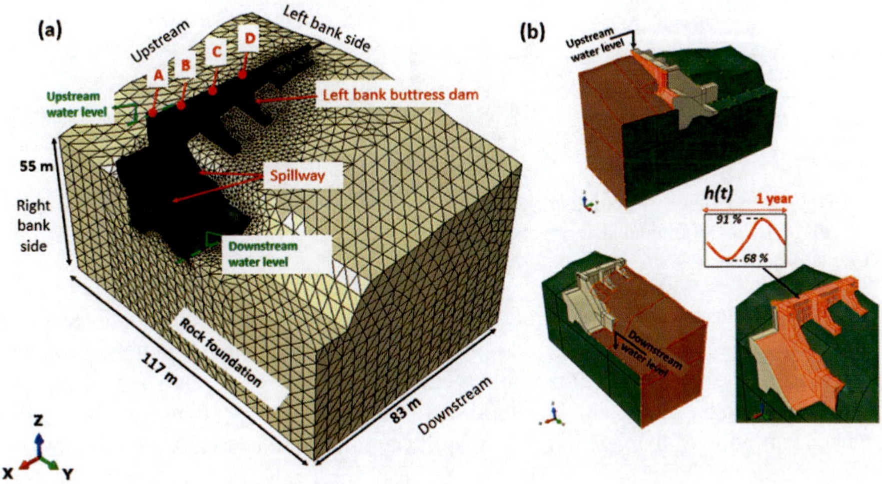

Fig. 6. FE model of the left bank side of the facility: (a) Geometry and mesh; (b) boundary conditions for the pore pressure analysis.

Figure 6b shows the boundary conditions considered in the pore pressure analysis. A sinusoidal variable relative humidity condition was specified for external concrete faces exposed to air by applying the equivalent variable capillary pressures on the nodes of the mesh. The following mechanical boundary conditions were applied for the final multi-physical analysis: (i) plane strain conditions are applied to all vertical boundary faces of the model; (ii) fixity condition is applied at the bottom face of the rock and full compatibility is assumed between concrete and rock. All the mechanical loads were applied at the beginning of the analysis. For reference, the computation time for this explicit analysis was around 6 h, using 12 parallel computing cores.

Two different options were considered for size effects. Firstly, size effects were neglected ($\alpha_{\tau_c} = \alpha_{\varepsilon_v} = 1.0$) and displacement results of the simulation were compared to the available monitoring data. In a second step, size effects are considered through fitting on the observed monitored vertical displacement of reference point A and using available results of residual expansion tests. This allows directly to assess the size factors: $\alpha_{\tau_c} = 7.2$ and $\alpha_{\varepsilon_{l\infty}} = 1.45$. Figure 7 presents comparison of numerical predictions with available vertical displacement monitoring data. When chemical size effects are not considered in the numerical simulation (Fig. 7a), AAR in the model reaches its asymptote around year 40, which is contradictory to the quasi-linear behavior observed for the vertical displacement of the structure. Additionally, the asymptotic AAR expansion appears to be under-estimated in the numerical simulation. When chemical size effects are included in the model (Fig. 7b), all structural vertical displacements are simulated with good accuracy.

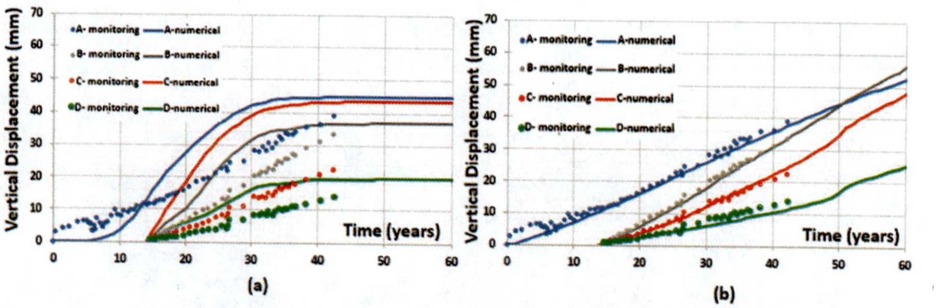

Fig. 7. Comparison of vertical displacements results: (a) without including chemical size effects; (b) with including chemical size effects.

Figure 8 shows comparison of observed cracking and predicted damage pattern for $t = 42$ years. The numerical results correspond to the model including chemical size effects. A good correlation can be generally seen between predicted and existing patterns. Due to the change in the lateral stiffness of the facility (along the X-axis), there is a concentration of damage in the dam/spillway intersection area. A compression strut plunges near the dam/spillway intersection resulting in the inclined cracks visible in both facility and model (Figs. 8a and b).

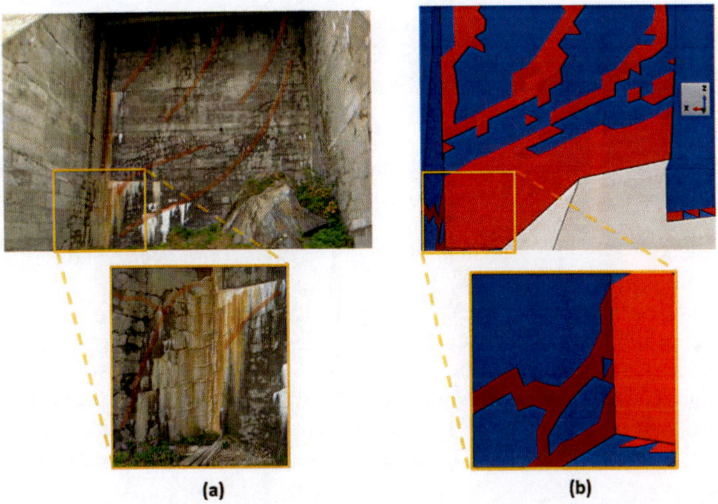

Fig. 8. Comparison of (a) existing cracking and (b)predicted damage pattern (t = 42 years)

5 Conclusions

This work presented numerical and experimental investigations on existing hydraulic structure affected by alkali-aggregate reaction (AAR). Experimental characterization campaign conducted on the reconstructed mass concrete mixture of the facility highlighted the importance of both mechanical and size effects, related respectively to the fracture energy of concrete and the free AAR expansion.

A new enhanced chemo-mechanical modelling approach has been developed, validated and applied to the case of an existing hydraulic facility affected by AAR. Though the simplicity of the coupling scheme, predictions of the model were generally in good agreement with available experimental results and allowed to understand the complex behavior of the facility under study. Future works are planned to include a numerical multi-scale analysis at the level of the laboratory specimens, in order to assess the AAR free expansion curve, without chemical size effects. The long-term objective is to be able to predict numerically the behavior of hydraulic structures affected by AAR using a feasible and adequate characterization campaign at the material level, without a calibration step at the structural level.

References

1. CDA: Dams in Canada. 41, Ottawa (2019)
2. Ben Ftima, M., Lemery, J.: Asymptotic fracture energy for nonlinear simulation of mass concrete structures. Construct. Build. Mater. **271**, (2021)
3. Trunk, B., Wittmann, F.H.: Influence of size on fracture energy of concrete. Mater. Struct. **34**(239), 260–265 (2001)
4. Guan, J.F., et al.: Minimum specimen size for fracture parameters of site-casting dam concrete. Constr. Build. Mater. **93**, 973–982 (2015)

5. Bakour, A., Ben Ftima, M.: Experimental investigations on the asymptotic fracture energy for large mass concrete specimens using wedge splitting test. Construct. Build. Mater. **279** (2021)
6. Duan, K., Hu, X.Z., Wittmann, F.H.: Explanation of size effect in concrete fracture using non-uniform energy distribution. Mater. Struct. **35**(250), 326–331 (2002)
7. 50-FMC, R.: Determination of the fracture energy of mortar and concrete by means of three-point bend tests on notched beams. Mater. Struct. **18**(106), 285–290 (1985)
8. Lindgard, J., et al.: Alkali-silica reaction (ASR)-performance testing: Influence of specimen pre-treatment, exposure conditions and prism size on alkali leaching and prism expansion. Cem. Concr. Res. **53**, 68–90 (2013)
9. Multon, S., Sellier, A.: Multi-scale analysis of alkali-silica reaction (ASR): impact of alkali leaching on scale effects affecting expansion tests. Cem. Concr. Res. **81**, 122–133 (2016)
10. Takahashi, Y., et al.: Scale-dependent ASR expansion of concrete and its prediction coupled with silica gel generation and migration. J. Adv. Concr. Technol. **14**(8), 444–496 (2016)
11. Abdalla, H.M., Karihaloo, B.L.: Determination of size-independent specific fracture energy of concrete from three-point bend and wedge splitting tests. Mag. Concr. Res. **55**(2), 133–141 (2003)
12. Lemery, J., et al.: The disturbed fracture process zone theory for the assessment of the asymptotic fracture energy of concrete. Eng. Fract. Mech. **231** (2020)
13. Ben Ftima, M., Joder, M., Yildiz, E.: Creep modelling for multi-physical simulation of mass concrete structures using the explicit finite element approach. Eng. Struct. **212** (2020)
14. Ben Ftima, M., et al.: Concrete hydraulic structures affected by AAR: numerical and experimental investigations. Submitted to Engineering Structures (2022)
15. Saouma, V., Perotti, L.: Constitutive model for alkali-aggregate reactions. ACI Mater. J. **103**(3), 194–202 (2006)
16. Massicotte, B., M. Ben Ftima: EPM3D v3.0, a user-supplied constitutive model for the nonlinear finite element analysis of concrete structures. École Polytechnique de Montréal, Montréal, QC (2017)
17. Multon, S., Toutlemonde, F.: Effect of moisture conditions and transfers on alkali silica reaction damaged structures. Cem. Concr. Res. **40**(6), 924–934 (2010)
18. Poyet, S., et al.: Influence of water on alkali-silica reaction: experimental study and numerical simulations. J. Mater. Civ. Eng. **18**(4), 588–596 (2006)
19. Multon, S., Seignol, J.F., Toutlemonde, F.: Structural behavior of concrete beams affected by alkali-silica reaction. ACI Mater. J. **102**(2), 67–76 (2005)
20. Larive, C.: Apports combinés de l'expérimentation et de la modélisation à la compréhension de l'alcali-réaction et de ses effets mécaniques [Thèse de doctorat]: École nationale des ponts et chaussées (France) (1997)
21. Baroghel-Bouny, V., Mainguy, M., Lassabatere, T., Coussy, O.: Characterization and identification of equilibrium and transfer moisture properties for ordinary and high-performance cementitious materials. Cem. Concr. Res. **29**, 1225–1238 (1999)

Biaxial Interaction Diagrams of a RC Section at Elevated Temperatures

Sabine Boulvard(✉) and Duc Toan Pham

Centre Scientifique et Technique du Bâtiment (CSTB), 84 avenue Jean Jaurès, Champs-sur-Marne, 77447 Marne-la-Vallée Cedex 2, France
sabine.boulvard@cstb.fr

Abstract. Based on the lower bound static approach of the yield design (or limit analysis) theory, this paper proposes and develops a simple method for deriving the biaxial interaction diagrams of a reinforced concrete section at both ambient and elevated temperatures. Such a method is an extension of the work of Pham *et al.* [5] (*A straightforward procedure for deriving the biaxial interaction diagrams of RC sections in fire*) where the previous complex 3D failure surfaces are simplified here to a superposition of simple 2D interaction diagrams for different orientations of the resulting bending moment in the section. In addition to the proposed theoretical solutions, their validation is provided by their favourable comparison with some experimental results available in the literature.

Keywords: Biaxial interaction diagrams · Reinforced concrete section · Fire loading · Yield design

1 Introduction

Normal force - bending moment interaction diagrams are useful tools for the design of reinforced concrete (RC) sections under the combined action of a tensile-compressive load and a bending moment. At ambient temperature, these interaction diagrams may be established within the framework of ultimate limit state design (ULSD) where the strain limitations of concrete and reinforcing steel are prescribed (see for example EN 1992–1-1 [1]). However, such a calculation procedure is rather complex for a RC section in fire conditions, essentially due to the influence of elevated temperatures on the constituent material properties (see for example Pham *et al.* [2]). The situation becomes much more complex in the case of a heated RC section under the combined action of an axial load and two bending moments about their two axes, since 3D failure surfaces (called here after biaxial interaction diagrams) are now required instead of classical 2D interaction diagrams.

This contribution is based on the lower bound static approach of the yield design (or limit analysis) theory (Salençon [3], de Buhan [4]) for deriving the biaxial interaction diagrams of a RC section. This is an extension of the work of Pham *et al.* [5] where the previous complex 3D failure surfaces are simplified here to a superposition of simple 2D interaction diagrams for different orientations of the resulting bending moment in

P. Rossi and J.-L. Tailhan (Eds.): SSCS 2022, RILEM Bookseries 38, pp. 61–68, 2023.
https://doi.org/10.1007/978-3-031-07746-3_6

the section. Indeed, while the classical methods are usually based on conventional limitations of strains of concrete and steel reinforcing bars (EN 1992–1-1 [1]), the yield design approach only requires that their stress (and not strain) limitations be prescribed, with no reference to other mechanical characteristics. In the present paper, material characteristics may be introduced in two steps: (i) a preliminary heat transfer analysis for evaluating the temperature distribution on the RC section and (ii) the introduction of reduced factors as functions of temperature into both the concrete and steel strength properties (see Pham *et al.* [2] for more details). As a result, a semi-analytical solution of axial load-bending moment interaction diagrams could be obtained for any prescribed temperature distribution and any prescribed orientation of bending moment. Such proposed theoretical solutions will be compared with some experimental results available in the literature.

2 Material Strength Properties at Elevated Temperature

In the same way as in the work of Pham *et al.* [5], in this section, failure criteria for concrete as well as steel reinforcing bars are defined under the assumption that spalling of concrete does not occur and the bonding between the reinforcing bars and the surrounding concrete is perfect. In addition, the tensile strength of concrete is neglected due to the brittleness of concrete under tensile stresses.

Following the approach proposed by Chen [6], Nielsen and Hoang [7], Bleyer and de Buhan [8] or more recently Pham *et al.* [9], a modified Mohr-Coulomb failure criterion is adopted for ascertaining the strength properties of concrete material. In the case where the tensile strength of concrete is neglected, the expression of this failure criterion may be written as follows:

$$f^c(\underline{\underline{\sigma}}) = \sup_{i,j=1,2,3} \left\{ \sigma_i(1+\sin\varphi) - \sigma_j(1-\sin\varphi) - f_c(1-\sin\varphi) \right\} \leq 0 \tag{1}$$

where σ_i, $i = 1,2,3$, are the principal stress components, f_c is the uniaxial compressive strength and φ the internal friction angle.

Neglecting the shear and bending resistance of steel reinforcing bars which are modeled as linear one-dimensional elements, the strength properties of a reinforcing bar may be characterized by the following condition on the axial force n:

$$|n| \leq n_0 = A_s f_y \tag{2}$$

where n_0 is the tensile-compressive resistance of the reinforcing bar, equal to the product of the steel yield strength f_y by its cross-sectional area A_s.

At elevated temperature, the internal friction angle of concrete is assumed to remain constant while its compressive strength f_c and the yield strength f_y of steel are reduced through the introduction of the non-dimensional multiplicative factors:

$$f_c(\theta) = k_c(\theta) f_c; f_y(\theta) = k_s(\theta) f_y \tag{3}$$

where k_c and k_s are decreasing functions of the temperature θ. These factors are equal to one for ambient temperature.

3 Yield Design Auxiliary Problem

The structure under consideration is a reinforced concrete (RC) beam modeled as a parallelepiped of thickness h along the Oz-axis, width b along the Oy-axis and length L along the Ox-axis.

Given any value β, comprised between 0 and $\pi/2$, the beam is subjected to mechanical loading conditions as follows:

- Body forces (self-weight) are neglected.
- The left section ($x = 0$) is in smooth contact with a fixed vertical plane.
- The right section ($x = L$) is in smooth contact with a plane in horizontal translation of velocity $\dot{\delta}$ at $z = 0$ and rotation of angular velocity $\dot{\alpha}$ about the Oy_β-axis (see Fig. 1).
- Lateral sides are stress free.

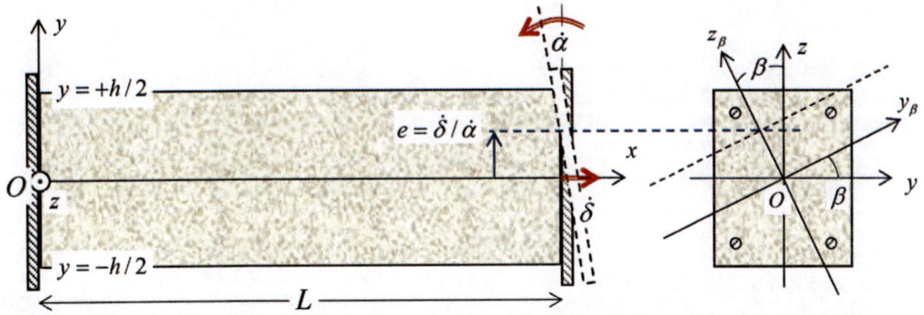

Fig. 1. Reinforced concrete beam subjected to an axial force and a bending moment of direction about β.

In these conditions $\dot{\delta}$ and $\dot{\alpha}$ may be interpreted as the generalized kinematic variables associated by duality to the generalized stress variables (N, M_β).

Considering any kinematically admissible (K.A.) velocity field $\underline{U}$ (which complies with the velocity boundary conditions), the work of the external forces in any such field may be written as follows:

$$W_e(\underline{U}) = N\dot{\delta} + M_\beta\dot{\alpha} \tag{4}$$

where N and M_β may be interpreted as the axial force along Ox and bending moment about Oy_β exerted on the section $x = L$ of the beam, which is thus subjected to a two-parameter loading mode.

According to the yield design theory (Salençon [3], de Buhan [4]), the so-called domain K of potentially safe loads (N, M_β) is the set of loads which can be equilibrated by a stress distribution in the beam (stress tensor fields in the concrete, tensile force distributions along the reinforcements), verifying the respective strength conditions (1) and (2) on any point of the beam. The boundary of this domain is called the interaction diagram of the RC beam subjected to combined axial load and bending moment of direction β.

4 Lower Bound Static Approach

Given any direction β and any value e, the two following uniaxial stress distributions are considered (see Fig. 2). This means that in configuration (a) both concrete and reinforcing bars reach their positive tensile (resp. negative compressive) strengths when located below (resp. above) the plane of equation $z_\beta = e$. The opposite applies in configuration (b).

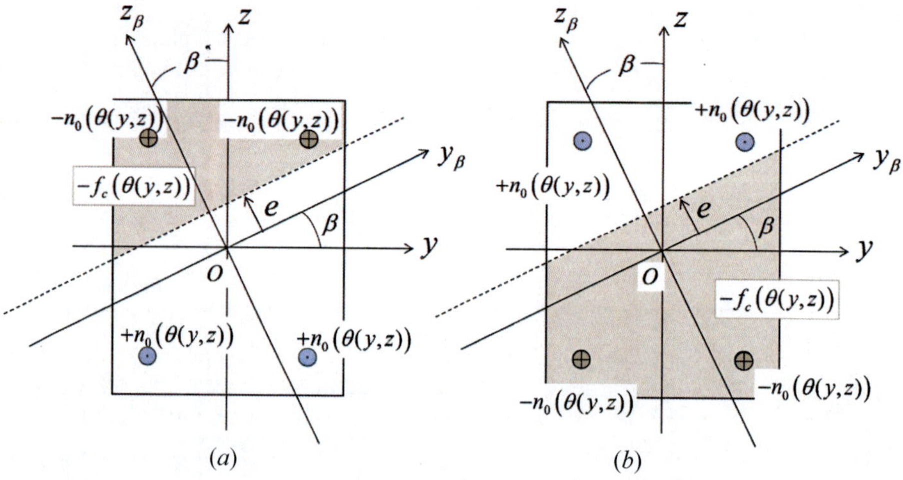

Fig. 2. Stress distributions in RC beam used in the lower bound static approach of yield design.

The corresponding values of the two loading parameters N and M_β in equilibrium with such stress distributions may be calculated:

$$N = \int \sigma(y_\beta, z_\beta) dy_\beta dz_\beta$$
$$M_\beta = \int -z_\beta \sigma(y_\beta, z_\beta) dy_\beta dz_\beta \tag{5}$$

leading to the determination of the interaction diagram of the RC beam under the combined actions of axial load and bending moment of direction β.

5 Numerical Examples

For illustrative purpose, the following set of data has been selected:

- Rectangular section $b \times h = 100 \times 200\ \text{mm}^2$,
- Normal weight concrete with siliceous aggregates, $f_c = 30$ MPa,
- Four hot rolled reinforcing steel bars of diameter 10 mm, with 18 mm of concrete cover, $f_y = 500$ MPa,
- Reduction factors k_c and k_s are considered temperature dependent referred to EN 1992–1-2 [10].

- A lateral face exposed to ISO 834 fire (EN 1991–1-2 [11]).

Figure 3 shows the temperature distributions across the beam section obtained by means of the SAFIR computer program [12], corresponding to 120 min fire duration.

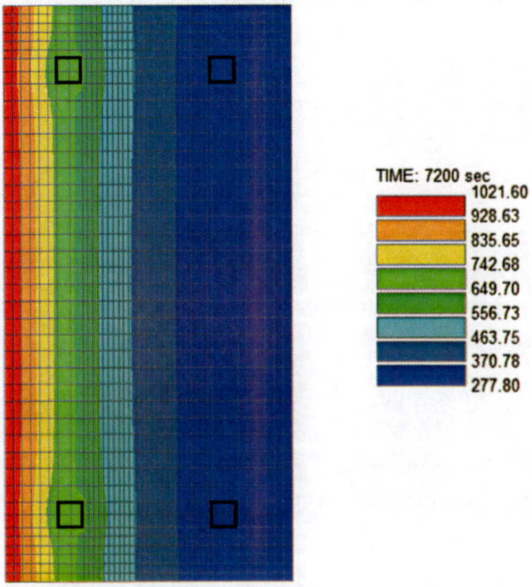

Fig. 3. Temperature distribution across the beam section for 120 min of fire exposure.

Introducing this temperature distribution into the above-described calculation procedure, the corresponding interaction diagrams for different bending moment orientation could be determined as shown in Fig. 4.

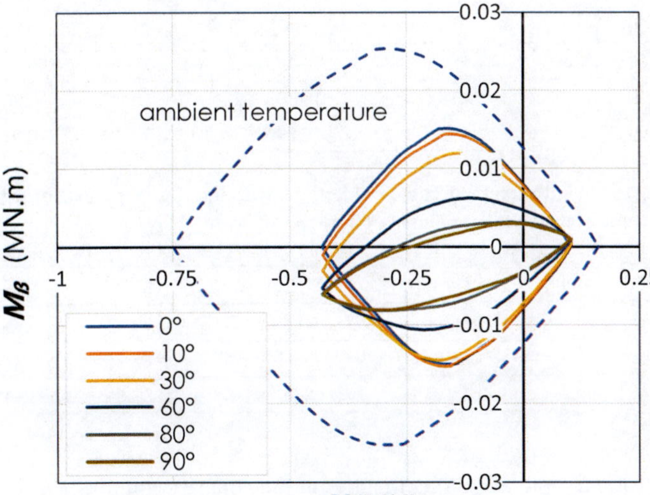

Fig. 4. Interaction diagrams for different bending moment orientations.

6 Comparison with Available Experimental Results

Validation of the proposed theoretical solutions is provided in this section by their favourable comparison with some experimental results available in the literature (see Table 1 and Table 2 for tests at ambient temperature performed by Kim and Lee [13] and Table 3 for tests at elevated temperature performed by Tan and Nguyen [14]).

Table 1. Comparison between theoretical predictions with the results of tests of square columns (100×100 mm^2) at ambient temperature of Kim and Lee [13].

Test	Orientation (°)	Ntest (kN)	Mtest (kN.m)	Mpred	Mtest/Mpred
SS-0-1	0°	119	6.664	6.165	1.08
SS-0-2	0°	126	7.3206	6.152	1.19
SS-30-1	30°	112	5.822	5.140	1.13
SS-30-2	30°	104	5.436	3.780	1.44
SS-45-1	45°	103	5.500	4.983	1.10
SS-45-2	45°	106	5.514	4.975	1.11

Table 2. Comparison between theoretical predictions with the results of tests of rectangular columns (200 x 100 mm^2) at ambient temperature of Kim and Lee [13].

Test	Orientation (°)	Ntest (kN)	Mtest (kN.m)	Mpred	Mtest/Mpred
RS-0-1	0°	204	11.342	11.580	0.98
RS-0-2	0°	206	11.804	11.606	1.02
RS-30-1	30°	208	11.241	14.806	0.76
RS-30-2	30°	217	11.609	14.921	0.78
RS-45-1	45°	266	13.279	19.208	0.69
RS-45-2	45°	239	12.342	19.086	0.65
RS-60-1	60°	313	15.074	22.360	0.67
RS-60-2	60°	295	14.189	22.554	0.63
RS-90-1	90°	418	18.393	22.867	0.80
RS-90-2	90°	443	19.846	21.775	0.91

Table 3. Comparison between theoretical predictions with the results of tests of square columns (300 x 300 mm^2) at elevated temperatures of Tan and Nguyen [14].

Test	Orientation (°)	Ntest (kN)	Mtest (kN.m)	Mpred	Mtest/Mpred
C3-1-25	45°	860	74.27	79.82	0.93
C3-2-40	45°	680	73.83	85.28	0.87
C3-3-60	45°	530	68.29	88.64	0.93

Focusing for example more specifically on the square column C3–1-25 at elevated temperatures (Tan and Nguyen [14]), Fig. 6 displays the interaction diagram calculated according to the proposed yield design procedure on account of the calculated temperature distribution of Fig. 5 (obtained by Safir program [12]), and the corresponding experimental point. Due to the axial compressive force, the resulting bending strength is increased by approximately 25% as compared to the ultimate strength under pure bending.

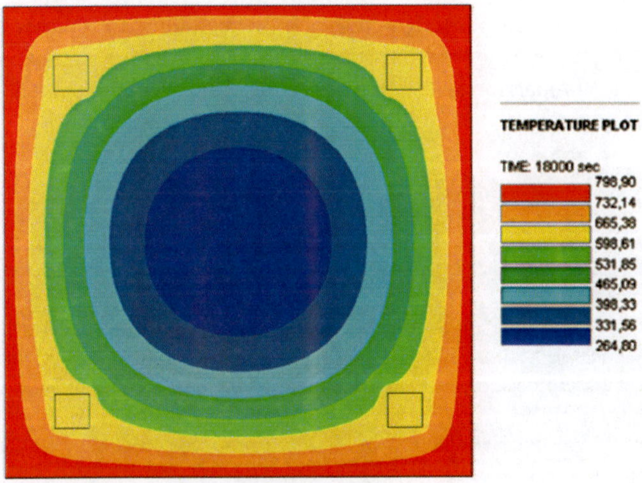

Fig. 5. Temperature distribution across the C3–1-25 column section.

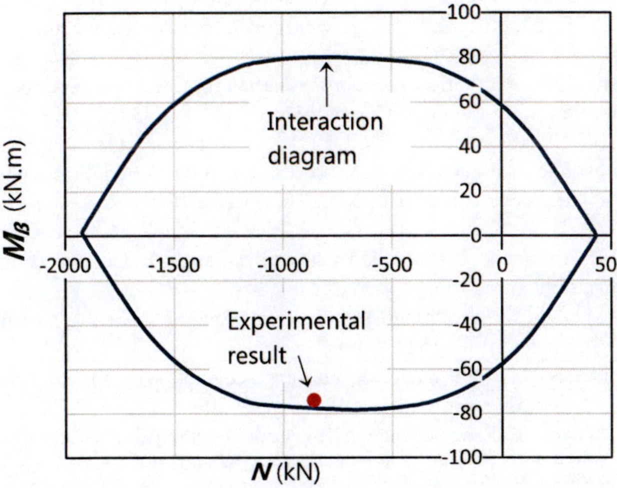

Fig. 6. Comparison of the predicted interaction diagram with the result of test C3–1-25 performed by Tan and Nguyen [14].

7 Conclusion

Relying on the lower bound static approach of the yield design, this paper has (proposed) a semi-analytical solution for the biaxial interaction diagrams of a RC section. This is an extension of the work of Pham *et al.* [5] where the previous complex 3D failure surfaces are simplified here to a superposition of simple 2D interaction diagrams for different orientations of the resulting bending moment in the section. Due to its simplicity, the proposed method allows performing parametric studies in a rather quick way. In addition to the proposed theoretical solutions, their validation has been provided by their favourable comparison with some available experimental results at both ambient and elevated temperatures.

References

1. EN 1992–1–1. Eurocode 2: Design of concrete structures - Part 1–1: General rules and rules for buildings (2004)
2. Pham, D.T., de Buhan, P., Florence, C., Heck, J.V., Nguyen, H.H.: Interaction diagrams of reinforced concrete sections in fire: A yield design approach. Eng. Struct. **90**, 38–47 (2016)
3. Salençon, J.: Yield design. ISTE Ltd and John Wiley and Sons Inc, Great Britain and the United States (2013)
4. de Buhan, P.: Plasticité et calcul à la rupture. Presses de l'Ecole Nationale des Ponts et Chaussées, Paris (2007)
5. Pham, D.T., Nguyen, H.H., Boulvard, S.: A straightforward procedure for deriving the biaxial interaction diagrams of RC sections in fire. In: IOP Conference Series: Material Science and Engineering, Prague, Czech Republic (2021)
6. Chen, W.F.: Plasticity in reinforced concrete. McGraw-Hill, Newyork (1982)
7. Nielsen, M.P., Hoang, L.C.: Limit analysis and concrete plasticity. 3rd edn. Taylor and Francis Group (2010)
8. Bleyer, J., de Buhan, P.: Yield surface approximation for lower and upper bound yield design of 3D composite frame structures. Comput. Struct. **129**, 86–98 (2013)
9. Pham, D.T., de Buhan, P., Florence, C., Heck, J.V., Nguyen, H.H.: Interaction diagrams of reinforced concrete sections in fire: A yield design approach. Eng. Struct. **90**, 38–47 (2015)
10. EN 1992–1–2. Eurocode 2: Design of concrete structures-Part 1–2: General rules - Structural fire design (2004)
11. EN 1991–1–2. Eurocode 1: Actions on structures-Part 1–2: General actions. Actions on structures exposed to fire (2002)
12. Franssen, J.M.: SAFIR: a thermal/structural program for modeling structures under fire. Engineering Journal **42**, 143–158 (2005)
13. Kim, J.-K., Lee, S.-S.: The behavior of reinforced concrete columns subjected to axial force and biaxial bending. Eng. Struct. **23**, 1518–1528 (2000)
14. Tan, K.-H., Nguyen, T.-T.: Experimental behaviour of restrained reinforced concrete columns subjected to equal biaxial bending at elevated temperatures. Eng. Struct. **56**, 823–836 (2013)

Probabilistic Modelling of Containment Building Leakage at the Structural Scale: Application to the PACE Mock-Up

M. Briffaut[1] (✉), M. Ghannoum[3], J. Baroth[2], H. Cheikh Sleiman[2], and F. Dufour[2]

[1] CNRS, Centrale Lille, UMR9013—LaMcube—Laboratoire de mécanique multiphysique et multiéchelle, Université de Lille, 59000 Lille, France
matthieu.briffaut@centralelille.fr
[2] Univ. Grenoble Alpes, CNRS, Grenoble INP, 3SR, 38000 Grenoble, France
[3] Univ. d'Angers, Pays de la Loire, France

Abstract. This work follows studies conducted in the framework of the French research program MaCEnA (PIA), aiming to predict air leakage through a reinforced (and prestressed) concrete structure. As mentioned by the international Benchmark VeRCoRs, only very few teams were able to predict them and variations of at least one order of magnitude between participants were observed.

Reinforced concrete tightness estimation is of the utmost importance for confinement vessels but also to assess concrete structures durability. One of the reasons for these difficulties lies in the fact that air leakage prediction is the last step of complex, multiphysics and coupled simulations. On the one hand, to predict concrete permeability, saturation evolution of the porous network needs to be correctly addressed. On the other hand, cracks predictions (numbers, openings, and appearance time) is essential but not sufficient since roughnesses, tortuosities and connectivities of these latter also strongly influence the leakage rate.

After a first part showing the capacity of the used model to mimic the behavior of a structural representative volume, this contribution quantifies the effect of the use of autocorrelated random fields modelling the tensile strength on the concrete structural leakage prediction. The results highlight that the air leakage prediction can easily vary by one order of magnitude for the same random field parameters. Moreover, during mechanical loading, observation of the cracks evolutions (numbers, openings and positions) allows for quantifying the prevalence of material and structural heterogeneity and explains the sudden evolution of leakage rate.

Keywords: Reinforced concrete · Leakage · Stochastic finite elements · Autocorrelated random fields

1 Introduction

Despite the recent progress of the last decade and especially during the French research program MaCEnA (PIA), accurate prediction of air leakage through a reinforced (and

Grenoble INP—Institute of engineering Univ. Grenoble Alpes.

P. Rossi and J.-L. Tailhan (Eds.): SSCS 2022, RILEM Bookseries 38, pp. 69–78, 2023.
https://doi.org/10.1007/978-3-031-07746-3_7

prestressed) concrete structure remains a difficult question. Indeed, as mentioned by the international Benchmark VeRCoRs (phase 2) restitution report, only very few teams are able to predict them and variations of at least one order of magnitude between participants are observed [1]. Even when global leakage is quite well predicted, the repartition between localized leakage (through cracks for instance) and diffuse one (through the porous network of cementitious material) is poorly estimated. Reinforced concrete tightness estimation is of the utmost importance for confinement vessels but also to assess concrete structures durability.

One of the reasons for these difficulties lies in the fact that air leakage prediction is the last step of complex, multiphysics and coupled simulations. On the one hand, to predict concrete permeability (i.e. flow through the connected porous media), saturation evolution (in time and space) of the porous network needs to be correctly addressed as well as the description of the porous network. On the other hand, cracks predictions (numbers, openings, and appearance time) are essential but not sufficient since roughnesses, tortuosities and connectivities of these latters also strongly influence the leakage rate. Moreover, localized phenomena (cracking) and continuous ones (stresses, strains, saturation) are strongly coupled. Indeed, saturation evolution (generating shrinkage), creep strains, early age behavior, mechanical stresses could lead to localized cracks due to reinforced concrete heterogeneity (material or structural ones). The most frequent technique employed to account for the heterogeneous nature of concrete at the structural scale is random field generation applied to one (or more) concrete behavior parameter.

This contribution aims to quantify the effect of the use of autocorrelated random fields modelling the tensile strength on the concrete structural leakage prediction. On the contrary to structural tests that are very expensive and cannot be reproduced, numerical simulations can be performed for many realizations of the same random field. In this contribution, the results of more than 50 simulations (for 2 different random fields) of a RSV (representative structural volume) involving rebars and prestressed tendons and based on a real structural experiment [2] will be analyzed. For this calculation, regularized damage model is used and the leakage rate results from the mechanical fields (damage and strains) post-processing through the recently developed coupling law between damage and leakage that could consider leakage rate evolution during cracks closure [3]. The first part of this contribution presents a deterministic calculation which considers the ageing of the concrete structure whereas the second part aims at studying the effect of heterogeneity (structural ones i.e. prestressed cables tubes and material ones i.e. concrete heterogeneous nature).

2 Deterministic Leakage Prediction of a Structural Representative Volume

2.1 PACE Mock-Up Brief Description

The PACE mock-up is a facility to study the Representative Structural Volume (RSV) in the standard zone of a nuclear power plant (PACE: "PArtie Courante de l'Enceinte" in French) and was built in a collaboration between the EDF R&D department and the MPA Karlsruhe (Materials testing and Research institute of the Karlsruhe Institute of Technology (KIT)). With realistic dimensions (see Fig. 1), the specimen is loaded similarly to a closed ring under internal pressure.

The reinforcement layout of the specimen mainly consists of rebars meshes near the intrados and extrados surfaces and four pre-stressing cables in the tangential direction. During the tests, the specimen is under a pressure of up to 6 bars (absolute term) that simulates the overpressure scenario in an accidental condition. The resulting pressure (Chaudronnier's equivalent pressure) is applied to the circumferential direction by mean of 128 GEWI's rebars connected to ears and tensioned by hydraulic jacks. As mentioned before, the specimen is post-tensioned by four tendons. The pre-stressing level was decreased over time in order to accelerate the pre-stressing losses and thus to simulate the aging of the structure. In addition, one pre-stressing cable in the original vertical direction is placed before concrete's pouring but without any pre-stressing force. Therefore, in order to simulate the pre-stressing of the containment in the original vertical direction, steel cushions, which were set under a pressure of up to 1 MPa, are placed on the original top and bottom surfaces of the specimen (please refer to [4–6] for more details about the mock-up).

(a)

(b)

Fig. 1. Pictures of PACE mock-up in Karlsruhe Institute of Technology: global view (a), side view before adding the loading device (b)

Regularly, pressure tests were performed to check of the leakage tightness of containment in France. These lasts were reproduced on the experimental mock-up and each pressure test is called RUN. As a result of some of these RUNs, especially RUNs 4, 5 and 6, cracking patterns appear on the extrados side (see Fig. 2). The black cracks, propagating horizontally, results from flexure loading, whereas the red ones, propagating vertically, are crossing cracks and are supposed to be the preferential paths for most of the leakage.

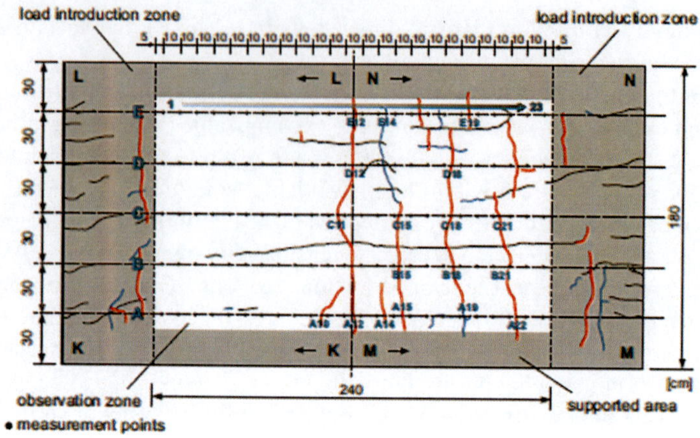

Fig. 2. Cracking extrados pattern after the RUN 4 [5]

2.2 Finite Element (FE) Model

The strategy described in Fig. 3 is based on a weakly coupled thermo-hygro-mechanical FE model. In this study, early age behavior is not considered and there is no experimental temperature evolution so the thermal part is not considered. The hydric model is based on phenomenological macroscopic model [7–9] with Neumann boundaries conditions type and the relationship between the relative humidity and the water content is the one defined by Thiery *et al.* [10].

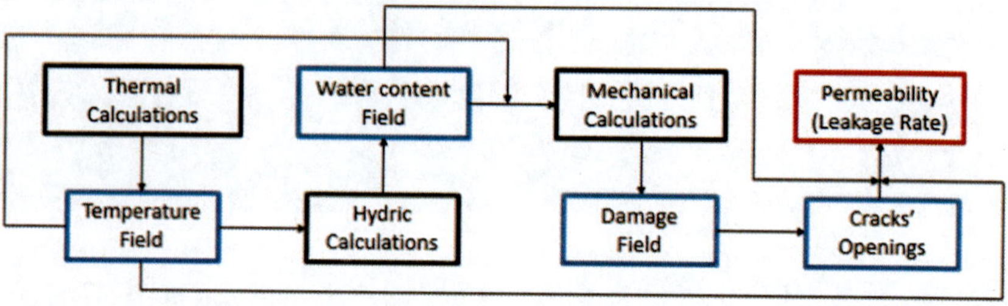

Fig. 3. Numerical THM weakly coupled (staggered) strategy to compute the global leakage

The mechanical model is based on strains partitioning and the total strain is the sum of thermal strains, desiccation strains, creep strains and elastic strains. This latter is used to compute the damage variable based on Mazars model. Please note that a part of the creep strains is also considered to calculate the Mazars equivalent strains. Finally, since the damaged model used is a local one regularized by fracture energy, the cracks (discontinuities) can be post-processed from the damage field using the approach of [11]. A reduction of the tensile strength of the concrete is also applied using a size effect law [12]. A more detailed description of the model used can be found in [13].

The global leakage rate can be seen as a post-processing of both hydric state of the porous media and the flow through the cracks. Indeed, in absence of cracks (or between two cracks) the flow is governed by the apparent permeability highly dependent of the saturation rate (the relation defined by [14] is used in this study). Through the cracks, the matching law defined by [3] based on an original idea of [15] is used to allow the possible decrease of the flow when the crack opening decreases.

2.3 Crack and Leakage Comparison

With this model and after the simulation of the mock-up life, including ageing through the drying and the decrease of prestressing, the comparison between predicted and experimental leakages for the run 6 is displayed on Fig. 4a. The predicted leakage slightly overestimates the experimental one. The comparison between numerical flow in and flow out also shows that even if the comparison is made on the experimental pressure plateau, the steady-state is not numerically completely reached.

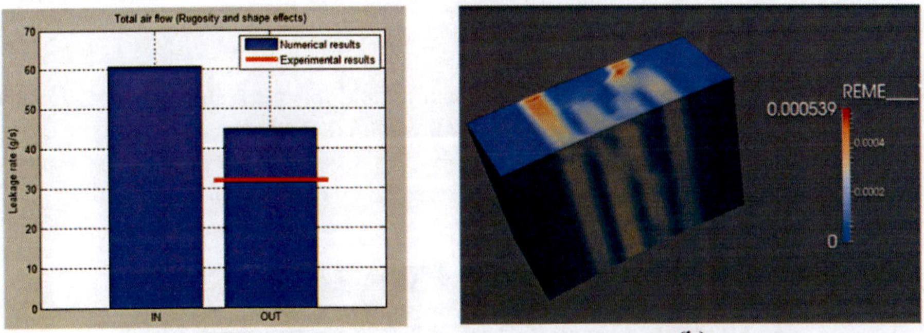

(a) (b)

Fig. 4. Comparison between numerical results in terms of global leakage (a) and final damage pattern (REME = crack opening in meters).

Looking at the cracking pattern (Fig. 4b), the crack position is clearly different with respect to the experimental crack pattern but the number of damaged zones (that represent in a more diffuse manner the cracks) is comparable. It is nevertheless worth noting that simplifications were made on the geometry (empty vertical prestressing tube was not represented) and that the mesh is rather coarse (to decrease computation time). Moreover, a random field was applied to the damage strain threshold to avoid damage diffusion but the field variability is very small and is not representative of the heterogeneous material behavior.

3 Probabilistic Leakage Prediction of the RSV

3.1 Stochastic Finite Element (SFE) Model

To study the impact of the heterogeneous nature of the concrete and to consider the structural heterogeneity (namely the presence of an empty prestressing vertical and horizontal tubes), a Gaussian random field (RF) is arbitrarily chosen to model the tensile strength of the concrete. Gaussian autocorrelation function and two coefficients of variation are considered (Table 1). SFE simulations were performed on a thinner mesh but with a simplified loading path. Indeed, the numerical PACE mock up is only submitted to a tangential tension and permeability evolution during loading (quasi-instantaneous one) is computed. As for the deterministic simulation presented in the part 2, the reinforcements are modelled using bars elements that coincide with the concrete mesh. Only a half of the total specimen is meshed. A total of 56 realizations of both RFs were performed.

Table 1. Parameter of the random fields used.

Random field number	Tensile strength mean value [MPa]	Autocorrelation length [m]	Coefficient of variation [%]	Number of realizations
1	2.5	30	10	30
2	2.5	30	15	26

Figure 5 displays the obtained crack opening field for one realization just after the first crack. It should be highlighted here that the first crack always appears at the center of the RSV due to the presence of the empty prestressed tube which creates a strong heterogeneity (at least stronger than the one created by the random field).

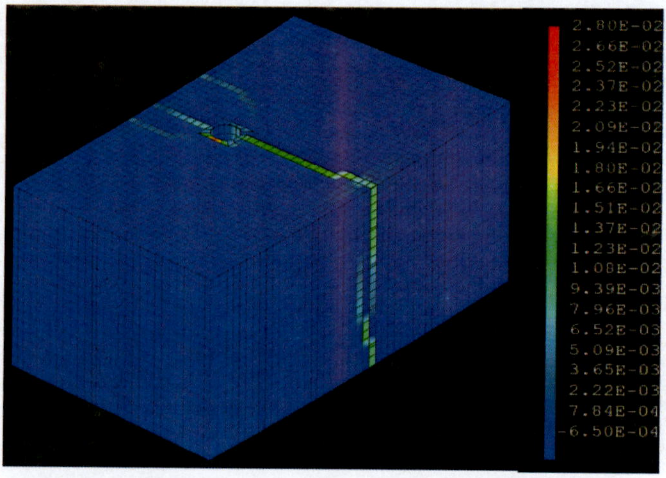

Fig. 5. Crack opening [mm] field just after the damage of the first finite elements

3.2 Crack and Leakage Results

Figure 6 gathers the results of permeability evolution for the 26 realizations with the 2nd random field (similar results are obtained with the first one). One can notice that the global trend is the same but that the final permeability (presented here as the ratio with respect to the initial one) can differ from at least one order of magnitude. The total flow cannot be directly compared to the experimental one because this probabilistic study only considers the evolution of flow through the cracks (permeability evolution of the porous media due to drying is not considered) but interesting feature can be deduced from this graph.

Firstly, some permeability sharp decrease during the loading progress. This decrease is attributed to secondary cracks that relax stress state of the rebars close to the location on previous cracks. Consequently, even if the total applied displacement is close to be the same between two successive time steps, the total crack opening is divided into two (or more) crack openings. As the flow through the crack is directly proportional to the cube of the crack opening (following Poiseuille law), the total flow decreases. In other words, in term of tightness, two cracks slightly open is preferable to one crack largely open.

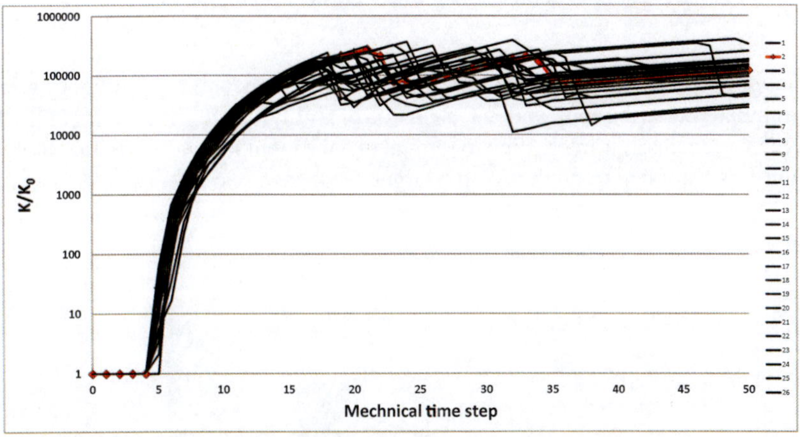

Fig. 6. Evolution of relative permeability (initial permeability $K_0 = 1.e{-}17$) versus the mechanical loading.

Crack pattern analysis also brings important information. Indeed, the crack number and crack position strongly evolve from one realization to another. For most of the realizations, we can consider that we have obtained the steady state cracking regime (i.e. the spacing between two cracks is not sufficient to create a new crack and increasing the loading will only lead to an increase of cracks opening) and the crack number varies from 3 to 5 (for most of them 14 under 26 realizations, 4 cracks appears). The case n°2 (in red on the Fig. 6) is a special one. Indeed, the experimental crack pattern shows four cracks on the same side of the mock-up raising immediately the issue of the real symmetry of the boundary's mechanical conditions.

Figure 7 presents the damage pattern and the crack opening field obtained on the PACE extrados for this special case. The damage field is slightly diffuse (especially close to the rebars) but the crack opening field clearly shows four cracks on the same specimen side. Cracking along the rebars also started indicating that we have reached the cracking steady state regime. Although this special case is not a proof that the symmetry of experimental boundaries conditions was achieved, it shows that with a relatively small number of realizations, with a random field applied only on tensile strength and for a reasonable coefficient of variation, complete dissymmetry of the cracking pattern can be observed.

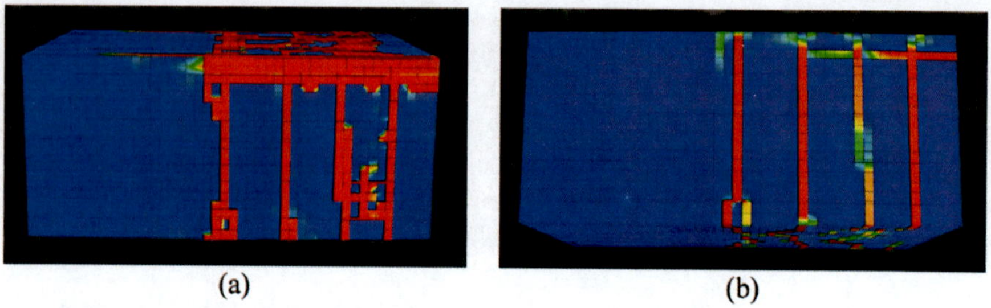

Fig. 7. Damage field (a) and crack opening field (b) for the case number 2

Figure 8 displays histograms of the final relative permeability, highlighting relative and cumulated distributions and a Gausian probability density function (PDF) fitting relative frequencies. It is interesting to remind that the input RF was Gaussian. It probably explains this Gaussian trend, but with a larger dispersion (CoV ~53%).

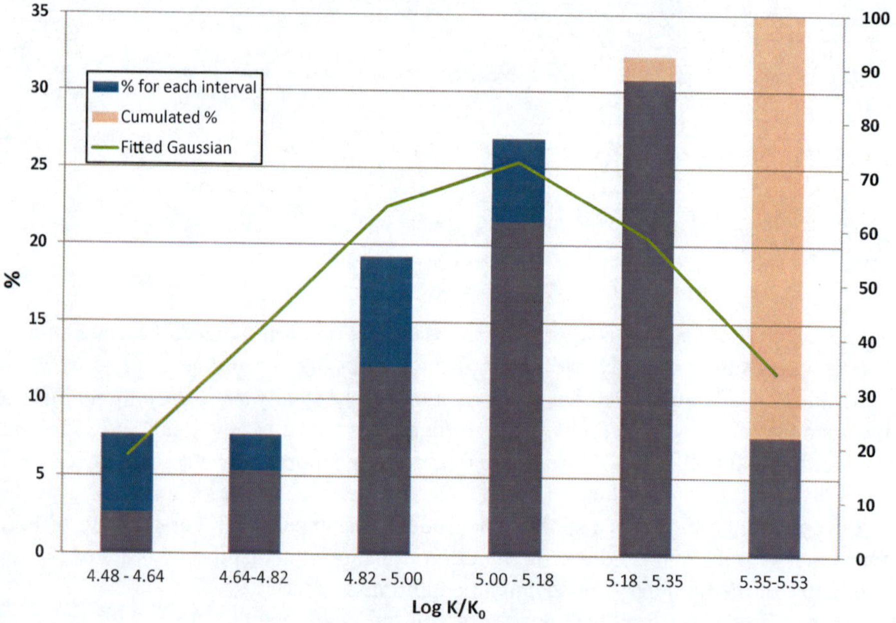

Fig. 8. Histogram of the relative final permeability

4 Conclusion

In this paper, simulations of the PACE mock-up are presented with two different approaches and objectives. During the first part, a HM simulation aiming at reproducing the complete life of the mock was conducted with boundaries conditions and loading as close as possible to the real one. The results show that considering the ageing effect (creep and drying of the porous media) allows to reproduce cracking and that numerical and experimental leakage are in the same order of magnitude.

The second part of this contribution allow to present the results of a probabilistic study of the same mock-up but with a thinner mesh and simplified loading (only mechanical one) in order to both decrease the calculation time and increase the representativity of the crack by damage band. The results highlight that the experimental results which can be a priori seen as a particular case (non-symmetry of the crack position) can be sometimes be retrieved but above all, the simulations show a larger uncertainty spread from mechanical random field (CoV = 15%) to leakage rate (variation of more than one order of magnitude on the prediction of the leakage rate).

References

1. Corbin, M.: Overview, synthesis and lessons learnt from the International Benchmark VeRCoRs (2018)
2. Niklasch, C., Herrmann, N.: Nonlinear fluid-structure interaction calculation of the leakage behaviour of cracked concrete walls. Nucl. Eng. Des. **239**, 1628–1640 (2009)
3. Bouhjiti, D.E.M., et al.: New continuous strain-based description of concrete's damage-permeability coupling. Int. J. Numer. Anal. Meth. Geomech. **42**(14), 1671–1697 (2018)
4. Herrmann, N., et al.: PACE-1450 – experimental investigation of the crack behavior of pre-stressed concrete containment walls considering the prestressing loss due to aging. In: 20th International Conference on Structural Mechanics in Reactor Technology (SMiRT 20), Espoo, Finland, SMirt 20-Division 1, Paper 1671, 9–14 August 2009
5. Herrmann, N., Muller, H.S., Niklasch, C., Michel-Ponnelle, S., Le Pape, Y., Bento, C.: PACE-1450 the crack and leakage behavior of a prestressed concrete wall considering the prestressing loss due to aging. In: SMiRT 22, San Francisco, California, USA, 18–23 August 2013
6. Herrmann, N., Muller, H.S., Niklasch, C., Michel-Ponnelle, S., Masson, B.: The PACE-1450 test campaign – leakage behaviour of a pre-stressed concrete containment wall segment. In: Key Engineering Materials (2016)
7. Bazant, Z.P., Najjar, L.J.: Non-linear water diffusion in non-saturated concrete. Mater. Struct. **5**(1), 3–20 (1972)
8. Mensi, R., Acker, P., Attolou, A.: Séchage du béton: analyse et modélisation. Mater. Struct. **21**, 3–12 (1988). https://doi.org/10.1007/BF02472523
9. Granger, L.: Comportement différé du béton dans les enceintes de centrales nucléaires. Thèse ENPC, France (1995)
10. Thiery, M., Baroghel-Bouny, V., Bourneton, N., Villain, G., Stefani, C.: Modelisation du séchage des bétons. Analyse des differents modes de transfert hydrique. Revue européenne de Génie civil. Modeling of concrete drying. Analyse of different hydric transfert modes **11**, 541–577 (2007)
11. Matallah, M., La Borderie, C., Maurel, O.: A practical method to estimate crack opening in concrete structures. Int. J. Numer. Anal. Meth. Geomech. **34**(15), 1615–1633 (2010)

12. Sellier, A., Millard, A.: Weakest link and localisation WL^2: a method to conciliate probabilistic and energetic scale effects in numerical models. Eur. J. Environ. Civ. Eng. **18**(10), 1177–1191 (2014)
13. Bouhjiti, D.E.M., Baroth, J., Briffaut, M., Dufour, F., Masson, B.: Statistical modeling of cracking in large concrete structures under thermo-hydro-mechanical loads: application to nuclear containment buildings, Part 1: random field effects (reference analysis). Nucl. Eng. Des. **333**, 196–223 (2018)
14. Verdier, J.: Contribution à la caractérisation de l'évolution du taux de fuite des enceintes de confinement du parc nucléaire. Thèse de doctorat, Université Paul Sabatier Toulouse (2001)
15. Pijaudier-Cabot, G., La Borderie, C.: Mechanical damage, chemical damage and permeability in quasi-brittle cementitious materials. Eur. J. Environ. Civ. Eng. **13**(7–8), 963–982 (2009)

Digital Twin for Modelling Structural Durability

Jan Cervenka(✉) and Jiri Rymes

Cervenka Consulting s.r.o., Prague, Czech Republic
jan.cervenka@cervenka.cz

Abstract. The reinforcement corrosion due to chloride ingress is an important deterioration mechanism, which may compromise the service life of reinforced concrete structures. This study presents a bridge monitoring system coupled with an advanced chemo-mechanical computational method for an estimation of chloride ingress and reinforcement corrosion. Based on-site measurements, a digital replica of the bridge is calibrated and by applying the degradation models, the reduction of structural performance is simulated. Apart from the structural analysis, the data from the monitoring system can be used to deduce information about the daily traffic crossing the bridge. Pilot applications of the proposed coupled framework are shown for two concrete bridges, where a 150-years-long chloride attack was assumed for the assessment of the long-term structural performance. The structural resistance was evaluated by the methods based on fib MC 2010, namely the estimation of a coefficient of variation method (ECOV) and partial factor method (PFM).

Keywords: Reinforcement corrosion · Chloride ingress · Fracture mechanics · Nonlinear simulation

1 Introduction

The role of civil engineering for the 21st century is to ensure safe and reliable infrastructure while keeping both financial and ecological costs low. In the EU region, most of the infrastructure was built during the rapid economic growth after World War II implying that the age of many structures now well exceeds 50 years and the maintenance const now represents a severe burden for the public budgets. At the same moment, the concrete industry produced about 7% of the man-made CO_2 mainly due to the production of the cement clinker, which is required for the construction processes [1]. One of the means how to reduce the negative environmental impact is to ensure the optimal structural lifespan. Apart from regular structural inspections, online monitoring systems can be installed on the structure to monitor structural performance characteristics in real-time. Such systems contribute to better structural safety and, through early diagnosis, help to reduce the cost of repair works.

Digital twin refers to a digital replica of a real structure. Based on the feedback from measurement data on the real structure, the most significant model properties are identified. These appropriate model properties are consequently used for assessments of

P. Rossi and J.-L. Tailhan (Eds.): SSCS 2022, RILEM Bookseries 38, pp. 79–89, 2023.
https://doi.org/10.1007/978-3-031-07746-3_8

safety, reliability, durability, and sustainability of the investigated structure under service as well as limit state conditions.

In this study, we show two examples of this methodology. The monitoring system was installed on the Wonka Bridge, Pardubice, Czech Republic and Vogelsang Bridge, Esslinger, Germany. Non-linear finite element (FE) models were subsequently developed for both structures. Upon calibration, the numerical models were used for accessing the long-term structural performance. By applying advanced chemo-mechanical models, namely for the chloride ingress and reinforcement corrosion, the degradation of the structure due to chloride attack was evaluated.

2 Bridge Monitoring System

Advanced measuring systems can be used to monitor structural behaviour in real-time. In this study, the iBWIM (Bridge-Weigh-In-Motion) technology (PEC – Petschacher Consulting, ZT-GmbH) was installed on the bridges. The system consists of strain gauges for deflection monitoring coupled with a laser rangefinder for detection of the vehicles crossing the bridge. The strain gauges are installed in both longitudinal and transversal directions on the underside of the bridge's deck, therefore, the installation can take place without any traffic disruption. The sensitivity of the used measuring system is suitable for the detection of vehicles with a gross weight exceeding 3.5 t. Before starting the measurements, the system is calibrated by crossing the bridge with trucks of known weights.

For instance, the coupled system of a laser rangefinder and strain gauges can be used to deduce data about the vehicle's speed, weight, or load distribution over the vehicle's axles. The data recorded during a single event are shown in Fig. 1 (left) and the deduced information is summarised in Table 1. The relationship between the load and the structure's response is needed for structural evaluations, namely for calibration and validation of the numerical models. Figure 1 (right) shows the relationship between

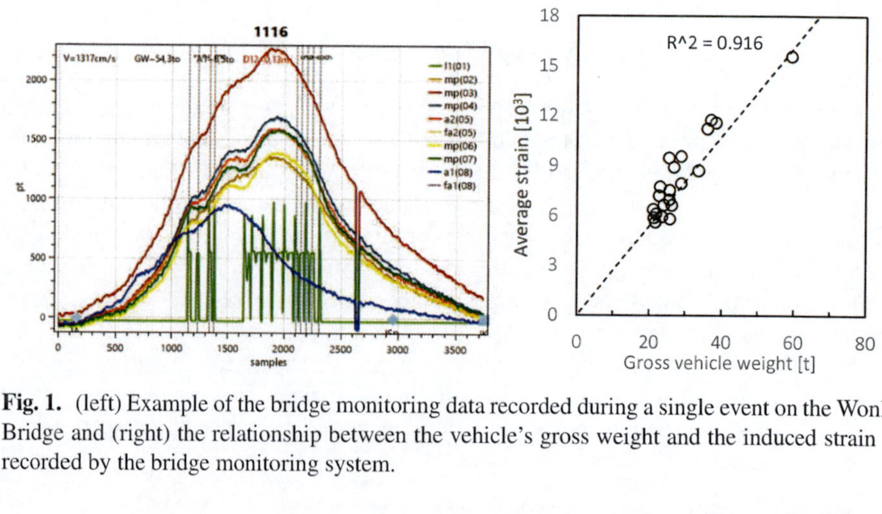

Fig. 1. (left) Example of the bridge monitoring data recorded during a single event on the Wonka Bridge and (right) the relationship between the vehicle's gross weight and the induced strain as recorded by the bridge monitoring system.

the vehicle's gross weight and the strain they induced while crossing the bridge. The generally good correlation of this data confirms the reliability of the measurement.

Table 1. Example of events and the measured values Fig. 1 (left).

TST	*ID*	*V*	*AC*	*VC*	*GW*	*Axles*	*Length*	*A2A*	*Q*	*T*	*S*	*y*
20180825 07:00:50	1116	47.4	4	80	59.4	18.6 25.4 9.0 6.4	19.96	3.91 12.50 3.55	7	23.7	15.590	4.477

TST: time and date, *ID*: event number, *V*: velocity [km/h], *AC*: axle count, *VC*: vehicle class, *GW*: gross vehicle weight [t], *Axles*: axles weights [t], *Length*: first-to-last axle distance, *A2A*: axle-to-axle distance [m], *Q*: fit quality [%], *T*: temperature, *S*: average strain [10^{-3}], *y*: lateral distance [m].

Furthermore, the obtained data allows for analysing the traffic load distribution during a week or day thus the traffic temporal patterns can be deduced. Figure 2 (left) shows that approximately 250 vehicles above 3.5 t cross the bridge during weekdays while the traffic falls to approximately 100 vehicles during the weekdays. The results of the speed monitoring showed that the traffic crosses the bridge at approximately 90 km/m; however, mainly during peak hours, the speed decrease to approximately 65 km/h as documented in Fig. 2 (right).

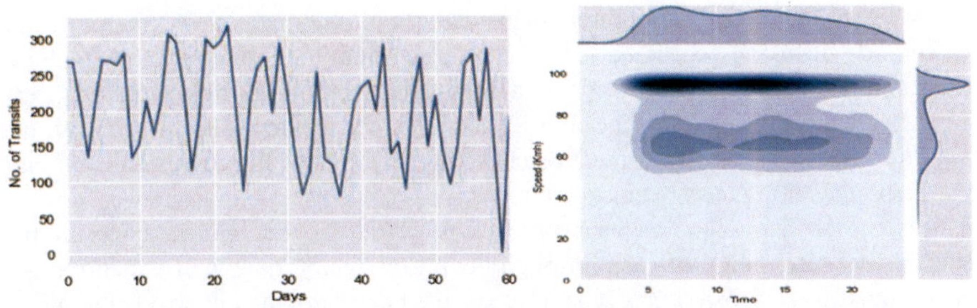

Fig. 2. Example of the daily traffic data crossing the Wonka Bridge showing the number of trucks above 3.5 t crossing the bridge (left) and how their speed varies during the day (right).

3 Numerical Simulation Methods

3.1 Non-linear Analysis

A digital twin refers to a computational model including its feedback with monitoring system, which is used to replicate the behaviour of the real process or object. In the structural engineering field, the digital twin often represents a structure or its section.

Such a computational model should be capable of simulating all important aspects of the real structure. In the case of durability assessments, on top of the usual static analysis outcomes such as structural resistance, the model should predict the deterioration mechanisms and their impact on the structural performance.

For the presented study, the numerical part of the digital twin was developed by the means of FE method within the framework of the ATENA software. The mechanical behaviour of concrete is simulated using the elasto-fracture-plastic model of Červenka et al. [2], and Červenka and Papanikolaou [3]. The applied material model can simulate the real concrete behaviour, including the compressive crushing described by the plasticity approach and the tensile cracking described by the smeared crack approach with a crack band.

3.2 Ageing Modelling

Ageing management is a term, which comprises various tools for ensuring the long-term, safe, and reliable service life of civil structures. In this sense, an online monitoring system coupled with a non-linear numerical model can be used together with the regular structures inspections to monitor the structure's health. Furthermore, the numerical model can be used to evaluate various scenarios, that might occur during the service life, and their impact on the structure's performance. In this study, we focused on the reinforcement corrosion induced by the chloride ingress as concrete transport infrastructure is vastly subjected to de-icing agents during winter.

The chloride ingress and reinforcement corrosion models applied in the analysis assume two phases of the process. First, during the induction phase, the chlorides penetrate the concrete microstructure. This process is often described by the diffusion equation [4]. The rate of chloride ingress depends both on the material behaviour through the chloride diffusion coefficient and the chloride-binding ability, and the structure's exposure conditions determined by suitable boundary conditions. Furthermore, in the presence of mechanical cracks, the chloride ingress is accelerated. Once the chloride concentration around the reinforcement reaches the critical level, the corrosion is initiated, and the propagation phase begins.

At first, the rate of the corrosion process is driven by the chloride concentration, temperature, corrosion time, and, through the pitting factor, how well the corrosion is localized. As the corrosion proceeds, it is assumed that only the uncorroded portion of the reinforcement cross-section transfers the mechanical stresses. Since the corrosion products have a larger volume than the steel, internal pressure builds up in the concrete cover. The implemented model assumes that once the critical corrosion depth is reached, spalling of the concrete cover occurs. This critical corrosion depth depends on the concrete strength, the initial reinforcement diameter and the concrete cover. After the spalling of the concrete cover, it is assumed that the corrosion continues with the rate given by the structure's exposure conditions.

Furthermore, in this study, we show an example of an analysis, where a sudden leak into the protective duct of the pre-stressed tendon was considered for the Wonka Bridge. This was done explicitly by reducing the tendons' cross-sectional area based on the prescribed corrosion rate.

Further details about the model can be found for example in reference [5]. For the study presented here, the input data were based mainly on the DuraCrete project report [4], long-term measurements of chloride ingress in concrete by the RISE Research Institutes of Sweden [6], and the corrosion rates of tendons published by the US Federal Highway Administration [7]. Further details can be found in reference [8] for the Wonka Bridge and reference [9] for the Vogelsang Bridge.

By varying the duration of the chloride attack, the different extent of the reinforcement corrosion is obtained in the numerical model. Since the chloride diffusion model takes into account the impact of mechanical cracks, the analysis results depend on the chosen load level for the application of the chloride ingress. In this study, two methods were used. For the Wonka Bridge, the chloride attack was assumed under characteristic dead loads while the ultimate design load level was used for the Vogelsang Bridge. The duration of the chloride attack was assumed up to 150 years. For the Wonka Bridge, which was constructed in 1959, it was assumed that a sudden leak occurs to the protective ducts of the unbonded cables after 63 years of its service (i.e., in 2022). The corrosion of the unbonded pre-stressed cables was assumed for 5 and 8 years. Finally, for both structures, the overloading was simulated to obtain the design structural resistance at different moments of their service life.

3.3 Structure Resistance Assessment

General structural design requirements specify that the design structural resistance (R_d) should be greater than the effects of the design loads (E_d):

$$E_d < R_d. \tag{1}$$

According to fib Model Code 2010 [10], three kinds of methods are admissible for non-linear analyses, i.e., the full probabilistic method, global resistance methods, and the partial factor method (PFM).

The global resistance method is represented by the ECOV method, which is based on the assumption that the structural resistance follows the log-normal distribution. The parameters of the distribution can be estimated by two analysis runs; one with characteristic (R_k) and one with mean (R_m) material properties. From this viewpoint, the ECOV method can be categorized as a semi-probabilistic method, whose coefficient of variation V_R is calculated as:

$$V_R = \frac{1}{1.65} \ln\left(\frac{R_m}{R_k}\right), \tag{2}$$

and the global resistance factor γ_R:

$$\gamma_R = \exp(\alpha_R \, \beta \, V_R), \tag{3}$$

where α_R is the sensitivity factor for the reliability of resistance and β is the reliability index. For general design practice, $\alpha_R = 0.8$ and $\beta = 3.8$ can be assumed.

Finally, the design resistance according to the ECOV method is expressed as:

$$R_{d,ECOV} = \frac{R_m}{\gamma_R \, \gamma_{Rd}}, \tag{4}$$

where γ_{Rd} is the uncertainty of the numerical model, which was for this study $\gamma_{Rd} = 1.16$ according to the previous research [11].

Furthermore, for the Wonka Bridge, the structural resistance obtained by the ECOV method was compared with the results of the partial factor method ($R_{d,PFM}$), which relies on design material characteristics. For their estimation, partial safety factors were based on Eurocode [12]. Excluding the model uncertainty from the material parameters, the partial safety factors give $\gamma_c = 1.46$ and $\gamma_s = 1.20$ for concrete and steel reinforcement, respectively.

4 Examples of Application

4.1 Pilot Applications

Two pilot applications of the proposed system are presented in this section for the Wonka Bridge, Pardubice, the Czech Republic over the Elbe River and the Vogelsang Bridge, Esslinger, Germany over the Neckar River. The in-situ data were previously collected within the framework of the European cyberBridge project (www.cyberbridge.eu).

The Wonka Bridge in the Czech Republic is a pre-stressed box-girder concrete bridge consisting of three arches with spans of 50 + 70 + 50 m. The cross-section depth is up to 3.5 m. The bridge was constructed between 1956 and 1959. During the service life, the bridge is loaded by road transport and pedestrians. Furthermore, the bridge is subjected to the deterioration mechanisms originating from the external environment, such as penetration of the de-icing agents and carbonation of the concrete cover. The data from the monitoring system were collected for 60 days from August until October 2018.

The Vogelsang Bridge in Germany consists of eight partial structures built in three different construction types. The bridge was built between the years of 1971 and 1973. The total length is approximately 595 m and it has a total area of 9 744 m^2 including ramps. For the monitoring, two spans of 13.8 + 13.2 m were chosen. From the structural point of view, this section is a continuous non-prestressed beam with a height of 0.6 m. The bridge monitoring ran for 61 days from Jan. until Mar. 2019.

4.2 Calibration of Digital Twins

Upon development of the model for FEM analysis, its ability to capture the behaviour of the real structure needs to be checked and, eventually, the unknown parameters in the model are calibrated using the feedback from the bridge monitoring data. This calibration is conducted by comparison of the measured and computed strains. For the Wonka Bridge, single strain measurements together with mid-span deflection data from a static load test were used while, in the case of the Vogelsang Bridge, data from two groups of strain gauges were used for the calibration. The results of the calibration are summarized in Table 2 for both bridges.

Table 2. Summary of calibration results for two digital twin models.

	Measured data	Numerical results
Wonka Bridge, Czech Republic		
Strain [μ]	8.35	8.88
Load test mid-span deflection [mm]	14.36	14.23
Vogelsang Bridge, Germany		
Strain (group 203) [μ]	77	74
Strain (group 204) [μ]	30	43

4.3 Results of Durability Assessment

As shown in Fig. 3, in the case of the Wonka Bridge, the predicted failure mode is compression/shear crushing above the support of the bridge (see Discussion section). Accounting for the brittle nature of this failure mechanism, the ultimate load-bearing capacity was assumed as the moment when compressive strains in concrete reach the value −0.0035. In the case of the Wonka Bridge, owing to the robustness of the structural design, the corrosion of the unbonded pre-stressed cables does not result in a significant reduction of the resistance, although the structural performance is compromised due to an apparent increase of the mid-span deflection.

The Vogelsang Bridge collapsed due to bending. First, reinforcement yielding occurred followed by expansion of the region with concrete crushing as shown in Fig. 4.

Fig. 3. Longitudinal view of the Wonka Bridge shows the stress in the bonded cables and cracks in the box girder at the peak load. The detail shows the development of the crushing zone above the support of the bridge.

Typical load-deflection curves are shown in Fig. 5. The data are given for the analyses with mean material parameters giving the mean structural resistance (R_m).

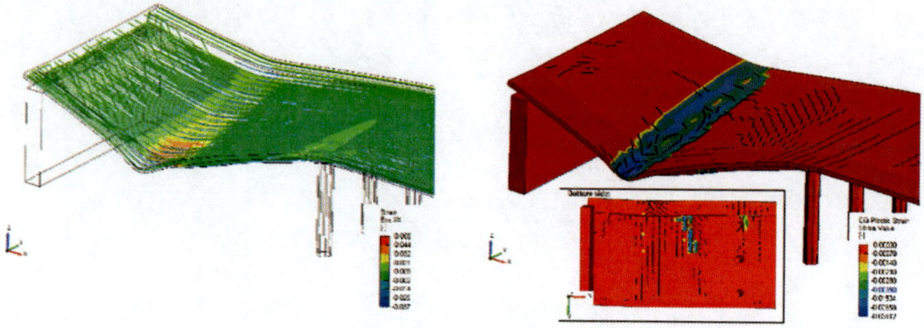

Fig. 4. (left) Deformed shame and plastic strains in the main reinforcement and (right) zone of concrete crushing at the peak load for the Vogelsang Bridge.

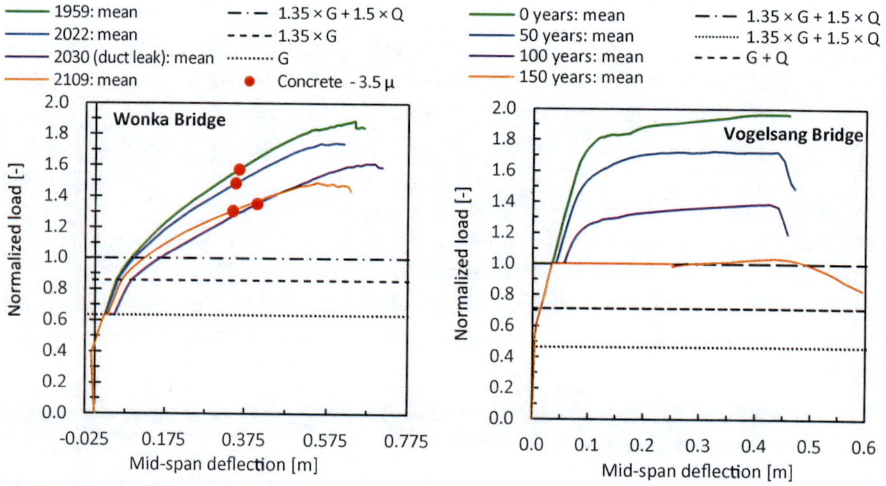

Fig. 5. Typical load-displacement curves for the Wonka Bridge (left) and Vogelsang Bridge (right) for the analyses with mean material properties. For the Wonka Bridge, the load when the compression strain of −3.5 µ is reached in concrete is indicated.

Figure 6 shows the reduction of the load-bearing capacity in time due to chloride ingress and subsequent reinforcement corrosion. The curves are given for the mean (R_m) and characteristic (R_k) structural resistance, which were used to calculate the design structural resistance according to the ECOV method ($R_{m,ECOV}$). For comparison, the development of the design structural resistance according to the PFM ($R_{m,PFM}$) is plotted for the Wonka Bridge. Both structures show good resistance against the deterioration up to 100 years of service life.

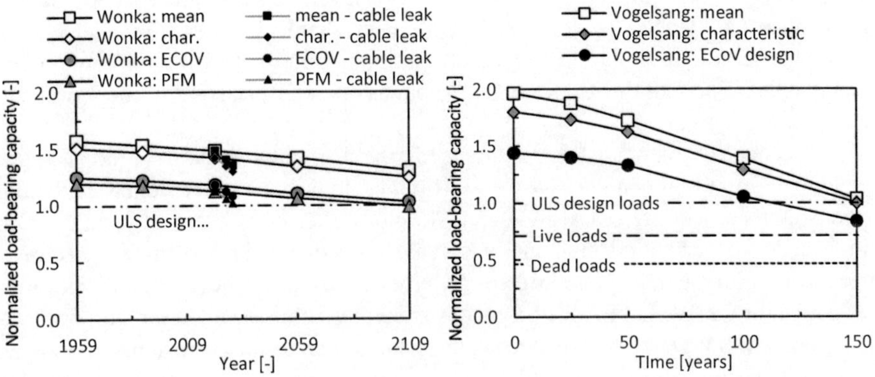

Fig. 6. Reduction of the load-bearing capacity in time due to chloride-induced reinforcement corrosion for the Wonka Bridge (left) and Vogelsang Bridge (right). For the Wonka Bridge, the black data points between the years 2022 and 2030 show a scenario, where a leak into the protective ducts of the unbonded pre-stressing cables occurs.

4.4 Discussion on the Application Results

At the peak load, the Wonka Bridge collapses due to a combination of shear/compression stresses in the concrete of the box girder above the support. The bonded pre-stressing cables efficiently transfer the shear load in the box girder due to their optimal inclination with respect to the direction of the shear stress while the unboned cables mainly contribute to the transfer of the bending moment. Both types of pre-stressing cables are bonded at the upper section of the box girder. Due to this mechanism, the concrete above the support is compressed in the vertical direction. At the maximum load, a longitudinal crack forms in the web of the box girder, which leads to the splitting of the concrete above the support. Although this failure mechanism is brittle by nature, the collapse of the bridge is preceded by a significant increase of the mid-span deflection exceeding 0.4 m due to the strain accumulation in the long external cables.

In the case of the Vogelsang Bridge, the bending failure mode develops in the mid-span at the peak load. Before the peak load is reached, reinforcement yielding occurs, which increases the deflection of the bridge. If the load is further increased, ultimate compression strain is reached in the compression zone and a region with concrete crushing forms.

Based on Fig. 6, the chloride-ingress reduces the structural performance at a faster rate for the Vogelsang Bridge than for the Wonka Bridge. This is mainly due to two factors: different load levels at the application of chloride attack and different failure mechanisms. For the Vogelsang Bridge, the chloride attack was simulated when the structure was subjected to the ULS design load level. Therefore, the rate of the corrosion process might be faster than in reality since the computed crack width, which accelerates the chloride diffusion process, maybe overestimated at this load level. This can be seen as a conservative approach. On the other hand, in the case of the Wonka Bridge, the chloride attack was assumed at the characteristic dead load level, which might be more realistic. Under lower load levels, the computed width of mechanical cracks is smaller thus the rate of chloride diffusion is lower. In addition, the main pre-stressing reinforcement is located in the middle of the walls further from the concrete surfaces thus more protected from chloride ingress.

The second reason behind the difference in the degradation rate in Fig. 6 originates from the different failure modes for the two bridges. In the case of the Vogelsang Bridge, at the peak load, a concrete crushing zone develops after reinforcement yielding in the mid-span due to the bending moment. In this case, the reduced reinforcement area directly impacts the ability to transfer the stresses in the tensile zone. On the other hand, the failure mechanism of the Wonka Bridge originates from the compression/shear stresses above the support. Although the failure mechanism is facilitated by the corrosion of the stirrups in the web of the box girder, the ability of concrete to transfer compression loads plays a key role.

For the Wonka bridge case, two safety formats are compared in Fig. 6. It shows that PFM method provides slightly more conservative results than ECOV. This can be expected as the semi-probabilistic ECOV method can be considered more advanced and with higher accuracy.

5 Summary

The paper presents a digital twin approach combining advanced monitoring system with non-linear numerical model to predict the time development of the reliability of two reinforced concrete bridges. It demonstrates the importance of the monitoring system feedback for the calibration of the model. The numerical simulation was used to predict the reinforcement corrosion and its impact on the structural resistance using the new safety formats for non-linear analysis introduced by fib model code 2010.

Acknowledgement. This paper was prepared with financial support of the Czech Technological Agency under the project CK03000023 "Digital twin for increased reliability and sustainability of concrete bridges". The financial support is greatly acknowledged.

References

1. Barcelo, L., Kline, J., Walenta, G., Gartner, E.: Cement and carbon emissions. Mater. Struct. **47**(6), 1055–1065 (2013). https://doi.org/10.1617/s11527-013-0114-5

2. Červenka, J., Červenka, V., Eligehausen, R.: Fracture-plastic material model for concrete, application to analysis of powder actuated anchors. In: Proceedings of FRAMCOS (3), pp. 1107–1116 (1998)
3. Červenka, J., Papanikolaou, V.K.: Three dimensional combined fracture–plastic material model for concrete. Int. J. Plast. **24**, 2192–2220 (2008)
4. The European Union–Brite EuRam III: Probabilistic performance based durability design of concrete structures: final technical report of Duracrete project (2000)
5. Hájková, K., Šmilauer, V., Jendele, L., Červenka, J.: Prediction of reinforcement corrosion due to chloride ingress and its effects on serviceability. Eng. Struct. **174**, 768–777 (2018)
6. Luping, T., Boubitsas, D., Utgenannt, P., Abbas, Z.: Chloride Ingress and Reinforcement Corrosion - After 20 years' field exposure in a highway environment, Borås (2018)
7. Hartt, W.H., Lee, S.-K.: Corrosion Forecasting and Failure Projection of Post-Tension Tendons in Deficient Cementitious Grout (2018)
8. Rymeš, J., Červenka, J., Pukl, R.: Advanced modelling of concrete structures for improved sustainability. In: Central Europe towards Sustainable Building (CESB22) (2022)
9. Červenka, J., Jendele, L., Žalský, J., Pukl, R.: Digital twin approach for durability and reliability assessment of bridges. In: 17th fib Symposium Proceedings (2020)
10. International Federation for Structural Concrete: fib Model Code for Concrete Structures 2010 (2013)
11. Červenka, V., Červenka, J., Kadlec, L.: Model uncertainties in numerical simulations of reinforced concrete structures. Struct. Concr. **19**, 2004–2016 (2018)
12. European Committee for Standardization: Eurocode 2: Design of concrete structures - Part 1-1: General rules, and rules for buildings (2004)

Numerical Modeling of Water Transport in Ultra-High-Performance Fiber-Reinforced Concrete

Xuande Chen[1], Abdoul Salam Bah[1], David Conciatori[1](✉), Luca Sorelli[1], Brahim Selma[2], and Mohamed Chekired[2]

[1] Université Laval, Quebec G1V 0A6, Canada
xuande.chen.1@ulaval.ca
[2] Institut de Recherche d'Hydro-Québec, Quebec J3X 1S1, Canada

Abstract. Chloride induced corrosion has been an important issue for many reinforced concrete structures. Over the past decades, numerous transport models has been developed to simulate the process. Most existing transport models focused and are validated only on well-known traditional materials such as regular concrete and cement paste. New materials such as Ultra-High-Performance Concrete (UHPC) and High-Performance Fiber-Reinforced Concrete (UHPFRC), however, are less studied from the numerical perspective. The objective of this study is to propose a new transport model for UHPC and UHPFRC to simulate the water transport process in these less permeable materials. The model has been implemented in an in-house finite element method (FEM) software TransChlor2D and validated with experimental results from Dynamic Vapor Sorption (DVS) test on crushed UHPFRC samples subjected to dry-wet cycles. The research provides a 2D numerical model allowing to estimate durability of UHPC and UHPFRC structures under realistic boundary conditions.

Keywords: UHPFRC · Transport model · Finite element method · DVS · Durability

1 Introduction

Over the past decades, concrete structures have suffered from chloride induced corrosion, especially in cold regions. During its lifetime, concrete structure's member's degradation due to corrosion will significantly decrease its durability and sustainability, thus increasing the risk of management for project managers or engineers. Amongst many major mechanisms involved in steel-bar corrosion, water transport is undoubtably the predominant one. Currently, most traditional predictive models are using 1D simplified Fick's law of diffusion to model the water transport process and the simulation objects are usually limited to regular concrete. However, with more and more uprising new materials such as Ultra-High-Performance Concrete (UHPC) and Ultra-High-Performance Fiber-Reinforced Concrete (UHPFRC) being introduced in the industry, the existing transport models can barely predict the water transport in these new materials. Therefore,

P. Rossi and J.-L. Tailhan (Eds.): SSCS 2022, RILEM Bookseries 38, pp. 90–99, 2023.
https://doi.org/10.1007/978-3-031-07746-3_9

a numerical model allowing to estimate durability of UHPC and UHPFRC structures under realistic boundary conditions is in need.

In the literature, the water transport in concrete has been modeled by many using the traditional Fick's law of diffusion [1–3], which, however, failed to capture the physical nature and kinetics of capillarity. More elaborated transport models [4–6] used multiphase equations to explain the capillarity effect by using the Darcy's law in the constitutive equations. The latter better described the transport of water vapor in concrete porosity, but the expressions are limited to 1D cases and simplified laboratory boundary conditions. A recent work from Chen et al. [7] has put efforts in extending these multiphase equations in 2D and applying them on more laboratory tests on regular concretes, but similar work can rarely be found for UHPC/UHPFRC.

On the other hand, water vapor sorption isotherm (WVSI) is an important aspect of a material's hygric property. It demonstrates the material's water sorption/desorption capacity and provides key functions to help solving the water transport equations [6, 8]. Many researchers have studied the regular concrete's WVSI of various types in search of the sorption isotherm curves/functions, but it was until recently that the similar work [9] has been done on UHPC/UHPFRC to discover the sorption isotherm curves of these new materials.

In this study, we are focusing on developing the water transport equations in UHPC/UHPFRC in a 2D framework. The work is based on the existing 1D FEM software for durability prediction, TransChlor [7, 10], which considered the coupling between temperature, carbonation, water transport, ion diffusion and chloride isothermal adsorption during the concrete corrosion process [11]. In the new transport model (TransChlor2D), not only new in-housed algorithms and equations were designed for a coupled hygrothermal problem in two dimensions, the major hygric characteristics of UHPC/UHPFRC, i.e. the water vapor sorption isotherm curves at controlled temperature, were also modeled with dynamic vapor sorption (DVS) experiments [9]. The mathematical developments of water transport equations in TransChlor2D will be presented in detail and an example of simulation will also be given. The aim of TransChlor2D is to provide interesting and useful information for the estimation of residual durability of structures made of UHPC/UHPFRC.

2 Modeling Water Transport in Porous Medium

2.1 Hygrothermal Transport Equations Implemented in TransChlor2D

The hygrothermal transport equations in TransChlor2D are [7]:

$$\begin{cases} \frac{\partial T}{\partial t} = \nabla.\left(\frac{\lambda_T(T,S)}{c_T(S)}\nabla T\right) \\ \varnothing\frac{\partial S}{\partial t} = -\nabla.(D_S(T,S)\nabla S) \end{cases} \tag{1}$$

where T is the temperature [K], t is the time [s] and S is the saturation degree in concrete pores [−]. The transport coefficients are: λ_T the thermal conductivity and c_T the heat capacity.

By doing the integration by parts, the weak form of thermal equation is thus obtained (using 2D Cartesian coordinates):

$$\iint \delta T.\frac{\partial T}{\partial t}d\Omega - \iint \nabla\delta T.\left(\frac{\lambda_T(T,S)}{c_T(S)}\nabla T\right)d\Omega + \iint_{\partial\Omega}\delta T.\overrightarrow{q_T}.\vec{n}ds = 0 \tag{2}$$

Same for the hygric transport equation:

$$\iint \delta S.\frac{\partial S}{\partial t}d\Omega - \iint \nabla\delta S.(D_S\nabla S)d\Omega + \iint_{\partial\Omega}\delta S.\overrightarrow{q_S}.\vec{n}ds = 0 \tag{3}$$

where δT and δS are the extended terms of "virtual displacement" from the definition of finite element method, $\overrightarrow{q_T}$ and $\overrightarrow{q_S}$ are the flux vectors representing heat and water flow throughout boundaries and $\vec{n}$ is the elemental unit vector perpendicular to the closed element boundary.

For mathematical discretization, let us consider using a square element with four nodes and four Gauss points as the elemental computation unit. Its coordinates (ξ,η) located in a Cartesian coordination are shown in the figure below (Fig. 1):

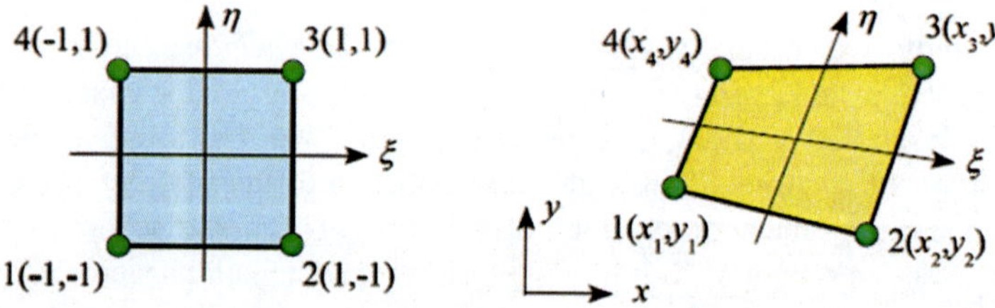

Fig. 1. Basic element for finite element shape function transformations (left: local Cartesian coordinates and right: global Cartesian coordinates).

Which has a shape function be like:

$$[N]^T = \frac{1}{4}\begin{bmatrix}(1-\xi)(1-\eta)\\(1+\xi)(1-\eta)\\(1+\xi)(1+\eta)\\(1-\xi)(1+\eta)\end{bmatrix} \tag{4}$$

And its derivative form on two orthodox axes are:

$$[\frac{\partial N}{\partial \xi}]^T = \frac{1}{4}\begin{bmatrix}-(1-\eta)\\(1-\eta)\\(1+\eta)\\-(1+\eta)\end{bmatrix} \tag{5}$$

and

$$[\frac{\partial N}{\partial \eta}]^T = \frac{1}{4}\begin{bmatrix}-(1-\xi)\\-(1+\xi)\\(1+\xi)\\(1-\xi)\end{bmatrix} \tag{6}$$

The derivative matrix for the shape functions then becomes:

$$[B(\xi,\eta)] = \frac{1}{4}\begin{bmatrix} -(1-\eta) & (1-\eta) & (1+\eta) & -(1+\eta) \\ -(1-\xi) & -(1+\xi) & (1+\xi) & (1-\xi) \end{bmatrix} \tag{7}$$

And the corresponding Jacobian matrix is:

$$[J(\xi,\eta)] = \begin{bmatrix} \frac{\partial N_1}{\partial \xi} & \frac{\partial N_2}{\partial \xi} & \frac{\partial N_3}{\partial \xi} & \frac{\partial N_4}{\partial \xi} \\ \frac{\partial N_1}{\partial \eta} & \frac{\partial N_2}{\partial \eta} & \frac{\partial N_3}{\partial \eta} & \frac{\partial N_4}{\partial \eta} \end{bmatrix} \cdot \begin{bmatrix} x_1 & y_1 \\ x_2 & y_2 \\ x_3 & y_3 \\ x_4 & y_4 \end{bmatrix}$$

$$= \begin{bmatrix} \frac{\partial N_1}{\partial \xi}x_1 + \frac{\partial N_2}{\partial \xi}x_2 + \frac{\partial N_3}{\partial \xi}x_3 + \frac{\partial N_4}{\partial \xi}x_4 & \frac{\partial N_1}{\partial \xi}y_1 + \frac{\partial N_2}{\partial \xi}y_2 + \frac{\partial N_3}{\partial \xi}y_3 + \frac{\partial N_4}{\partial \xi}y_4 \\ \frac{\partial N_1}{\partial \eta}x_1 + \frac{\partial N_2}{\partial \eta}x_2 + \frac{\partial N_3}{\partial \eta}x_3 + \frac{\partial N_4}{\partial \eta}x_4 & \frac{\partial N_1}{\partial \eta}y_1 + \frac{\partial N_2}{\partial \eta}y_2 + \frac{\partial N_3}{\partial \eta}y_3 + \frac{\partial N_4}{\partial \eta}y_4 \end{bmatrix} \tag{8}$$

Therefore, for the thermal diffusion equation along (Eq. (8)), the integrated matrix form of the balance equation becomes:

$$\iint \delta S.\frac{\partial s}{\partial t}d\Omega - \iint \nabla\delta S.(D_S \nabla S)d\Omega + \iint_{\partial\Omega} \delta S.\overrightarrow{q_S}.\vec{n}ds = 0 \tag{9}$$

On one element, we use the Gauss' principal of integration for four points:

$$\int_{-1}^{1}\int_{-1}^{1} f(\xi,\eta)d\xi d\eta = \sum_{i=1}^{4} f(\xi_i,\eta_i)w_i \tag{10}$$

where $f(\xi,\eta)$ is the function or the variable to be integrated and w_i is the weight function.

Then the discretized thermal equation can be written as (explicit scheme):

$$\{A\}^e\{\overline{T}\}^e + \{b\}^e \cdot \frac{1}{\Delta t}\left(\{T\}^e - \left\{T^{t-1}\right\}^e\right) = 0 \tag{11}$$

with:

$$w_i = 1, (\xi_i,\eta_i) = (\pm\frac{1}{\sqrt{3}}, \pm\frac{1}{\sqrt{3}}) \tag{12}$$

Then to simplify the integrated equation, we define two nominal matrices:

$$\{A\}^e = \sum_{i=1}^{4}\left([B(\xi_i,\eta_i)]^T \frac{\lambda_T(T(\xi_i,\eta_i), S(\xi_i,\eta_i))}{c_T(S(\xi_i,\eta_i))}[B(\xi_i,\eta_i)]detJ(\xi_i,\eta_i) * 1\right) \tag{13}$$

and

$$\{b\}^e = \sum_{i=1}^{4}\left([N(\xi_i,\eta_i)]^T[N(\xi_i,\eta_i)][J(\xi_i,\eta_i)] \cdot 1\right) \tag{14}$$

The thermal system of heat transfer equation finally leads to:

$$\left(\frac{\{A\}^e}{2} + \frac{\{b\}^e}{\Delta t}\right)\{T\}^e = \left(\frac{\{b\}^e}{\Delta t} - \frac{\{A\}^e}{2}\right)\left\{T^{t-1}\right\}^e \tag{15}$$

Following the similar steps, the hygric system of water transport equation yields:

$$\left(\frac{\left\{A'\right\}^e}{2}+\frac{\varnothing\left\{b'\right\}^e}{\Delta t}\right)\{S\}^e=\left(\frac{\varnothing\left\{b'\right\}^e}{\Delta t}-\frac{\left\{A'\right\}^e}{2}\right)\left\{S^{t-1}\right\}^e \tag{16}$$

where:

$$\left\{A'\right\}^e=\iint_{\Omega}B^T D_S B d\Omega=\sum_{i=1}^{4}\left([B(\xi_i,\eta_i)]^T D_S(\xi_i,\eta_i)[B(\xi_i,\eta_i)]detJ(\xi_i,\eta_i)*1\right) \tag{17}$$

and

$$\left\{b'\right\}^e=\varnothing\iint_{\Omega}N^T N d\Omega=\varnothing\sum_{i=1}^{4}\left([N(\xi_i,\eta_i)]^T[N(\xi_i,\eta_i)][J(\xi_i,\eta_i)]\cdot 1\right) \tag{18}$$

To solve the two explicit equations simultaneously, a Gauss elimination method has been adopted and the in-housed algorithms are developed in the software TransChlor2D.

2.2 Water Vapor Sorption Isotherm in UHPC/UHPFRC

To model the water transport in UHPC/UHPFRC, a set of dynamic vapor sorption (DVS) experiments have already been carried out by Chen et al. [9]. Based on the experimental sorption curves, formulas for both main isotherm curves and scanning curves are developed in this study. In this paper, the empirical hysteretic models initially developed for ordinary concrete [12–14] was adopted with proper modifications.

Introducing the definition of moisture capacity ξ, which is the slope of scanning curves in sorption hysteresis (Fig. 2), the scanning curves can be analytically determined by using the empirical model proposed by Pedersen [15]:

$$\xi_{hys,a}=\frac{0.1(w-w_a)^2\xi_d+(w-w_d)^2\xi_a}{(w_d-w_a)^2} \tag{19}$$

$$\xi_{hys,d}=\frac{(w-w_a)^2\xi_d+0.1(w-w_d)^2\xi_a}{(w_d-w_a)^2} \tag{20}$$

where ξ_{hys} is the moisture capacity (or the slope of scanning curves) at a specific RH and the subscripts a and d represents for absorption and desorption, respectively. w is the estimated pore water content (water mass/dry sample mass) at current RH, while w_a and w_d are corresponding values derived from the main absorption and desorption curves. ξ_a and ξ_d are the moisture capacities on corresponding mains isotherm curves.

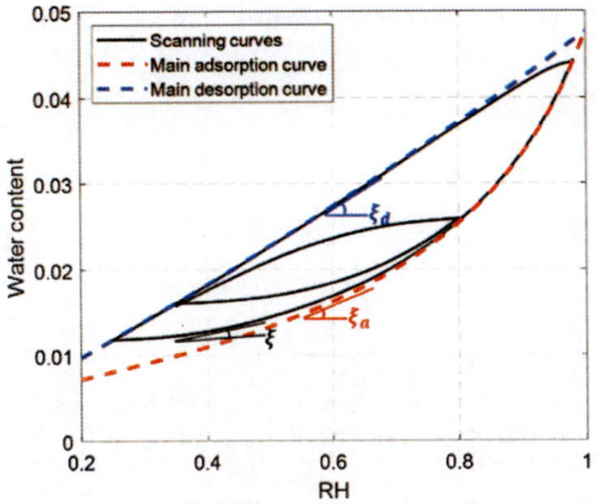

Fig. 2. Schematic representation of WVSI hysteresis with the main isotherm curves, the drying-wetting scanning curves, and the moisture capacity ξ.

The original empirical expressions from Ozbolt et al. [16] for main absorption and desorption isotherms were intended to apply on ordinary concrete (OC) with water-to-binder ratio of 0.48:

$$w_{a,OC}(H) = 4.79 \cdot e^{-\frac{1}{1.13}\cdot\ln(1-\frac{\ln(H)}{0.214})} \tag{21}$$

$$w_{d,OC}(H) = 4.76 \cdot e^{-\frac{1}{0.18}\cdot\ln(1-\frac{\ln(H)}{4.85})} \tag{22}$$

To extend its use on UHPFRC, the mathematically fitted parameters were recalibrated with the main isotherms measured in this study. Unlike ordinary concretes, the main isotherm of tested UHPFRC samples had a triangular shape. For the main adsorption curve, the expression of water content w_a depending on the humidity H is:

$$w_{a,UHPFRC}(H) = 5.14 \cdot e^{-\frac{1}{0.63}\cdot\ln(1-\frac{\ln(H)}{0.289})} \tag{23}$$

As for the main desorption curve, to account for the triangular shape of UHPFRC's sorption isotherm, two different expressions are given for low RH (RH $\leq$ 45%) and high RH (RH $\geq$ 45%), respectively:

$$w_{d,UHPFRC}(H) = \begin{cases} 5.14 \cdot e^{-\frac{1}{0.18}\cdot\ln\left(1-\frac{\ln(H)}{20.5}\right)} (0.45 \leq H \leq 1) \\ 8.26 \cdot e^{-\frac{1}{0.22}\cdot\ln\left(1-\frac{\ln(H)}{4.95}\right)} (0 \leq H < 0.45) \end{cases} \tag{24}$$

The water content during scanning between the main adsorption and desorption isotherms is determined by the following relation:

$$w_{j+1} = w_j + \xi_{hys} \cdot \Delta H \tag{25}$$

in which w_j is the water content at previous RH step, $\Delta H = H_{j+1} - H_j$ is the RH step in loading cycles, where the subscript j represents the previous RH step and j + 1 represents the current RH step.

Knowing that the water content and the saturation degree have the following relation:

$$S = \frac{w}{\phi} \tag{26}$$

Combining Eq. (23)–(24) and Eq. (26), it allows us to build an explicit isotherm expression between saturation degree and relative humidity:

$$S(H) = \frac{w(H)}{\phi} \tag{27}$$

3 Simulation of Water Transport in UHPFRC Samples

3.1 Water Vapor Sorption Isotherm in UHPC/UHPFRC

The hysteretic isotherm model described in Sect. 2.2 was then applied on data collected from drying test samples [9] to test its validity in modelling UHPFRC's sorption-desorption behavior. In Fig. 3, the simulation results using both models (for normal concrete and for UHPFRC) are reported. The lines represent the empirically modelled adsorption-desorption isotherm curves using adapted formula for UHPFRC, while the dot lines represent empirically the modelled adsorption-desorption isotherm curves using formula for normal concrete. The scanning curves are modelled with the scanning hysteretic formulas with combination of the Euler forward formula.

It can be clearly observed that the original hysteretic sorption isotherm model for normal concrete, is not compatible with the sorption isotherm of UHPFRC, while the newly adapted model (Eq. (25)), on the contrary, shows a good performance in capturing both the main isotherm curves and the hysteresis tendances, since a rather satisfying correlation between the numerical prediction and the experimental data can be observed. Thanks to the switching conditions applied in the desorption function, the experimentally measured unusual triangular sorption isotherm is correctly reproduced by the numerical model as well.

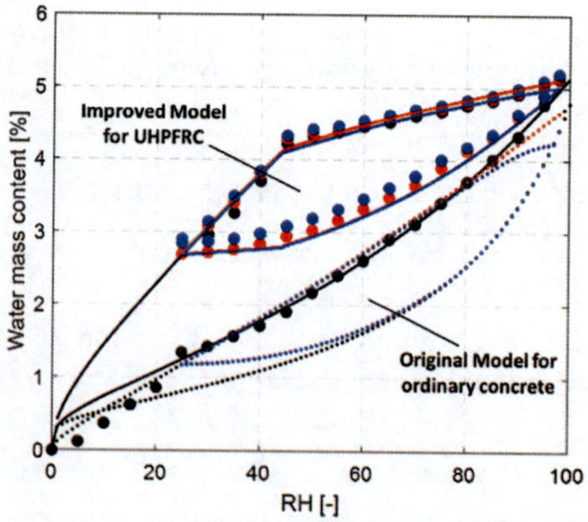

Fig. 3. Modelling the sorption hysteresis in UHPFRC, with traditional model (dot lines) and the adapted model for UHPRFC (lines)

4 Simulation of UHPFRC's Drying Test with TransChlor2D

Having the isotherm functions calibrated, the complete water transport model developed in Sect. 2.1 was then validated by running a simulation of a drying test on a real UHPFRC beam of dumbbell shape (Fig. 4 left). The tested UHPFRC samples were fabricated with the ASTM C494 standard [17]. The composition was: UHPC premix (Portland Cement 712 kg/m^3, Fine sand 1020 kg/m^3, Silica Fume 231 kg/m^3, Ground Quartz 211 kg/m^3), superplasticizer (35 kg/m^3) of type A&F [17], steel fibers of 2 mm long (156 kg/m^3) with a volume fraction of 2%, and water (125 kg/m^3). The water-to-binder ratio was about 0.17 [18]. The compressive strength of the tested UHPFRC was 120 MPa at 28 days. A flow-table test was conducted with the apparatus from ASTM C230 [19] and the slump diameter for cylinder was about 210 mm [20]. The UHPFRC samples had capillary pores between 2 and 50 nm and an overall porosity of 5%, according to the preliminary water absorption and mercury intrusion porosimetry (MIP) test results [20]. After casting, the original Ductal® beams were cured in a humidity room (100% RH) for 4 months and then saturated in laboratory for 1 year before drying tests.

Numerically, the whole sample's initial relativity was considered at 99% RH (H = 0.99) and the boundary conditions applied to the samples were 25% RH (H = 0.25). Most of the sample's surfaces were sealed with resin to have only unidimensional drying in a depth of 55 mm. By combining the isotherm functions (Eq. (25)) with the water mass conservation equations (Eq. (16)–(17)), we obtained the relative humidity profiles in the UHPFRC sample after a drying of 1 month (720 h). As a comparison, same simulation was also performed on regular concrete with exactly same dimensions (profiles in dot lines). It can be clearly observed that after one month of drying, the water loss on UHPFRC is quite limited to the first few layers (2–3 mm) whereas the water loss on regular concrete reaches almost 10 mm from the boundary.

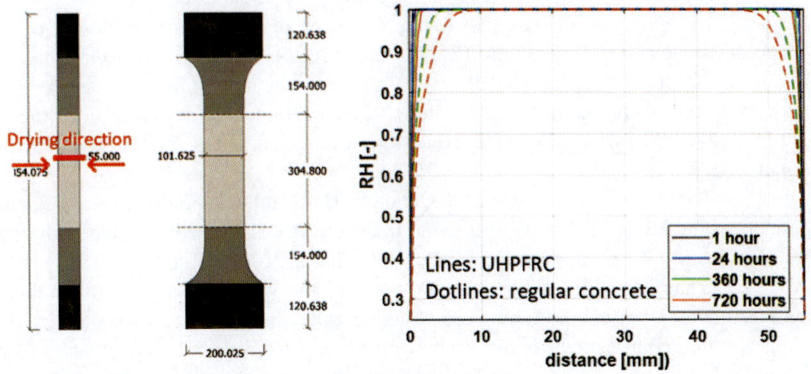

Fig. 4. Simulation of an UHPFRC beam of full scale, right: relative humidity profiles after drying at 1 h, 24 h, 360 h and 720 h.

5 Conclusions

This study demonstrates the successful implement and solution of the water transport problem in ordinary concrete, UHPC and UHPFRC, based on a modified diffusion equation coupled with a water vapor sorption isotherm function. A numerical example showed the FEM solution of the coupled transport equations with the effect of sorption isotherm and the excellent performance of the numerical model. The following conclusions can be drawn:

- The finite element approach developed in this paper is appropriate to simulate the water transport process in UHPFRC materials.
- The 2D water transport model, TransChlor2D, appropriately replicates the relative humidity profiles for drying test at constant temperature and environment humidity, showing promising simulation results, and can be potentially used as a powerful tool to estimate the long-term water penetration depth in real-life structures.
- Extend the model to a 3D framework to better simulate real structures.
- Corporate the water transport model with wider range of UHPC/UHPFRC materials.

References

1. Bazant, Z.P., Najjar, L.J.: Drying of concrete as a nonlinear diffusion problem. Cem. Concr. Res. **1**, 461–473 (1971). https://doi.org/10/frtk4t
2. Xi, Y., Bažant, Z.P., Jennings, H.M.: Moisture diffusion in cementitious materials adsorption isotherms. Adv. Cem. Based Mater. **1**, 248–257 (1994). https://doi.org/10/b229wz
3. Xi, Y.: Modeling chloride penetration in saturated concrete. J. Mater. Civ. Eng. **11**, 8 (1999). https://doi.org/10/bz9rhp.
4. Coussy, O., Coussy, O.: Poromechanics, 2nd edn. Wiley, Chichester (2004)
5. Mainguy, M., Coussy, O., Baroghel-Bouny, V.: Role of air pressure in drying of weakly permeable materials, J. Eng. Mech. **127**, 582–592 (2001). https://doi.org/10/bgt6hn
6. Zhang, Z., Thiery, M., Baroghel-Bouny, V.: Numerical modelling of moisture transfers with hysteresis within cementitious materials: verification and investigation of the effects of repeated wetting–drying boundary conditions. Cem. Concr. Res. **68**, 10–23 (2015). https://doi.org/10/gc4pg6
7. Chen, X., Sanchez, T., Conciatori, D., Chaouki, H., Sorelli, L., Selma, B., Chekired, M.: Numerical modeling of 2D hygro-thermal transport in unsaturated concrete with capillary suction. J. Build. Eng. **45**, 103640 (2022). https://doi.org/10/gnm4vt
8. Wu, M., Johannesson, B., Geiker, M.: A study of the water vapor sorption isotherms of hardened cement pastes: possible pore structure changes at low relative humidity and the impact of temperature on isotherms. Cem. Concr. Res. **56**, 97–105 (2014). https://doi.org/10/f3pbqd
9. Chen, X., et al.: An experimental study on the sorption in UHPFRC: adaptation of the DVS measurement procedure. In: Pellegrino, C., Faleschini, F., Zanini, M.A., Matos, J.C., Casas, J.R., Strauss, A. (eds.) EUROSTRUCT 2021. LNCE, vol. 200, pp. 1278–1285. Springer, Cham (2022). https://doi.org/10.1007/978-3-030-91877-4_145
10. Conciatori, D., Sadouki, H., Brühwiler, E.: Capillary suction and diffusion model for chloride ingress into concrete. Cem. Concr. Res. **38**, 1401–1408 (2008). https://doi.org/10/b3gvv4

11. Chen, X., Sanchez, T., Conciatori, D., Chaouki, H., Sorelli, L., Selma, B., Chekired, M.: Effect of the temperature on the water transport by capillarity into the concrete porosity. In: 4th International Rilem Conference on Microstructure Related Durability of Cementitious Composites, p. 8 (2021)
12. Mualem, Y.: A new model for predicting the hydraulic conductivity of unsaturated porous media. Water Resour. Res. **12**, 513–522 (1976). https://doi.org/10/bb7g6c
13. Mualem, Y.: Modified approach to capillary hysteresis based on a similarity hypothesis. Water Resour. Res. **9**, 1324–1331 (1973). https://doi.org/10/frt9wh
14. Mualem, Y.: A conceptual model of hysteresis. Water Resour. Res. **10**, 514–520 (1974). https://doi.org/10/drztnk
15. Cr, P.: Combined Heat and Moisture Transfer in Building Constructions. Technical University of Denmark, Thermal Insulation Laboratory (1990)
16. Ožbolt, J., Oršanić, F., Balabanić, G.: Modeling influence of hysteretic moisture behavior on distribution of chlorides in concrete. Cem. Concr. Compos. **67**, 73–84 (2016). https://doi.org/10/gf27kn
17. ASTM C494.C494M - Chemical Admixtures for Concrete. ASTM International (2017)
18. Graybeal, B.A., Tanesi, J.: Durability of an ultrahigh-performance concrete. J. Mater. Civ. Eng. (2007)
19. ASTM C230.C230M - Flow Table for Use in Tests of Hydraulic Cement. ASTM International (2014)
20. Provete Vincler, J., Sanchez, T., Turgeon, V., Conciatori, D., Sorelli, L.: A modified accelerated chloride migration tests for UHPC and UHPFRC with PVA and steel fibers. Cem. Concr. Res. **117**, 38–44 (2019). https://doi.org/10/gfrjqj

Predicting Early Age Temperature Evolution in Massive Structures from Non-standard Characterization Test

Arnaud Delaplace(✉), Regis Bouchard, and Paul O'Hanlon

Holcim Innovation Center, Saint-Quentin-Fallavier, France
arnaud.delaplace@holcim.com

Abstract. Two methods are proposed to characterize the thermal fingerprint of a binder from an insulated mockup test realized on a jobsite, rather than using a classic quasi-adiabatic test in lab condition. The first part introduces the model used to compute the temperature evolution at early age, using a Finite Element solver. The second part presents the two methods. The first method is based on the estimation of the heat loss of the mockup. Then, a methodology similar to a standard quasi-adiabatic test is done, by computing (i) the total heat generated during hydration, (ii) the hydration rate and (iii) the chemical affinity. The second method is based on an imposed form of the hydration affinity function, with three parameters. A Finite Element model of the mockup is used, and a minimisation between the computed temperature evolution and the experimental one allow to identify the three parameters. The two methods are validated on a purely numerical study. Then, a real example is presented, and the relevance of the two methods are discussed.

Keywords: Heat of hydration · Massive structures · Early age behavior · Temperature evolution

1 Introduction

In massive concrete structures, heat generated during the exothermic hydration reaction of the binder can lead to important temperature at core. The risk associated to an excessive temperature is nowadays well known and has been largely communicated in the concrete industry (see for example [1]). The consequence is that today, for all big construction projects, thresholds are imposed on the maximal temperature and on the maximal gradient temperature. It is therefore necessary to have an accurate prediction of early age temperature at any point and any time in the structure. Different methods and recommendations have been published in the last two decades for predicting temperature evolution [2,5,9]. These methods are usually based on the characterization of the binder hydration by using a standardized quasi-adiabatic test. Unfortunately, depending on the country and on the local standards, such a test can not be accessible, and just a temperature monitoring of a mockup is available. The objective of

P. Rossi and J.-L. Tailhan (Eds.): SSCS 2022, RILEM Bookseries 38, pp. 100–109, 2023.
https://doi.org/10.1007/978-3-031-07746-3_10

this study is to propose a general framework allowing to predict temperature evolution in massive structures, whatever the type of hydration characterization test.

2 Background

Predict temperature in concrete structure is usually done by solving the heat equation, using for example a Finite Element solver:

$$\rho C_p \frac{\partial T}{\partial t} - \lambda \nabla^2 T = Q_\infty \frac{\partial \xi(T)}{\partial t} \tag{1}$$

The source term on the right hand side, describing the heat release during the binder hydration, can be written:

$$\frac{\partial \xi(T)}{\partial t} = A(\xi) \exp\left(-\frac{E_a}{RT}\right) \tag{2}$$

where the following notations are used:

- t: time [s]
- T: temperature [K]
- ξ: degree of hydration ($0 \leq \xi \leq 1$)
- E_a: activation energy [J/mol]
- A: affinity function [s^{-1}]
- ρ: material density [kg/m^3]
- C_p: specific heat capacity [J/(kg.K)]
- $Q_\infty = \rho C_p (T_\infty - T_0)$: hydration heat [$J/m^3$]
- λ: thermal conductivity [W/(m.K)]
- $R = 8.314$: perfect gas constant [J/(mol.K)]

The computation is straightforward if and only if the material parameters are well defined. Some models allow to estimate the hydration parameters (E_a, A, Q_∞) based on the binder composition [6,9]. But the increasing usage and the wide varieties of alternative cementitious materials limit the applicability of these models. It is therefore preferred to characterize at lab scale the *thermal fingerprint* of the binder. It's achieved by using a quasi-adiabatic test, at one or better at two temperatures [3,4]. Unfortunately, such quasi-adiabatic test are not globally available, and for some projects the site environment doesn't allow to perform the test with the required specifications (temperature-controlled lab, identical initial temperature for reference sample, test apparatus and raw materials).

On the other hand, a characterization of the temperature evolution in an 1 m^3 insulated mockup, in ambient external conditions, is often required in the project specifications. In some regions, such mockup is also a standard to identify the maximum temperature reached during the hydration of a binder. Considering this source of experimental data, we propose in this study to identify the thermal fingerprint from such mockup. We will evaluate two procedures, described in the next section.

3 Identification of the Thermal Fingerprint from a 1 m^3 Mockup Test

In this section, two procedures are presented to identify the thermal fingerprint of a binder from the internal temperature evolution of a 1 m^3 mockup test.

3.1 Identification Using an Evaluation of the Heat Loss (Method A)

Global Procedure. In this approach, an initial estimation of the heat loss of the mockup is done. This approach is similar the the quasi-adiabatic tests, in which heat loss calibration parameters are characterized for each apparatus. Following the quasi-adiabatic standard test, the heat release is computed as:

$$q(t) = C_{\text{tot}}\Big(\theta(t) - \theta(0)\Big) + \int_0^t \alpha\Big(\theta(u)\Big)\theta(u) \quad du \tag{3}$$

where:

- $\theta(t) = T^{\text{sample}}(t) - T^{\text{ref}}(t)$,
- C_{tot} is the total heat capacity of the setup defined as $C_{\text{tot}} = C_{\text{conc}} + C_{\text{cal}}$,
- C_{conc} and C_{cal} are respectively the heat capacity of the concrete sample and of the calorimeter,
- $\alpha\Big(\theta(u)\Big)$ is the heat loss of the mockup.

The evolution of hydration is then computed from the following relation:

$$\frac{\xi(t)}{\xi_\infty} = \frac{q(t)}{q_\infty} \tag{4}$$

where ξ_∞ is the final degree of hydration and q_∞ is the total heat release. Knowing the water-binder ratio W/B, the ultimate hydration degree can be computed using the relation $\xi_\infty = 1 - \exp(1 - 3.25W/B)$ [10]. Finally, assuming an Arrhenius law for describing the temperature dependence of the hydration rate, the chemical affinity reads:

$$\tilde{A}(\xi) = \frac{\mathrm{d}\xi}{\mathrm{d}t} \exp\left(\frac{E_a}{RT}\right) \tag{5}$$

In the last equation, the activation energy could be identified from a second test at a different temperature. Because it's rarely the case, the value of the activation energy will be imposed (see for example [6] for reference values).

Identification of the Heat Loss Parameter. The heat loss parameter α can be easily identified following the procedure shared in [3]: in the steady state regime, the heat produced by a heat source of power P applied in the insulated cube is equal to the heat loss:

$$P = \alpha\theta_c \tag{6}$$

with θ_c the difference of temperature inside the cube and of the reference one (initial temperature of concrete if we assume a constant ambient temperature equals to this temperature).

3.2 Best Fitting Using a Finite Element Model of the Mockup (Method B)

For this procedure, an assumption will be done on the thermal fingerprint model. We will use the model proposed in [7,8], but any other relevant model can be used. The degree of hydration ξ reads:

$$\frac{d\xi}{dt} = \tilde{A}_{25}(\xi) \tag{7}$$

with $\tilde{A}_{25}$ the chemical affinity at 25 °C, defined as:

$$\tilde{A}_{25}(\xi) = B_1 \left(\frac{B_2}{\xi_\infty} + \xi \right) (\xi_\infty - \xi) \exp \left(-\eta \frac{\xi}{\xi_\infty} \right) \tag{8}$$

where ξ_∞ is the ultimate hydration degree, B_1, B_2 and η are coefficients to be identified. The affinity function at temperature T is then:

$$\tilde{A}_T = \tilde{A}_{25} \exp \left[\frac{E_a}{R} \left(\frac{1}{273.15 + 25} - \frac{1}{T} \right) \right] \tag{9}$$

The heat release $q(t)$ is linked to the hydration degree $\xi(t) = q(t)/Q_{pot}$ where Q_{pot} is the infinite heat release for $\xi = 1$.

The identification of the model parameters will be done with an inverse analysis. A Finite Element model of the mockup will be done. The temperature evolution in the cube center will be computed using the hydration model. The model parameters will be identified by minimising the difference between the numerical response and the experimental one.

3.3 Comparison of the Two Methods

Both methods have a drastically different approach. Table 1 gives a comparison of the two methods.

Table 1. Comparison of the two approaches

	Method A	Method B
Pro's	Free shape of the chemical affinity Very fast computation time (few seconds)	Bounded model avoiding any nonphysical thermal fingerprint
Con's	Simplified assumption used to represent the heat loss (not relevant for limited insulation)	Slow identification (up to 30 mns) Imposed shape of the chemical affinity

4 Numerical Analysis

In this section, a full numerical study is performed:

1. A synthetic adiabatic curve is generated for modelling the heat release in adiabatic conditions, allowing to create a thermal fingerprint.
2. A 1 m^3 mockup is modelled, and the temperature evolution is computed using the binder thermal fingerprint.
3. A reverse identification of the thermal fingerprint is performed, using Methods A and B.

4.1 Synthetic Adiabatic Curve

We consider a concrete with the component dosages given in Table 2. The concrete properties are assumed to be: density $\rho = 2400\,\mathrm{kg/m^3}$, thermal conductivity $\lambda = 1.9\,\mathrm{W/m/K}$, heat capacity $C_{conc} = 1000\,\mathrm{J/kg/K}$. The activation energy is $E_a = 42000\,\mathrm{J/mol}$.

Table 2. Concrete design used for the numerical analysis

Components	Dosage (kg/m^3)	Components	Dosage (kg/m^3)
CEMI	280	Fine aggregates	950
Fly ash	70	Coarse aggregates	1005
Water	140		

In order to model a temperature evolution representative of a real concrete, and to have a different model than the one presented in Sect. 3.2, the CIRIA hydration model is used [9]. The heat release reads:

$$q(t) = \frac{Q_{ult}}{2}\left(1 - \exp(-Bt^c)\right) + \frac{Q_{ult}}{2}\frac{t - t_2}{t - t_2 + D} \tag{10}$$

Assuming a heat generated after 41 hrs for the CEMI to be Q_{41}^{CEMI}=330 kJ/kg, and considering the 20% of fly ash in the binder, the model parameters are computed and given in Table 3.

4.2 1 m^3 Model

Thanks to the double symmetries, a quarter of a 1 m^3 cube is modeled (Fig. 1-left).

The initial temperature of concrete is 20 °C. We consider on all faces a 25 mm plywood formwork, plus an additional 10 cm-thick expanded polystyrene layer. It leads to an exchange coefficient $h = 0.32\,\mathrm{W/m^2/K}$. We consider a 24 h-period sinusoidal curve for ambient temperature evolution, varying from $T_{night} = 15\,°\mathrm{C}$ to $T_{day} = 25\,°\mathrm{C}$.

Table 3. Concrete model parameters

Parameter	Value	Unit	Parameter	Value
Q_{41}	270	kJ/kg	B	0.011724
Q_{ult}	382	kJ/kg	C	1.58
t_2	3.972	h	D	10.462

The temperature evolution at center, that will be used for identification, is represented in Fig. 1-right.

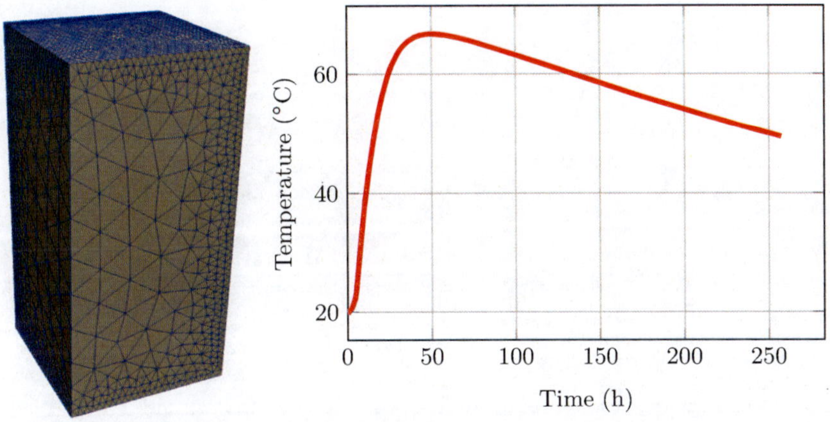

Fig. 1. Mesh used for the modeling (left) and reference temperature in the cube center (right).

4.3 Thermal Fingerprint Identification

Method A. For method A, the first step is to estimate the heat loss coefficient of the insulated cube (Eq. (6)). We assume a simple constant for α. Imposing a constant voluminal power of 50 W/m^3, the temperature increase in the steady state regime is $\theta_c = 27.87\,°\mathrm{C}$. It gives the heat loss parameter $\alpha = 6459$ J/h/K.

Using this coefficient in Eq. (3), the complete thermal fingerprint is directly obtained (Eqs. (4) and (5)). The computing time is negligible for this method.

Method B. In that case, the identification is done by minimizing the difference between the numerical model and the experiment. For this numerical study, the identified coefficients of Eq. (8) are:

$$B_1 = 3.61 \times 10^{-4} \qquad B_2 = 1.15 \times 10^{-3} \qquad \eta = 5.8$$

Without any specific optimization, the computation time is around 90 mns on a laptop.

4.4 Comparison of the Two Methods

In this specific numerical study, the exchange coefficient for the insulated mockup is known, so one can expect to have a good identification of the thermal fingerprint. In order to compare the two methods, once the thermal fingerprint is obtained, we simulate the $1\,\mathrm{m}^3$ mockup in adiabatic conditions, and with an exchange coefficient $h = 0.32\,\mathrm{W/m/K}$ used for the reference calculation. Results are shown respectively in Fig. 2 and Fig. 3. The agreement of the predictions are good, with a difference lower than $1\,^\circ\mathrm{C}$ all along the curves.

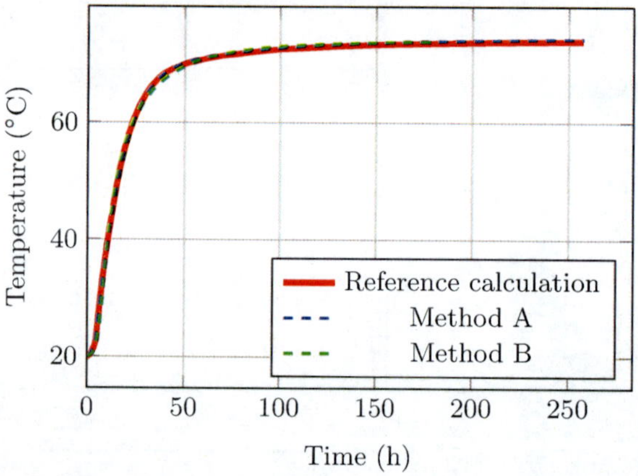

Fig. 2. Comparison of the temperature evolution for methods A and B with the reference solution, in adiabatic conditions.

5 Real Case Study

The previous numerical analysis allowed to validate the two methods and the associated algorithm. In real world, the main difference is that the exchange coefficient value is not known. It means that it has to be computed from the formwork and insulation characteristics, knowing that the computed theoretical value will never fit perfectly the real world one, with the hazards encountered on jobsite (non constant thickness of the insulation, different weather exposures on mockup faces...). We propose here to apply the analysis on a real example. The concrete design is given in Table 4.

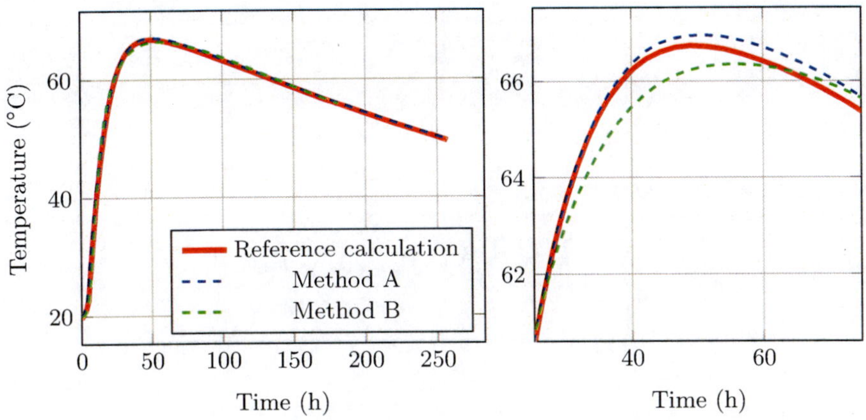

Fig. 3. Comparison of the temperature evolution for methods A and B with the reference solution, with an exchange coefficient $h = 0.32\,\mathrm{W/m/K}$. On the right hand side, a zoom around the temperature peak is shown.

Based on raw material properties, the concrete properties are computed as: density $\rho = 2364\,\mathrm{kg/m^3}$, thermal conductivity $\lambda = 1.8\,\mathrm{W/m/K}$, heat capacity $C_{conc} = 998\,\mathrm{J/kg/K}$. The activation energy is chosen to be $E_a = 40000\,\mathrm{J/mol}$.

Table 4. Concrete design

Components	Dosage (kg/m^3)
Binder (50% CEMI+20% FA+30% GGBS)	405
Water	183
Fine aggregates	778
Coarse aggregates	998

The evolution of temperature at the cube center, as well as the ambient temperature, are represented in Fig. 4. Considering that the insulation is not perfect, the exchange coefficient on all faces is assumed to be $h = 0.4\,\mathrm{W/m^2/K}$.

5.1 Thermal Fingerprint Identification

Method A. Because we consider a different exchange coefficient than for the numerical study, we have to recompute the heat loss for this configuration. Applying the same methodology, the heat loss parameter is $\alpha = 7944\,\mathrm{J/h/K}$. The thermal fingerprint is then computed using Eqs. (3), (2) and (5).

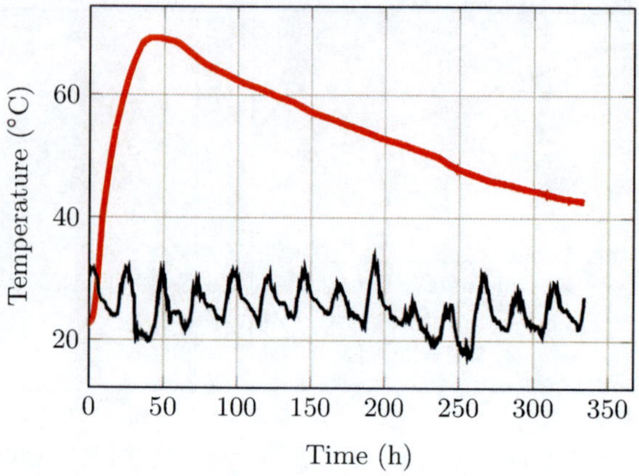

Fig. 4. Evolution of temperature at the center of the mockup (red curve), and ambient temperature (black curve).

Method B. For this case, the Q_{pot} is not known. Considering the blended binder, we assume a value of $Q_{pot} = 395\,\mathrm{J/g}$. The minimisation algorithm gives the following parameters:

$$B_1 = 2.60 \times 10^{-4} \qquad B_2 = 1.14 \times 10^{-3} \qquad \eta = 5.1$$

5.2 Results

We use the fingerprints identified with methods A and B to compute the temperature evolution in the mockup (Fig. 5). The fitting is for sure less good for

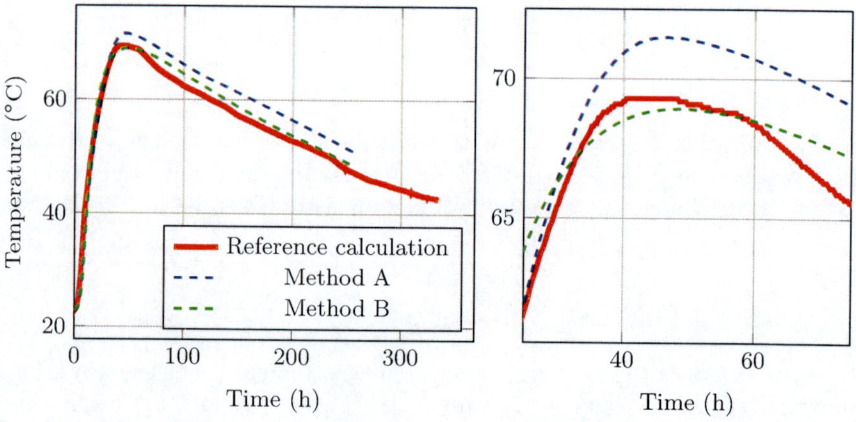

Fig. 5. Comparison of the temperature evolution for methods A and B with the experimental evolution. On the right hand side, a zoom around the temperature peak is shown.

this real example, but the two methods capture the main characteristics of the temperature evolution. Method A overestimates the maximum temperature by 2 °C, that can be explained by the fact that the total heat release is identified as well from the reference curve: if the exchange coefficient is under- or over-estimated, it can lead to an under- or over- estimation of the max temperature. On the other hand, method B is more flexible and can better fit the maximum temperature thanks to the minimisation algorithm. But if the exchange coefficient is not representative of the real experiment, it can lead to a good fitting of the experimental curve, but without a representative fingerprint of the binder.

6 Conclusion

Two methods are presented in this study, allowing to identify the thermal fingerprint of a binder from an insulated 1 m^3 mockup. It can be used for projects in which no standard quasi-adiabatic tests are available. Even if the mockup is less sensitive to the ambient conditions than a quasi-adiabatic test, it's important to have a good estimation of the exchange coefficient, that can be easily computed by considering the different layers of the mold. It means also that on the jobsite, high precautions must be taken in order to avoid any additional heat loss, that will affect the accuracy of the identification.

References

1. IFSTTAR: Recommendations for preventing disorders due to Delayed Ettringite Formation. Ifsttar, Technics and methods, GTI5-A, Marne-la-Vallée (2018)
2. Boulay, C., Torrenti, J.M., Andre, J.L., Saintilan, R.: Quasi-adiabatic calorimetry for concretes: influential factors. Bulletin des laboratoires des ponts et chaussees **278**, 19–36 (2010)
3. EN 196-9: Methods of testing cement - Part 9: Heat of hydration - Semi-adiabatic method
4. ASTM C186-17: Standard Test Method for Heat of Hydration of Hydraulic Cement
5. Schindler, A.K.: Effect of temperature on the hydration of cementitious materials. ACI Mater. J. **101**(1), 72–81 (2004)
6. Schindler, A.K., Folliard, K.J.: Heat of hydration models for cementitious materials. ACI Mater. J. **102**(1), 24–33 (2005)
7. Cervera, M., Oliver, J.: Thermo-chemo-mechanical model for concrete. I: hydration and aging. J. Eng. Mech. ASCE **125**(9), 1018–1027 (1999)
8. Faria, R., Azenha, M., Figueiras, J.A.: Modelling of concrete at early ages: application to an externally restrained slab. Cem. Concr. Compos. **28**(6), 572–585 (2006)
9. Bamforth, P.B.: Control of cracking caused by restrained deformation in concrete, CIRIA C760, London (2018)
10. Waller, V.: Relations entre composition des bétons, exothermie en cours de prise et résistance en compression, Collection Études et Recherches des Laboratoires des Ponts et Chaussées, OA35 (2000)

Numerical Approaches Aimed at a Sustainable Design: The Case of Wind Tower Foundations

Marco di Prisco(✉), Paolo Martinelli, Matteo Colombo, and Giulio Zani

Politecnico di Milano, P.za L. da Vinci 32, 20133 Milan, Italy
marco.diprisco@polimi.it

Abstract. The current effort towards the progressive switch from carbon-based to renewable energy production is leading to a relevant spreading of both on- and off-shore wind turbine towers. Regarding reinforced concrete shallow foundations of onshore wind turbine steel towers, possible reductions of reinforcement obtainable by employing steel fibre-reinforced concrete (SFRC) may increase their sustainability, speed of erection, and competitiveness. At the same time, there is a strong need to extend the life of foundations erected more than 15 years ago, originally designed for only 20 years. The paper presents a numerical investigation based on the results of a research programme in progress at Politecnico di Milano with ENEL, concerning the reinforcement with steel fibres of concrete shallow foundations embedded in a sandy soil subjected to both cyclic and monotonic loading. The approach takes advantage from a careful modelling of soil-structure interaction to highlight the safety margins correlated to conventional design and shows how the multi-directional resistance of SFRC allows a significant reduction of resources, preventing any brittle behaviour correlated to a reinforcement reduction.

Keywords: Onshore wind turbine · Shallow foundations · Reinforced concrete · Fibre-reinforced concrete · Soil-structure interaction (SSI) · Nonlinear FE analyses

1 Introduction

The need to switch from carbon-based to renewable energy production is leading to a relevant spreading of onshore and offshore wind turbine towers. Over time wind turbines have rapidly evolved in both height and rotor size to meet the demand for a greater energy production. With reference to onshore wind turbines, the increase in height implies a careful design of the foundations, whose response is expected to be significantly affected by its interaction with the soil. Onshore wind turbine foundations are massive reinforced concrete (RC) structures, gravity based, with some design peculiarities strictly related to their large geometrical sizes. Polygonal or circular shapes are typically adopted for these foundations, with (equivalent) diameters that can be larger than 20 m and thicknesses larger than 4 m. Concrete is typically cast in place in these foundations, and reinforcement is placed along the radial direction, with circumferential bars placed at different depths

P. Rossi and J.-L. Tailhan (Eds.): SSCS 2022, RILEM Bookseries 38, pp. 110–119, 2023.
https://doi.org/10.1007/978-3-031-07746-3_11

and stirrups connected to the radial bars. Steel anchor bolts ("anchor rod-type") or circular steel tubes ("anchor ring-type") are embedded into the RC foundation to transfer the wind loads acting on the superstructure.

The design of onshore wind turbine foundations is governed by the wind load, which acts mainly in one direction and induces load cycles with a frequency very different than the structure first mode of vibration. The speed of execution of the foundations is a relevant aspect to reduce construction costs: a reduction in the preparation and installation times of the steel reinforcement cages is thus essential. Moreover, soil-structure interaction (SSI) has to be accounted to correctly design onshore wind turbine foundation. Possible reductions of reinforcement, obtainable by employing steel fibre-reinforced concrete (SFRC), may increase sustainability, speed of erection, and competitiveness of these type of foundations: for this reasons reinforcement volume is here regarded as the only design parameter to be optimized. The elastic shell models with variable thickness usually adopted by designers cannot guarantee a reliable safety check even in the serviceability state and cannot at all be considered a reference to investigate shear failure and fatigue behaviour correlated to stress distributions.

To achieve this aim, a careful non-linear modelling taking into account a detailed geometry and materials non linearities for both R/FRC and soil is required. The paper presents a refined numerical approach validated with results of a wide experimental research programme carried out by Politecnico di Milano in cooperation with ENEL Green Power, concerning the use of hybrid reinforcement obtained by coupling steel fibres with conventional rebars in concrete shallow foundations embedded in a sandy soil, subjected to both cyclic and monotonic loading.

2 Experimental Programme and Materials

The tower-foundation-soil system was experimentally reproduced in a scale 1:15 and a schematic sketch together with a picture of the test set-up is shown in Figs. 1a and 1b. The system consisted of a truncated cone foundation (#3 in Fig. 1a), a flanged steel tube connected to the basket with the post-tensioned vertical bars via a bolted connection (#7 in Fig. 1a), and a tank filled with about 15.5 m^3 of Ticino sand (#1 in Fig. 1a). Four additional masses (#4, 5 and 10 in Fig. 1a) were introduced to reproduce the state of stress calculated in the real structure according to the similitude law: the mass at the top (#10 in Fig. 1a) reproduces the weight of the tower, rotor and blades of the wind tower, whereas the two steel coils (#5 in Fig. 1a) and the circular concrete crown (#4 in Fig. 1a) reproduce the stress due to the weight of the concrete foundation and are applied directly in contact with the soil. A picture of the concrete foundation is shown in Fig. 1c.

The foundation was positioned on a dry sand stratum at a depth of 0.185 m from the ground surface. A sand deposition device was adopted to deposit the soil layers guaranteeing a relative density of about 80% and test repeatability. The soil domain has a diameter of 3.5 m and a depth of 1.5 m. These dimensions were chosen after a series of ad hoc preliminary numerical analyses which showed that this geometry, under the considered loads, is not affected by boundaries. The vertical displacement of the foundation and indirectly the rotation of the foundation were measured through four vertical Linear Variable Displacement Transducers (LVDTs) inserted into the soil (#2 in

Fig. 1a). The top horizontal displacement was measured through a wire transducer (#9 in Fig. 1a), while a load cell measured the load, reproducing the wind action, applied by the horizontal jack.

Three solutions were adopted for the scaled (1:15) foundation, which mainly differ for the quantity of steel reinforcement. For each solution two nominally identical specimens were realized. The first solution consists of a reference prototype with the typical reinforcement layout of a real (full scale) foundation (specimens named FO1). In the second prototype solution the shear reinforcement was not included (specimens named FO2). The third solution (specimens named FO3) is a hybrid prototype in FRC without the shear reinforcement and with radial and circumferential reinforcement reduced by approximately 50% compared to the reference solution. The reinforcement introduced in the reference prototype of the concrete foundation (specimens FO1) is specified in Fig. 2. It is worth mentioning that the reinforcement ratios of the circumferential and radial reinforcement were accurately reproduced passing from the full-scale foundation to the scaled prototype foundation.

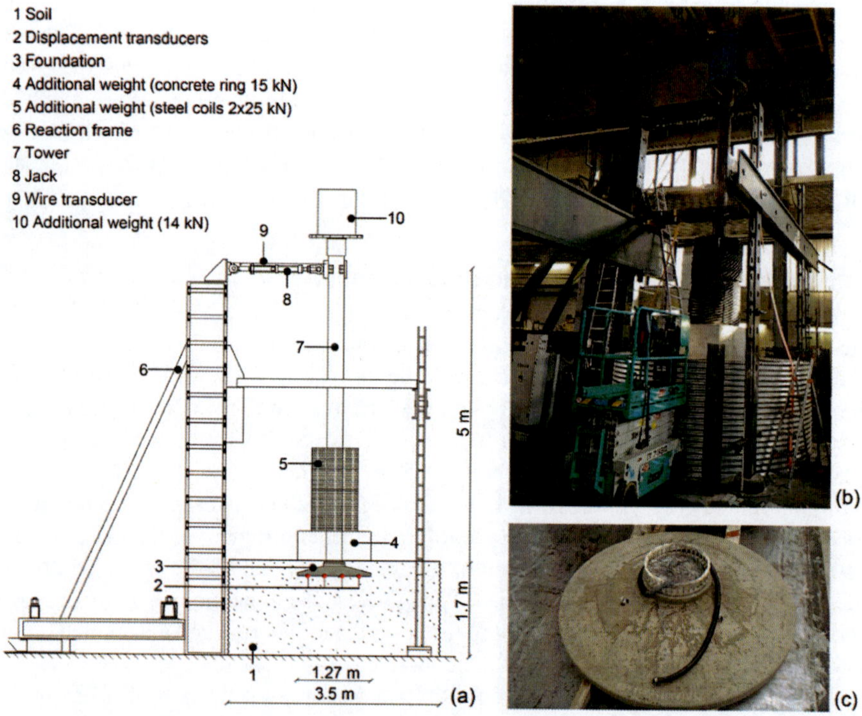

Fig. 1. Sketch (a) and lab photo (b) of the test set-up of the scaled tower-foundation-soil system; (c) top view of the RC foundation.

Several load phases were applied during each test: gravitational vertical loads, horizontal cyclic loads and a final phase with horizontal monotonic load up to the maximum stroke of the jack. A detailed description of the loading phases is reported in [1]. Vertical and horizontal cyclic loads have introduced in the RC foundations very light nonlinearities and therefore they are neglected in the numerical simulations. For this reason, only the final loading phase is analysed in this study.

The mix design of the materials used is described in [1]. Twenty-two cubes of 100 mm side were prepared in order to characterize the compressive strength of the concrete. The tests on the cubes and on the foundation prototypes were not performed on the same day. The approach suggested in *fib* Model Code 2010 [2] was used to estimate the cylindrical strength values at the foundation test date. The average cylindrical strengths for each specimen projected at the date of the monotonic test was 45.8, 40.6 and 40.5 MPa for specimens FO1, FO2 and FO3, respectively.

The post-cracking behaviour of the FRC material was characterized according to EN 14651 [3] by means of three-point bending tests (3PBTs) on six notched beams. For all the notched beams, the limit of proportionality f_{LOP} and the residual flexural tensile strengths f_{R1}, f_{R2}, f_{R3} and f_{R4} for a crack mouth opening displacement (CMOD) of 0.5, 1.5, 2.5 and 3.5 mm, respectively, were assessed. The average and standard deviation (SD) values of f_{R1}, and f_{R3} are equal 3.48 MPa (SD = 0.63 MPa) and 2.15 MPa (SD = 0.39 MPa), respectively. The complete curves in terms of nominal stress (σ_N) vs. CMOD together with the average response are reported in [4]. The bending behaviour identified according to EN14651 [3] allows to classify the materials as "2b" according to [2]. The original FRC target class for the full scale foundation was "3c", characterized by characteristic residual strengths $f_{R1k} \geq 3$ MPa and $f_{R3k} \geq 2.7$ MPa [2]. Nevertheless, the down-sizing of the 1:15 scaled foundations forced the use of a large amount of the smallest steel fibres available on the market ensuring the flowability through the reduced sizes of the cage but with a reduction of the FRC performance.

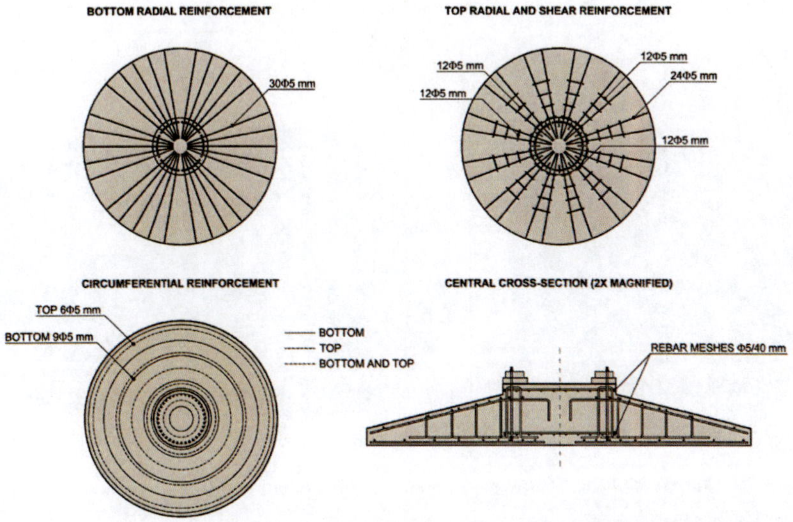

Fig. 2. Reinforcement introduced in the scaled foundation FO1.

3 Numerical FE Models

The design practice of shallow foundations of wind turbines based on the use of shell finite elements with variable thickness was validated in a preliminary phase which results are not reported here for the sake of brevity. After this initial check, a nonlinear finite element (FE) modelling of prototypes FO1–FO3 was performed by means of ABAQUS software [5].

A schematic drawing of the tower-foundation-soil system which reproduces the test setup of Fig. 1a, is shown in Fig. 3a. A close-up view of the numerical model of the RC foundation is shown in Fig. 3b. A non-structured mesh was adopted, with finer elements employed where large stress/strain gradients were expected both in the soil adjacent to the foundation and in the central core of the foundation. The FE models reproduced the steel rebars embedded in the scaled prototypes.

The steel tower was included in the FE models allowing to carried out displacement-controlled analyses during the application of horizontal loads. This choice also permitted to overcome numerical instabilities associated with the development of short softening branches in the response of the system.

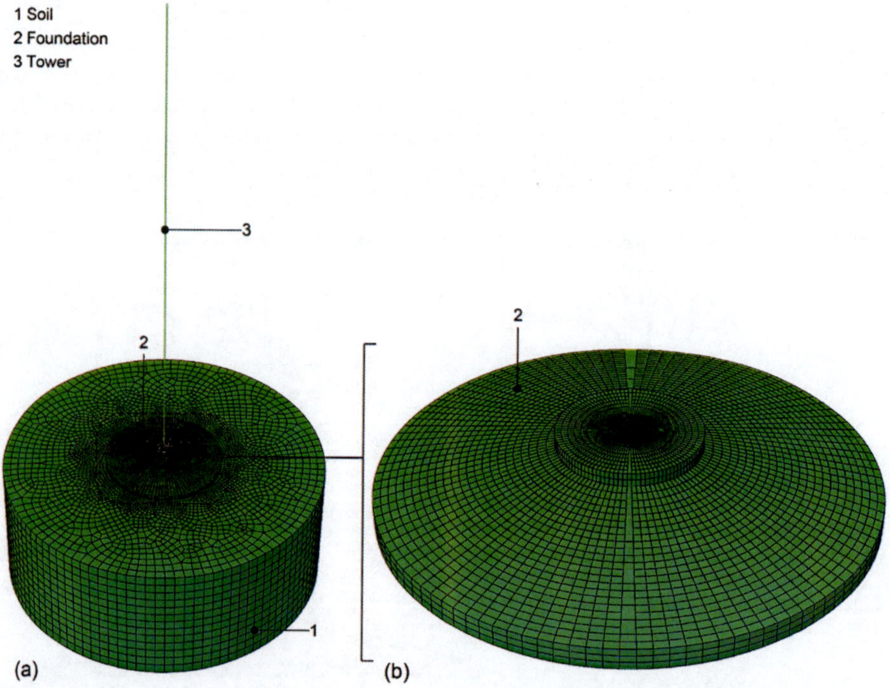

Fig. 3. (a) 3D view of the FE tower-foundation-soil system and (b) 3D view of foundation.

The concrete elastic properties were derived from the *fib* Model Code 2010 [2] for a concrete grade C40 (Table 1). The inelastic behaviour of concrete was simulated by using the concrete damaged-plasticity model (CDP) [6, 7]. Damage was not accounted for, meaning that the CDP model adopted was based on an elasto-plastic constitutive law with a non-associated flow rule. The adopted plasticity parameters, listed in Table 1, were derived from the literature [8]. The compressive strength values of concrete reported in Sect. 2 were used to determine the compressive and tensile constitutive laws of concrete. The nonlinear curve proposed by the *fib* Model Code 2010 [2] was used to describe the uniaxial compressive stress–strain law for plain and fibre-reinforced concrete.

Table 2 provides the values of cylindrical compressive strength f_{cm}, of strain corresponding to the maximum stress, ε_{c1}, and of the plasticity parameter k. A mixed stress–strain and stress–crack opening relation was adopted in uniaxial tension. It consists of an initial pre-peak linear segment and a post-peak nonlinear descending branch. For plain concrete, the tensile strength f_{ct} and fracture energy G_f were evaluated with the *fib* Model Code 1990 [9] with a bilinear descending branch (Table 2). For FRC specimens, the post-peak tension branch consists of a bi-linear law. The first branch was evaluated as for plain concrete following the prescriptions of *fib* Model Code 1990 [9], while the second branch was evaluated according to the proposal of di Prisco et al. [10]. It is worth mentioning that the first branch accounts for the contribution of the concrete matrix and it depends on tensile strength f_{ct} and fracture energy G_f, while the second branch accounts for the contribution of the fibres pull-out.

An elasto-J2 plastic material model was assumed in the analyses for the steel rebars. Elastic and inelastic steel material properties used in the numerical analyses are given in [11]. A non-associated elasto-perfectly plastic constitutive law with a Mohr-Coulomb failure criterion was adopted to simulate the mechanical behaviour of the soil. The elastic parameters were assumed to be constant with depth. Elastic and inelastic soil parameters (Table 1) were calibrated starting from the results of a series of triaxial tests [12].

Table 1. Mechanical properties used for the concrete (E_c: Young's modulus; ν: Poisson coefficient; ρ: density; Ψ: dilation angle; ε: flow potential eccentricity; f_{b0}/f_{c0}: ratio between the biaxial and the uniaxial compression strengths; K: Drucker-Prager surface modifier; ξ: viscosity parameter) and for the soil (E: Young's modulus; ν: Poisson coefficient; ρ: density; ϕ': friction angle; ψ: dilatancy angle).

Concrete		Soil	
Parameter	Values	Parameter	Values
E_c (MPa)	36300	E (MPa)	110
ν (–)	0.2	ν (–)	0.2
ρ (kg/m^3)	2500	ρ (kg/m^3)	1650
ψ (°)	13	ϕ' (°)	40
ε (–)	0.1	ψ (°)	12
f_{b0}/f_{c0} (–)	1.16	–	–
K (–)	0.7	–	–
ξ (–)	1×10^{-6}	–	–

Table 2. Concrete material parameters (f_{cm}: cylindrical compressive strength; ε_{c1}: strain at maximum compressive stress; k: plasticity number; f_{ct}: tensile strength; G_f: fracture energy).

Model	Compression			Tension	
	f_{cm} (MPa)	ε_{c1} (–)	k (–)	f_{ct} (MPa)	G_f (N/m)
FO1	48.0	0.0024	1.82	3.52	82.53
FO2	40.6	0.0023	1.92	3.07	71.34
FO3	40.5	0.0023	1.92	3.07	73.21

4 Results and Discussion

The 3D FE models have been validated in [11] by comparing the numerical analyses with the experimental results in terms of moment at the base vs. rotation at the base and crack patterns related to the monotonic behaviour of the specimens FO1, FO2 and FO3. The refined numerical models allowed reaching a reasonable reproduction of the experimental responses for all prototypes.

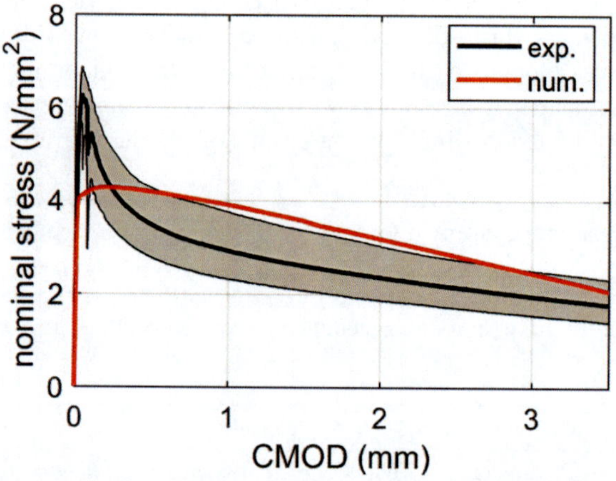

Fig. 4. Nominal stress vs. CMOD flexural responses of the six FRC notched beams: comparison between the experimental and the numerical results.

Figure 4 shows the average nominal stress σ_N vs. CMOD curve derived from six three-point bending tests superimposed to the solid-hatched dispersion of the results, together with the numerical FE results. The experimental response is characterised by a high peak stress compared to the numerical one, reflecting that the actual mean tensile strengths of concrete resulted higher than those expected from a concrete class C30/37. With reference to the post-peak branch associated with the fibre pull-out, the numerical analyses provide an overestimation of the experimental response, both at the serviceability and ultimate limit states ranges.

The numerical responses of the three types of foundations (i.e. models FO1, FO2 and FO3) expressed in terms of moment at the base vs. rotation at the base is proposed in Fig. 5. It is possible to observe as the curve of model FO3, corresponding to hybrid solution, exhibits a response very close to that exhibited by conventional RC models (curves FO1 and FO2) in the last monotonic load step. With respect to the RC foundations FO1 and FO2, for the FRC foundation FO3 the numerical model is slightly less accurate in reproducing the experimental results, with an overestimation of the experimental ultimate moment of about 15%. A careful analysis of the experimental data in previous load steps [1] has clarified that the small load jump in Fig. 5 corresponds to the first bending crack at the intrados. Filled circles in Fig. 5 correspond to the formation of the first flexural crack, to be intended in this context as a fully developed crack along a chord, while filled diamonds correspond to the first yielding in the reinforcement bars. In the numerical model FO3 plastic strains, rather than developing in narrow bands, are diffused in a much wider area. This leads to both (i) a delayed activation of the first cracking and the yielding of the reinforcing bars and (ii) an overestimation of the global behaviour of the foundation.

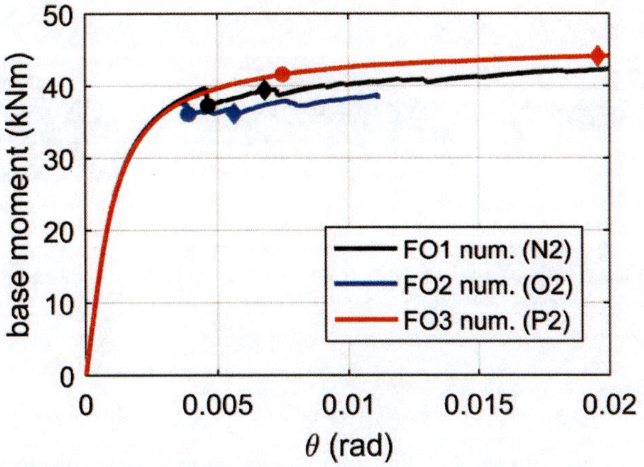

Fig. 5. Numerical comparison in terms of base moment vs base rotation for foundations FO1–FO3.

Figure 6 compares the tangential stress τ_{rz} along the radial direction for models FO1-FO3. The figure shows that stresses higher than 3 MPa (in grey in Fig. 6) are limited to a small region for the three structural solutions investigated. These results reveal (i) the absence of problems related to shear and (ii) solutions without shear reinforcement (i.e. models FO2 and FO3) have a comparable behaviour to the solution with shear reinforcement (i.e. model FO1). Numerical models FO1-FO3 do not show tensile plastic strains associable with shear cracks. Moreover, model FO1 is not characterized by the yielding of the shear reinforcement.

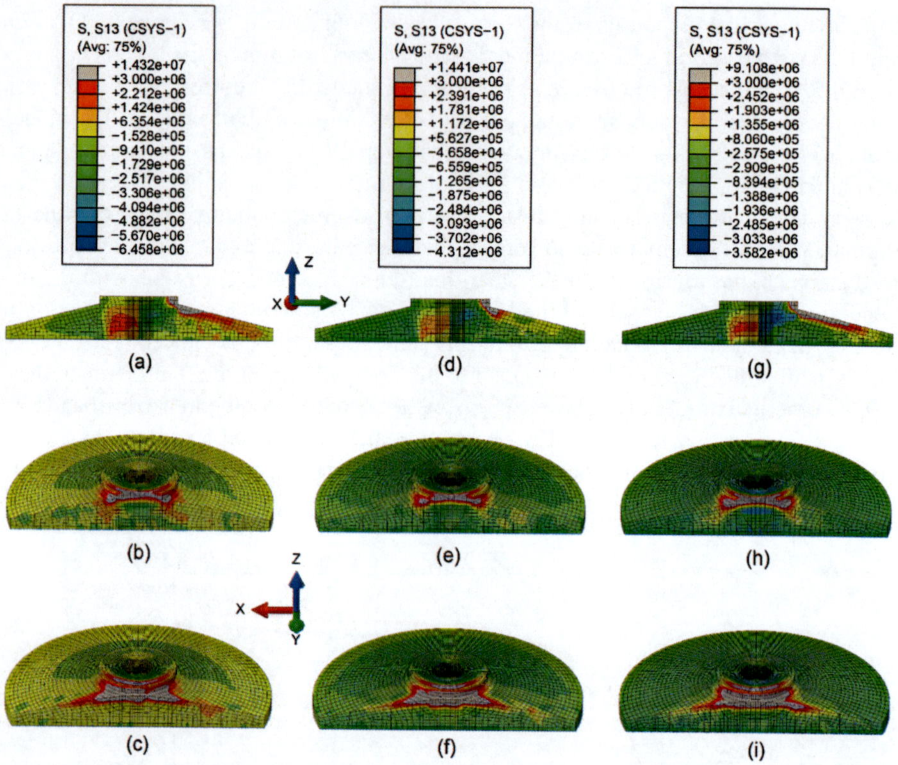

Fig. 6. Tangential stress τ_{rz} in the Y-Z plane (X = 0.0 m) and in the X-Y plane for cutting planes increasing along the Y axis (Y = 0.15, 0.2 m): (a-c) model FO1 ($\theta = 0.02$ rad), (d-f) model FO2 ($\theta = 0.011$ rad) and (g-i) model FO3 ($\theta = 0.02$ rad).

5 Conclusions

The numerical simulations on the wind tower foundation have highlighted the following conclusions:

- global response (moment-rotation curve) and crack patterns (tensile plastic strains) are similar through all the FE models, regardless the structural solution adopted. Cracking in hybrid solution (model FO3) is delayed and plastic strains are spread on a wider area, with respect to the conventional RC solutions (models FO1 and FO2).
- The FE models of the scaled 1:15 specimens exhibited a strong soil-structure interaction affecting the foundation bearing capacity. In particular, the failure mode for all the foundation models is characterized by the contemporary development of cracks in the foundation element with yielding of the radial reinforcement, and the attainment of the soil bearing capacity in a localized region relatively close to the edge of the foundation.
- The adoption of metallic fibres in partial replacement of traditional steel rebars (hybrid solution) for wind tower foundations appears a smart and promising option to be

pursued for the optimization of the reinforcing bars, thus reducing both construction times and costs.

- In agreement with the experimental findings, numerical results do not show yielding of transverse reinforcements and shear cracks developments.

Specific analyses on the size effect – not covered in this work – are in progress to understand whether the conclusions drawn from the 1:15 scaled models can be extended to real-scale foundation.

Acknowledgements. The authors thank the financial support of Enel Green Power Spa in the framework of the research contract between Enel Green Power Spa and Politecnico di Milano.

References

1. Dal Lago, B., Flessati, L., Marveggio, P., Martinelli, P., Fraraccio, G., di Prisco, C., et al.: Experimental tests on shallow foundations of onshore wind turbine towers. Struct. Concr. (2022). https://doi.org/10.1002/suco.202100655
2. fib. fib Model Code for Concrete Structures 2010, Lausanne (2013)
3. EN 14651. Test method for metallic fibered concrete – Measuring the flexural tensile strength (limit of proportionality (LOP), residual), Brussels (2004)
4. di Prisco, M., di Prisco, C., Fraraccio, G., Lago, B.D., Martinelli, P., Flessati, L., et al.: Wind tower FRC foundations: research and design. In: Serna, P., Llano-Torre, A., Martí-Vargas, J.R., Navarro-Gregori, J. (eds.) BEFIB 2021. RB, vol. 36, pp. 831–842. Springer, Cham (2022). https://doi.org/10.1007/978-3-030-83719-8_71
5. Dassault Systèmes. Abaqus Analysis User's Manual - Version 6.14 (2016)
6. Lubliner, J., Oliver, J., Oller, S., Oñate, E.: A plastic-damage model for concrete. Int. J. Solids Struct. **25**, 299–326 (1989). https://doi.org/10.1016/0020-7683(89)90050-4
7. Lee, J., Fenves, G.L.: Plastic-damage model for cyclic loading of concrete structures. J. Eng. Mech. **124**, 892–900 (1998). https://doi.org/10.1061/(ASCE)0733-9399(1998)124:8(892)
8. Alfarah, B., López-Almansa, F., Oller, S.: New methodology for calculating damage variables evolution in Plastic Damage Model for RC structures. Eng. Struct. **132**, 70–86 (2017). https://doi.org/10.1016/j.engstruct.2016.11.022
9. CEB-FIP. CEB-FIP Model Code 90 (1993)
10. di Prisco, M., Colombo, M., Dozio, D.: Fibre-reinforced concrete in fib Model Code 2010: principles, models and test validation. Struct. Concr. **14**, 342–361 (2013). https://doi.org/10.1002/suco.201300021
11. Martinelli, P., Flessati, L., Dal Lago, B., Fraraccio, G., di Prisco, C., di Prisco, M.: Role of numerical modelling choices on the structural response of onshore wind turbine shallow foundations. Structures **37**, 442–458 (2022). https://doi.org/10.1016/j.istruc.2022.01.002
12. Pisanò, F., Flessati, L., Di Prisco, C.: A macroelement framework for shallow foundations including changes in configuration. Geotechnique **66**, 910–926 (2016). https://doi.org/10.1680/jgeot.16.P.014

Numerical Modeling of New Conceptions of 3D Printed Concrete Structures for Pumped Storage Hydropower

Eduardo de M. R. Fairbairn(✉), Larissa D. F. Santos, Marina B. Farias, and Oscar A. M. Reales

Civil Engineering Department/COPPE, The Federal University of Rio de Janeiro, Rio de Janeiro, Brazil
{eduardo,larissa.santos,marina.farias,oscar}@coc.ufrj.br

Abstract. This paper explores the possibilities of pumped storage hydropower (PSH), checking the possibilities of using new engineering tools such as 3D printing. The use of such technology for concrete has gained rapid development in recent years due to the advantages in structural optimization and economy with formwork in conventional construction. Practical engineering applications have proven the applicability of 3D printing in large-scale construction of building components, and compared to conventional manufacturing methods, this advanced technique has several advantages and offers almost unlimited potential for geometric complexities. We explore new design conceptions with the help of numerical modeling in two ways: (i) during the early ages considering the phenomena of hydration; (ii) after hardening of concrete verifying the integrity of the structure. The results indicated that numerical modeling can point to new solutions that will help address the challenges posed by greater sustainability in energy generation and storage in the 21st century.

Keywords: Pumped storage hydropower · 3D printing

1 Introduction

Pumped storage hydropower (PSH) plants, also called "water battery", are storage energy systems consisting of two water reservoirs, a tunnel connecting these reservoirs and a powerhouse with turbines-pumps and motor-generators. It represents one of the most sustainable, economical, and efficient solutions for energy storage, being an excellent alternative to store energy from intermittent sources such as wind and solar. To store energy, water is pumped from the lower reservoir to the upper reservoir. On the other hand, to generate energy, the water from the upper reservoir flows through the turbine towards the lower reservoir. To reduce system costs, whenever feasible, turbines and generators are also used as pumps and motors [1], as illustrated in Fig. 1 (taken from reference [2]). PSHs respond today for about 160 GW installed and 9,000GWh of energy storage across the globe [3]. The distribution of these plants (existing and planned) around the world are shown in Fig. 2 taken from reference [4].

P. Rossi and J.-L. Tailhan (Eds.): SSCS 2022, RILEM Bookseries 38, pp. 120–129, 2023.
https://doi.org/10.1007/978-3-031-07746-3_12

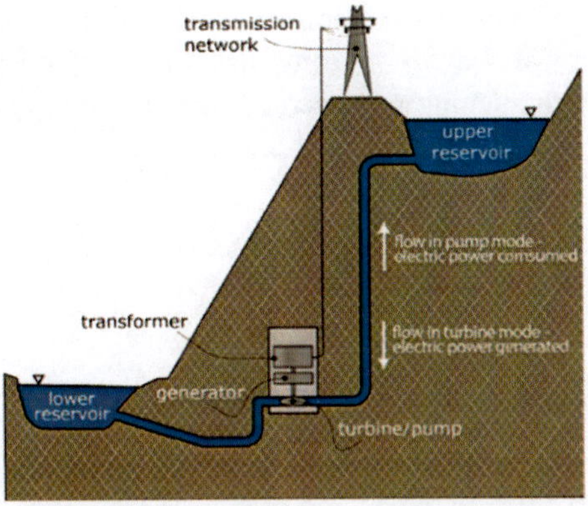

Fig. 1. Schematic of PSH (taken from reference [2])

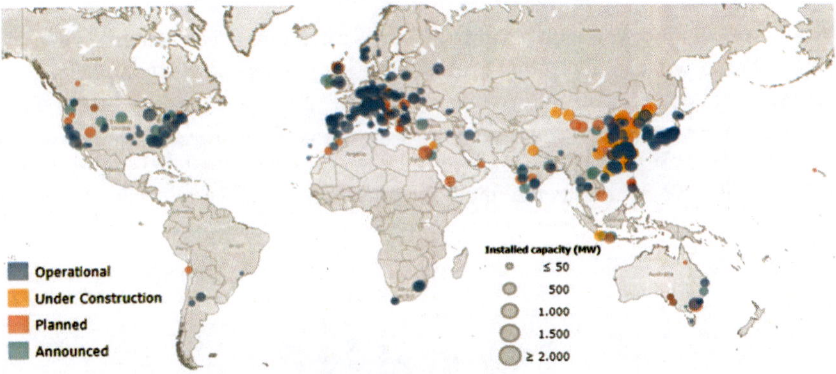

Fig. 2. Distribution of PSHs in the globe (taken from reference [4])

It can be seen in Fig. 2 that, although PSHs are relatively new energy storage systems, they are already widely used in some regions of the world. On the contrary, in Brazil, although the hydroelectric potential of rivers is greatly explored, the use of PSHs is practically non-existent. However, the recent expansion of solar and wind energy generation is imposing the installation of "water batteries" to ensure the constancy of energy supply. In a recent comprehensive study [1] it was demonstrated that Brazil has great potential for building PSHs with different types of storage arrangements and cycles, which result in different benefits for the Brazilian Electricity Sector.

This paper explores the possibilities of this new type of hydroelectric development, checking the possibilities of using new engineering tools such as 3D printing. The use of such technology for concrete has gained rapid development in recent years due to the advantages in structural optimization and economy with formwork in conventional construction [5]. Practical engineering applications have proven the applicability of 3D printing in the large-scale construction of building components and compared to conventional manufacturing methods [6], this advanced technique has several advantages and offers almost unlimited potential for geometric complexities [7]. We explore the potential of using 3D printing for the construction of PSH structures with the help of numerical modeling in two ways: (i) during the early ages considering the phenomena of hydration; (ii) after hardening of concrete verifying the possibility of introducing new design concepts. The results indicated that numerical modeling could point to new solutions that will help address the challenges posed by greater sustainability in energy generation and storage in the 21st century.

2 Theoretical Background

2.1 Modeling Parameters for 3D Printed Process

The process of 3D printing of cementitious material corresponds to the extrusion of this material through a tip with deposition of successive layers. The extruded material must be able to withstand its own weight and the weight of subsequent layers without significant deformations. This technique is capable of printing objects of great magnitude and complex geometries.

The physical model of a 3D printing that uses a cement matrix can be seen in Fig. 3. The model represents a 3D printing process through the deposition of successive layers and illustrates the parameters relevant to the printing process. Q represents the flow rate injected into the system, V represents the speed of the extruder nozzle and L represents the length to be expired. The total height is called Hm while the height of each layer is called h. Finally, W represents the thickness of the layer [8].

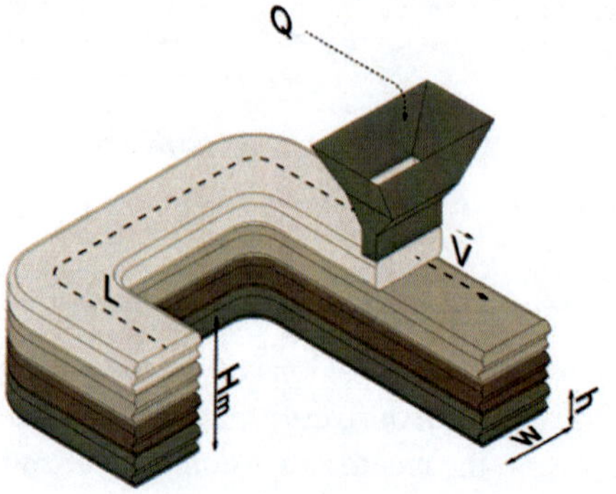

Fig. 3. Schematic of a layered extrusion process with concrete [8].

Some of the requirements that printable materials have is to verify the necessary structural strength that the building implies, given its geometric properties, and the correct consolidation of a layer before the deposition of the following, always maintaining a level of moisture that allows a good adhesion. Therefore, research carried out over time resulted in a series of equations that describe the behavior of fresh concrete submitted to the 3D printing process.

These materials, as well as any other cementitious materials, have the rheological behavior close to a Binghamian fluid, and two parameters are needed for their characterization: the flow stress and the plastic viscosity.

When the material is deposited, it displays an initial yield stress (τ_0). Below this yield stress the material exhibits an elastic behavior. Due to thixotropic phenomena and cement hydration, the yield limit of the cement matrix increases over time. The initial yield limit, that is, the one measured as soon as the matrix has been mixed, will determine the maximum height ($h_{m,max}$) that each layer can have through the following equation [8].

$$h_{m,max} = \frac{\sigma_{0,0}\sqrt{3}}{\rho g} \tag{1}$$

Thixotropy of these types of materials and their ability to build an internal structure at rest is the key to most printing applications [9]. Thus, another important rheological property for the 3D printing process is thixotropic gain (A_{thix}). This corresponds to the rate of increase of yield limit over time, that is, the higher the thixotropic gain, the faster the cement will gain yield limit.

After the deposition of the layers, the rheological parameters evolve, and the experimental results show that, while the flow stress and the shear module increase, the deformation of the layers decrease, as a function of time [10, 11].

Since it is essential that the lower layer has sufficient flow limit to support its own weight beyond the weight of the top layer, the structural kinetics of the object must obey its size and its time window. The relationship between the minimum waiting time required between the layers ($t_{h,min}$) and the thixotropic gain is given by Eq. 2 [8].

$$t_{h,min} = \frac{\rho g h}{A_{thix}\sqrt{3}} \tag{2}$$

The maximum print speed (V_{max}) can be obtained by dividing the print length (L), which is the path to be traveled, by the minimum waiting time. Printing at speeds higher than maximum will lead to the structure collapsing.

Quick structuring is one of the benefits of 3D printing. However, there is a limit to this growth rate from the rheological point of view. High times between one layer and another can lead to so-called "cold joints" that lead to weak interface zones between these layers, and yet, the higher the thixotropy, the weaker the interface will also be [12].

Therefore, this will occur if a critical rest time between successive depositions is exceeded, thus defining a maximum layer time (T_{max}) given by Eq. 3 [9].

$$t_{h,max} = \frac{\sqrt{\frac{(\rho g h)^2}{12} + \left(\frac{2\mu_0 L}{h t_{h,max}}\right)^2}}{A_{thix}} \tag{3}$$

3 Experimental Program

Concrete was prepared with 600 kg/m^3 of Portland CPII-E cement, 1379 kg/m^3 of fine aggregate, with grain smaller than 0.2 mm, 259 kg/m^3 of water and 0.8% of superplasticizer per cement mass. Mixed for 2 min the solid compounds for complete homogenization, and another 8 min after added the rest of the materials for the action of the superplasticizer.

Rheology of the concrete was measured at room temperature using a rotational rheometer equipped with a 0.69 cm diameter and 1.176 cm height Vane spindle. Material was placed in a glass beaker, densified using a glass stirring rod, covered with plastic film and left to rest. The resting time (t) was defined as the time between the adding water and the beginning of the rheological testing. At the end of t, the vane spindle was introduced in the paste using the vertical guide of the rheometer, taking care of not disturbing the paste. The spindle was positioned in the center of the beaker at middle height. No pre-shearing was applied before the rheological measurements. Similar to [13] the author adopted rheology test for cementitious material.

Static yield stress ($\tau_{0;t}$) after a given t was measured by shearing each sample from rest to a 0.2 s^{-1} shear rate in 180 s. The value of $\tau_{0,t}$ was determined as the maximum stress recorded in the shear stress versus shearing time curve [14]. Tests at different resting times (0, 23, 45, 68 and 90 min) were used to characterize the structural buildup of the pastes. A fresh sample was used for each resting time to guarantee undisturbed conditions for each experiment [15].

The static yield stress data was used to characterize the structural buildup process of each formulation trough the rate of thixotropic build up (A_{thix}) [9]. The rate A_{thix} was obtained by fitting the data to the linear equation below:

$$\tau_{0,t} = A_{thix}.t + \tau_{0,fit} \tag{4}$$

where A_{thix} is the slope of the linear equation, t is the resting time and $\tau_{0;fit}$ is the static yield stress at $t = 0$, obtained from the intercept of the fitted equation.

3.1 Results

The static yield stress ($\tau_{0,t}$) versus time (t) for all studied mixtures are presented in Fig. 4. Linear equations of the form (4) were adjusted to these experimental results and are presented in the same figure. The parameters for construction of the 3D structure (A_{thix}, $\tau_{0,fit}$) were extracted from these linear equations.

An initial static yield stress of 2585 Pa and A_{thix} of 0.85 Pa/s was obtained. Considering the density of 2162 kg/m^3 of the concrete and using the formulas (1) and (2), a layer of 20 cm height and 1 h of deposition time was adopted between the layers.

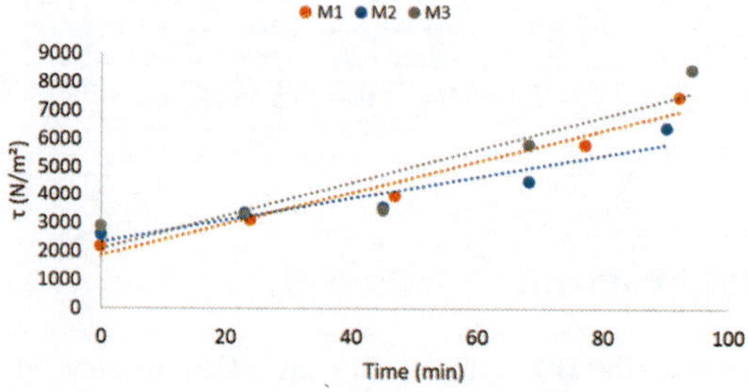

Fig. 4. Static yield stress ($\tau_{0;t}$) versus different mixtures and linear fits used to obtain A_{thix} and $\tau_{0;fit}$ for concrete

4 Numerical Model

From previous studies and the analysis of three shapes commonly used in 3D printing, the lattice shape was chosen to be modeled in this work. Among the three shapes shown in Fig. 5, the structure with a square lattice interior has a slightly better result of stress levels, according to the study from reference [16], but the triangular truss shape is lighter and easier to execute, being the chosen shape to be studied in greater depth.

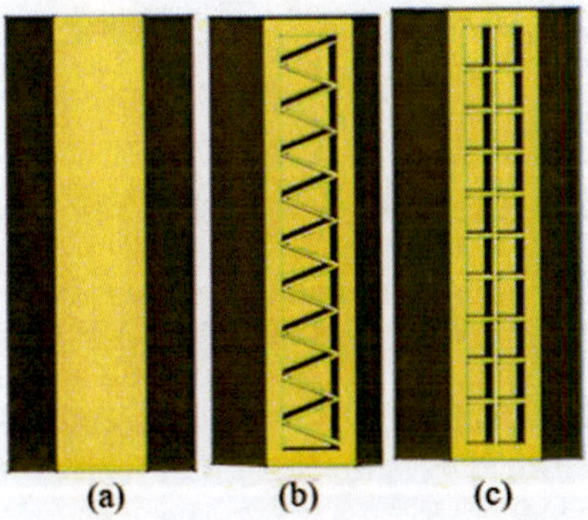

Fig. 5. Different dam wall shapes [16].

A numerical modeling was carried out using the finite element method and the design of the shape was made in the Autocad program to later be imported into the Diana FEA program, where the boundary conditions, material properties, mesh and, finally, the numerical analysis were established.

From the results of the tested concrete, a wall simulating a part of a wall of a PHS was modeled, built with 3D technology. This structure was simulated with 6 m high, 2.5 m thick and 6 m in length and it was divided into 30 layers with 20 cm in height each, as shown in Fig. 6.

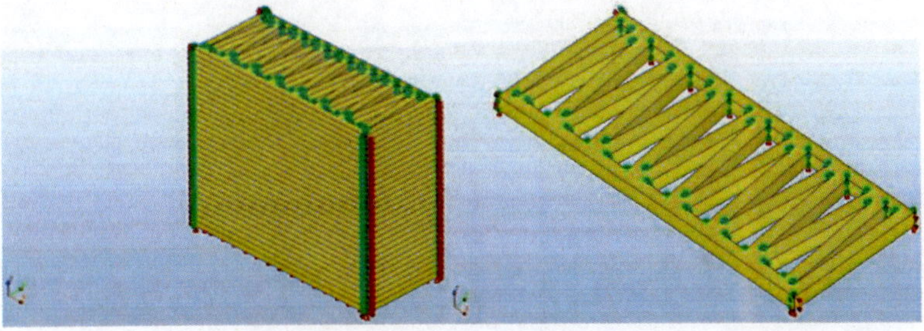

Fig. 6. Geometry of the numerical model

The model´s mesh, shown in Fig. 7, is formed by hexahedral elements with 10 cm of size. For the finite element analysis, quadratic elements were defined.

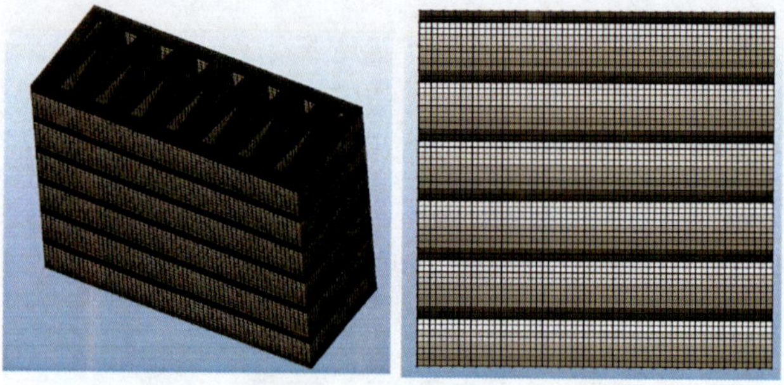

Fig. 7. Model's mesh

4.1 Numerical Modeling of the Structure in the Fresh State

The results were observed from the pouring of the first layer to the 16th layer, by limitations of the software and hardware, obtaining the highest possible degree of hydration, considering the placing temperature of 24 °C, constant ambient temperature of 22 °C, height of the layer of 20 cm and interval of 1 h between the layers. The usual thermal properties of concrete estimated by Fairbairn [13], together with the tested thermal data, were considered, using conductivity of 4.0 W/mK and capacity of 2.5 x 10^6 J/m^3K.

The boundary conditions considered the continuous wall in the direction of the longest length, heat flux between the layers, and concrete as an elastic material. The stresses are primarily influenced by the temperature field, and to a less degree by the dead load, while the effect of other loads was neglected.

Results. The result of the initial tension and tension increments due to the deposition of the layers both numerical and the experimental can be observed in Fig. 8. The setting process of the concrete starts 2:30 h after mixing, and ends at about 5 h, after which is considered as a hardened material.

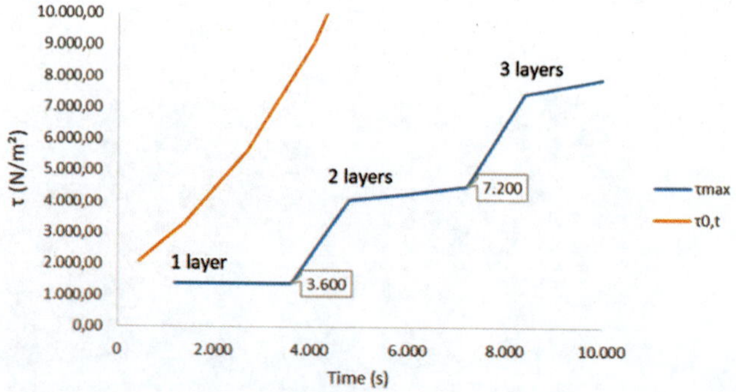

Fig. 8. Static yield stress (τ0,t) and maximum numerical shearing stress versus time

It can be seen in the above figure that the construction process is stable, given that the yield stress is always greater than the maximum shear stress during construction.

4.2 Numerical Modeling of the Structure in the Hardened State

As well as for the early age analysis, DIANA FEA was used to analyze the structure's performance at the hardened state. A structural nonlinear analysis of the walls was made to evaluate the stress distribution and integrity of the structure due to the acting forces, namely the dead weight and hydrostatic pressure. It is important to note that the program calculates the water pressure automatically, with the need to enter only the hydraulic head.

The following support restrictions were applied: the base was restricted in the x, y and z directions and the sides were restricted in the x direction. Between the layers, interface elements were used.

Material Properties. The material used was plain concrete and interface connections were applied between layers. the main properties of concrete are: young's modulus of 14 GPa, Poisson's rate of 0.3, mass density of 2062 kg/m^3, tensile strength of 3.0 MPa and a fracture energy (G_f) of 200 J/m. For the tensile behavior, the linear model was used with post-crack behavior based on fracture energy (linear-crack energy). for compression, the elastic model was used. For the interface material, the main properties are: normal and shear stiffness of 42 GPa, cohesion of 1.5 MPa, friction and dilatancy angle of 0.01 rad and tensile strength of 1.5 MPa. The interface opening model is the gapping model and the model for gap appearance is brittle.

Crack Index. To assess the capacity of the structure to resist the solicitant efforts, the crack index was used, which indicates how far the state of stresses is from the state of crack formation. This index is calculated by the ratio between the maximum principal stress and the tensile strength of concrete, as shown in Eq. (5) [17]:

$$F_{tu} = \frac{\sigma_1}{f_t} \tag{5}$$

where σ_1 is the maximum principal stress and f_t is the concrete tensile strength.

Results. The maps of normal stresses (S_1), shear stresses (T_{max}) and crack index (F_{tu}) were obtained, as shown in Fig. 9. As expected, there was a concentration of tensile stress at the extremes of the edges and base of the walls. Figure 9 present the results for the side view (where the water pressure is acting) of the model.

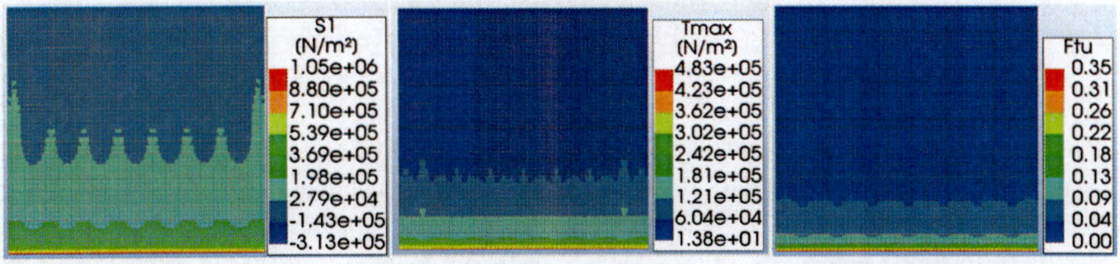

Fig. 9. Normal stress, shear stress and crack index maps at load step 1.0

To evaluate the behavior of the structure until failure, more load steps were taken, up to a load factor of 1.86, which multiplied by the water level would give a level of 11.16 m. From Fig. 10, it can be seen that there was a rupture between the first and second layers, evidencing that the structure reached stresses values greater than it can resist and, as consequence, there was a crack formation.

Fig. 10. Detail of crack between layers

5 Concluding Remarks

This paper presents an early analysis of an innovative design of structure to be used in new types of hydroelectric projects. The main idea is that a new type of hydroelectric plant is an interesting exploratory field for the development of innovative structural concepts that can make use of new technologies such as 3D construction. To this end, tools such as computer modeling using the finite element method are used, not only for hardened structures but also for the fresh concrete phase. The procedures presented here are still in an initial phase, but they intend to show the civil construction industry the advantages of using sophisticated simulation tools that may allow the development of new structural concepts.

References

1. Brandão, R., Nivalde, D. C., Hunt, J.: The Viability of Reversible Power Plants in the National Interconnected System. ANEEL (2021). (in Portuguese)
2. Viadero, R.C., Singh, A., Rehbein, M.: Hydropower on the Mississipi River (2017)
3. Pump it up : Recommendations for urgent investment in pumped storage hydropower to back the clean energy transition. International Forum on Pumped Storage Hydropower Policy and Market Frameworks Working Group. Global Paper (2021)
4. Pumped Storage Tracking Tool: IHA (International Hydropower Association) (n.d.). https://professional.hydropower.org/page/map-pumped-storage-tracking-tool. Accessed 10 Mar 2022
5. Ngo, T.D., Kashani, A., Imbalzano, G., Nguyen, K.T.Q., Hui, D.: Additive manufacturing (3D printing): a review of materials, methods, applications and challenges. Compos. Part B: Eng. **143**(December 2017), 172–196 (2018). https://doi.org/10.1016/j.compositesb.2018.02.012
6. Zhang, J., Wang, J., Dong, S., Yu, X., Han, B.: A review of the current progress and application of 3D printed concrete. Compos. Part A: Appl. Sci. Manufact. **125**(April), 105533 (2019). https://doi.org/10.1016/j.compositesa.2019.105533

7. Paul, S.C., van Zijl, G.P.A.G., Gibson, I.: A review of 3D concrete printing systems and materials properties: current status and future research prospects. Rapid Prototyp. J. **24**(4), 784–798. https://doi.org/10.1108/RPJ-09-2016-0154
8. Wangler, T., et al.: Digital concrete: opportunities and challenges. RILEM Tech. Lett. **1**, 67 (2016). https://doi.org/10.21809/rilemtechlett.2016.16
9. Roussel, N.: A thixotropy model for fresh fluid concretes: theory, validation and applications. Cem. Concr. Res. **36**(10), 1797–1806 (2006). https://doi.org/10.1016/j.cemconres.2006.05.025
10. Wolfs, R.J.M., Bos, F.P., Salet, T.A.M.: Early age mechanical behaviour of 3D printed concrete: numerical modelling and experimental testing. Cem. Concr. Res. **106**(January), 103–116 (2018). https://doi.org/10.1016/j.cemconres.2018.02.001
11. Mettler, L.K., Wittel, F.K., Flatt, R.J., Herrmann, H.J.: Evolution of strength and failure of SCC during early hydration. Cem. Concr. Res. **89**, 288–296 (2016). https://doi.org/10.1016/j.cemconres.2016.09.004
12. Zareiyan, B., Khoshnevis, B.: Interlayer adhesion and strength of structures in contour crafting - effects of aggregate size, extrusion rate, and layer thickness. Autom. Constr. **81**(June), 112–121 (2017). https://doi.org/10.1016/j.autcon.2017.06.013
13. Fairbairn, E.M.R., Silvoso, M.M., Ribeiro, F.L.B., Toledo-Filho, R.D.: Determining the adiabatic temperature rise of concrete by inverse analysis: case study of a spillway gate pier. Eur. J. Environ. Civil Eng. **21**(3), 272–288 (2017). https://doi.org/10.1080/19648189.2015.1112843
14. Mendoza Reales, O.A., et al.: Influence of MWCNT/surfactant dispersions on the rheology of Portland cement pastes. Cem. Concr. Res. **107**(August 2017), 101–109 (2018). https://doi.org/10.1016/j.cemconres.2018.02.020
15. Mendoza Reales, O.A., Duda, P., Silva, E.C.C.M., Paiva, M.D.M., Filho, R.D.T.: Nanosilica particles as structural buildup agents for 3D printing with Portland cement pastes. Constr. Build. Mater. **219**, 91–100 (2019). https://doi.org/10.1016/j.conbuildmat.2019.05.174
16. de Farias, M.B., Fairbairn, E.M.R., Reales, O.A.M.: Numerical modeling of 3D-printed concrete dams designed for pumped-storage hydropower (2021)
17. Diana Fea, B.V.: DIANA finite element analysis user's manual. Release 10.5. (2021). Delft, the Netherlands

Assessing 3D Concrete Structures at ULS with Robust Numerical Methods

Agnès Fliscounakis[1], Mathieu Arquier[2], Mohammed-Khalil Ferradi[3], and Xavier Cespedes[2](✉)

[1] Univ Rennes, INSA Rennes, LGCGM - EA 3913, 35000 Rennes, France
[2] Strains, 75012 Paris, France
xavier.cespedes@strains.fr
[3] Modelling Simulation and Data Analysis, Mohammed VI Polytechnic University, Benguerir, Morocco

Abstract. After a rapid state of the art, an innovative numerical method dedicated to reinforced concrete is presented. Constructed upon a strong mathematical basis, this method gives qualitative results on real-life projects. It illustrates the contribution dual analysis can have to the civil engineering world, bringing together robustness, readable results, and error bracketing.

Keywords: Concrete structures · ULS · Numerical methods · Optimization

1 Introduction

Assessing the real state of a concrete structure at ULS is often challenging. By definition, the ULS imposes to consider nonlinear law of material, and concrete is known to have complex heterogeneous behavior. Not only the reinforcements must obviously be taken into account, but the concrete itself can show scattered values of limit strength, especially in tension. Eurocodes conservatively recommend to consider a null tensile limit for the concrete, and to apply a safety factor to the compressive strength. This point of view is only one-dimensional, and some adaptations must be made when dealing with multi-axis stressing like confined concrete.

Hence the engineers find themselves with two possibilities for analyzing concrete structures at ultimate states:

- Applying the codes, generally by boiling down complex problems to easier ones. For instance, transversal sections of slender structures are classically calculated thanks to 2D tools, providing the ULS longitudinal reinforcement steel ratio. For more complex geometries, engineers can apply the strut-and-tie method (an illustration of limit analysis) to find compressive and tensile parts of concrete and then find the design of the needed reinforcements. Although this method is still used nowadays, it requires a certain level of expertise.

P. Rossi and J.-L. Tailhan (Eds.): SSCS 2022, RILEM Bookseries 38, pp. 130–139, 2023.
https://doi.org/10.1007/978-3-031-07746-3_13

- When the above methods cannot be easily applied, like on complex massive 3D structures, as a last resort, engineers turn to full 3D finite elements nonlinear elastoplastic analysis. This solution, which can be very time-consuming, is not so easy to put into practice, and often leads to more questions than answers: which software to use? What kinds of finite elements? What kind of 3D nonlinear law to use for concrete? How to take into account the reinforced bars (which are one-dimensional) into a 3D model? And even so, once the analysis has been made, how to efficiently post-process the 3D results?

  After a rapid state of the art, we present hereafter a new numerical method to assess the ULS of reinforced concrete. The presentation is illustrated all along with real project results, showing the ability of the method to cope with existing structures' complexity without oversimplifying them.

2 Current Methods and Their Limitations

2.1 When Standards Can't Solve Your Problem

In a civil engineering office, standards are always the first call. If by any chance, your problem complies with the standard framework, the study can be straightforward and, furthermore, you have the pleasant feeling to enforce an absolute law. Unfortunately, the situation is rarely as simple, mainly because the standard framework is narrow and its universality also has to be questioned.

As explained in [4] for design codes [5–9], regarding flexural slender beams the plan section hypothesis is universally accepted and codes have almost all the same predictions (less than 10% of standard deviation). This reassuring result is not the same for reinforced concrete beams subjected to shear for instance, where very large discrepancies can appear among standards (factor of more than 2). This particular example shows the difficulty to establish sound simplified computational methods as soon as there is no agreed basis for a rational theory. The present comparison, over different national codes, makes clear the limitations of standards universality as much regarding different countries at the time being as regarding standard evolution for the existing structures. On this subject, let's point out that the paradigm shift, from new constructions to aging existing structures, will increase the need for timeless/worldwide prescriptions.

On the other hand, beyond this non-universality, complex 3D RC structures suffer from a lack of completely general rules of design prescribed by standards. Thus, when dealing with complex problems, the strut-and-ties method arise to be the most usual way to bring out a reasonable design.

2.2 Strut-and-Ties Responses

When it comes to assessing ULS in reinforced concrete complex geometry, strut-and-ties method is one of the most widely used. This method stems from the Ritter [1] and Mörsh [2] work at the end of the nineteenth century, which introduced the truss analogy and became practical thanks to Schlaich and Shäfer simplifications presented in [3].

Roughly, in this method, an RC structure is seen as a truss made from the rebars in traction and the concrete in compression. Thus, a stress diagram must be designed, which can be done by hand or sometimes automatically. Although a yield capacity is sought, this stress diagram is often provided thanks to a load path resulting from an elastic stress distribution (Fig. 1).

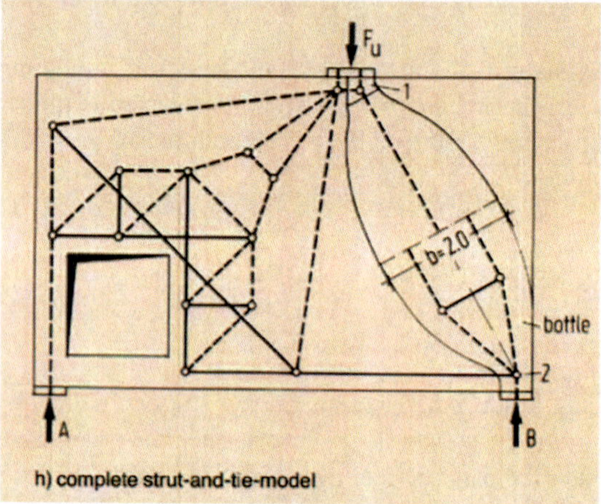

Fig. 1. Complete strut-and-tie-model of a deep beam with large hole [3]

Let's notice that the strut-and-ties method imposes compliance to the standard limit values on the traction stress in reinforcements and on the compression stress in concrete. As explained in [3], those limitations imply that the structure is designed according to the theory of plasticity lower bound theorem, and therefore, is in the spirit of the limit analysis formalism used in the presented work.

Dealing with simple structures and loadings, those diagrams can be straightforward, but they become really complex as soon as the problem is more unusual. In such cases, only experimented engineers can bring out a solution thanks to a deep knowledge of the theory. And what about an existing structure? When reinforcements are already laid out in the structure, the engineer is not free anymore to find a load path but has to trace back the design history and justify its viability. This is not the spirit of strut-and-ties that iterates on design to establish the verification.

The presented numerical method doesn't seek to replace strut-and-ties as a design tool, and neither does the understanding it gives at the first stage. However, in its application of the lower bound theorem in a more exhaustive way, it seems to be much more efficient when it comes to justifying, especially a posteriori, any kind of 3D reinforced structures.

2.3 3D Elastoplastic Analysis as the Last Solution

Let's come back to the standards discrepancies described in Sect. 2.1. A unified concept is often based on a strong physical basis. However, physical bases also mean laborious and time-consuming computations and standards can't prescribe unreachable computations to the operational world. Therefore, simplifications are made even if they imply unevaluated security margins and sometimes inconsistencies.

This fear of physical computation is largely caused by the difficulty to complete a relevant elastoplastic analysis on a 3D mesh, and then to bring out from them a readable result. While very general software, like Abacus or Ansys, offer a large number of different types of mesh and constitutive law, using them appears to be really hard for everyday life civil engineers. Indeed, those software tools are general mechanical software and don't offer proper tools for civil engineers. Moreover, uncontrolled convergence of non-linear models can prevent computation to succeed while hard work has already been done on the modelization part.

The presented below numerical method intends to solve those limitations regarding, at the same time, a suitable software tool and a robust numerical method. Let's add that several existing software already attempt to offer operational tools without over simplifying the problem, like for instance ATENA or DIANA (see [13] and [14]). However, to our knowledge, the dual analysis presented hereafter doesn't exist in any other commercial piece of software.

3 A General Optimization Formalism to Assess ULS for Reinforced Concrete Structures

Let's look into the following RC structure, a member of an externally and internally prestressed box girder subjected to cracking (Fig. 2):

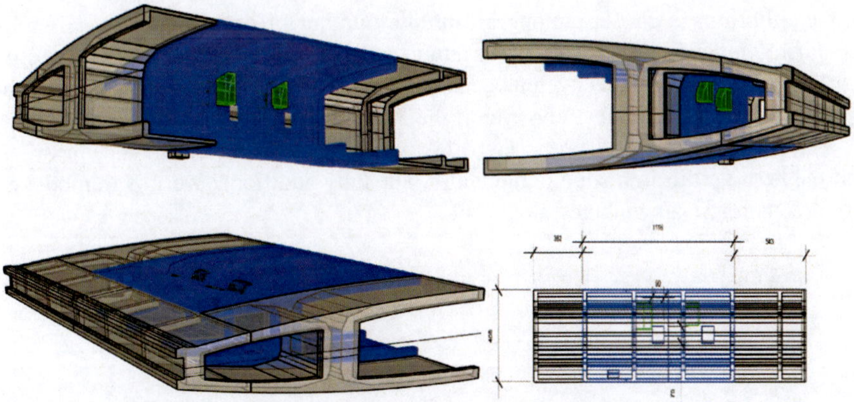

Fig. 2. CAD model of the prestressed box girder subjected to cracking. Half of the box girder is modeled due to longitudinal vertical plan symmetry. Only a few panels in the vicinity of supports are computed. External prestressing anchorage are in green.

Then, consider a civil engineer responsible for answering the following question: “Do we have to strengthen the structure or can we let it this way?”.

For this example, standards are poor help once the material limit values are determined. Moreover, since it concerns an existing structure, they have evolved since the first design whose story is hard to trace back. Therefore, a 3D finite element model is needed, either to help the strut-and-ties stress diagram, or to justify the structure full part.

3.1 An Inhouse IPM[1] Solver

Non-linear mechanical problems generally mean, mathematically speaking, constrained optimization problems (which are much more difficult than unconstrained ones). Therefore, reliable computation tools for civil engineering are necessarily based on strong mathematical foundations. An answer to this reliability need is the primal-dual interior point method, which offers great robustness as much as important calculation speed. In a simple way, this method, adopted in the present work, consists in applying Newton's method to a set of perturbed non-linear equations.

There exists a large number of available optimization solvers that can handle a wide variety of problems and one can find a good review of them in [18]. However Strains's will to exploit at most the mechanical insight lead the company to develop its own optimization software based on the primal-dual interior point method. Therefore, it must be pointed out that the strength of the whole presented numerical method results widely from this inhouse solver development and, in the same way, the quality of the presented results largely benefit from this demanding work.

3.2 The Static Approach or How to Find the Best Equilibrium State

As explained before, the strut-and-ties method would have tried to find an equilibrium state according to an elastic load path and elastoplastic materials strength limitations. This equilibrium is chosen among an infinite number of possibilities and is entirely dependent on the engineer's ability to intuit the best one. Moreover, to find the struct-and-tie diagram the engineer often needs to work out an elastic 3D computation which leads to a time-consuming study, even if the 3D model is only an intermediate step.

The presented method proposes to take advantage of the 3D model, in order to find the best equilibrium state numerically. The only additional work is to model each reinforcement, as shown below (Fig. 3):

[1] Interior Point Method.

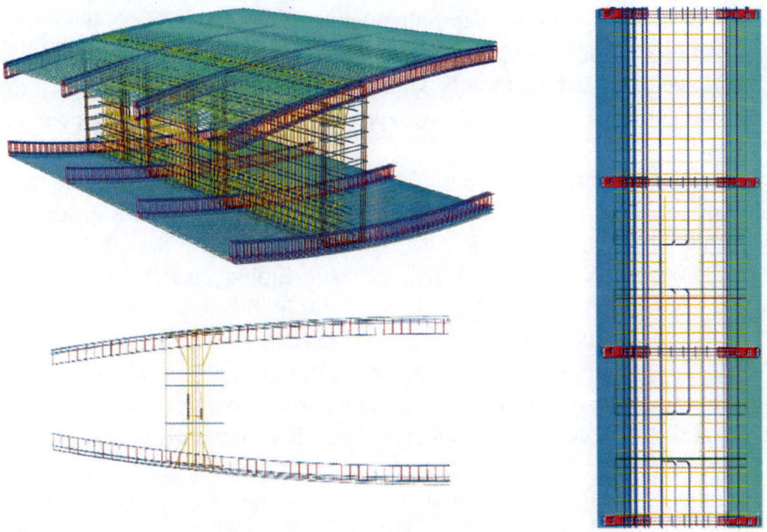

Fig. 3. CAD Model of the reinforcements embedded in concrete needed for the nonlinear analysis

A founded remark could be that drawing all the reinforcements geometry in a CAD model can be very time-consuming and that the everyday life engineer can't cope with this task. And, indeed, it would have been the case if the last decades' great improvements concerning CAD Software would have not occurred. Thanks to those powerful tools, this work only takes nowadays a few hours (once the data are collected) which doesn't seem prohibitive in front of the whole study time span.

Once the CAD model is performed, a 3D mesh is generated automatically, as well as the intersection of all the volumic mesh's tetras with the reinforcement cad lines. Then the algorithm tries to maximize the complementary energy, which only depends on the stress field, among stress values complying with the materials limit values. Different kinds of behavior can be prescribed: elastic, elastoplastic, or rigid-plastic (each with frictional contact if needed).

First of all, the used variational formulation is innovative, in the way it links concrete to reinforcement dealing only with stress variables, which can't be explained in the present paper. Let's only precise that this formulation is based on the Hybrid Equilibrium Element introduced in [16] and extended in [17], and which inherits from the work of [15]. Thus the general energy functional to be minimized is of the form:

$$\Pi_c(\sigma, \lambda) = U_c(\sigma) - V_c(\sigma) \pm \oint_{\Gamma_e \backslash \Gamma_u} \lambda^T (\mathcal{N}^T \sigma - \bar{t}) d\Gamma \quad (1)$$

The first term $U_c(\sigma) - V_c(\sigma)$ is the classical complementary energy. Imposing strongly the equilibrium on the whole structure and minimizing this first term one can retrieve an equilibrium stress field solution of the approximated mechanical problem. However, the stress field approximation used in the hybrid equilibrium formulation ensures only the equilibrium inside each finite element and does not consider equilibrium between elements. This fact explains the need to enforce *a posteriori*, and so in a weak form, the equilibrium on each finite element boundary. This is the reason of the more unusual second term $\oint_{\Gamma_e \setminus \Gamma_u} \lambda^T (\mathcal{N}^T \sigma - \bar{t}) d\Gamma$, where the Lagrange multipliers λ can be seen as generalized displacements. Under some conditions on the λ's discretization, which are respected in our case, a strong equilibrium is retrieved making the stress field codiffusive.

The second major advance is the mathematical formulation of the problem. It inherits from the work of [10] and [11] which brought out the interest of conic constraints in limit analysis. This formulation gives access to a robust solver (see Sect. 3.1) and allows to deal with all kinds of analysis (elastic, elastoplastic, limit analysis) in the same extended framework. This numerical robustness is enhanced by the ability to lead limit analysis. Indeed, limit-analysis is on its own a very direct and robust approach to assess a ULS state. While it is commonly used in geotechnical studies, structural studies hardly take advantage of this method, mainly for traditional reasons. For a civil engineer confronted with a complex RC structure, limit analysis can be seen as a very efficient tool giving access to the largest scaling value accepted for a particular loading. This scaling value is all the more interesting as the associated stress state is provided, illustrating the equilibrium scheme. Thus, static limit analysis can be seen as the optimal strut-and-ties scheme according to an automatically found safety factor (Fig. 4).

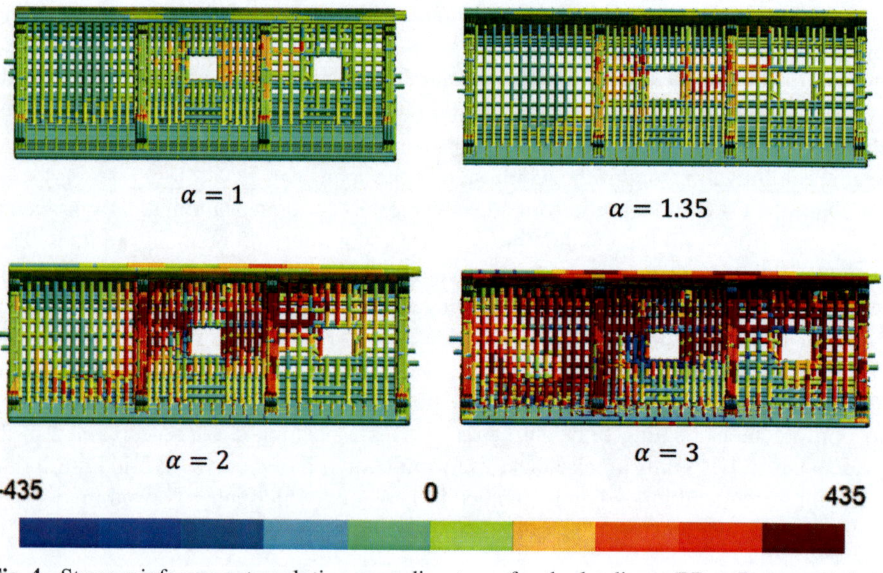

Fig. 4. Stress reinforcement evolution according to α, for the loading α(PP + P_ext + P_int). Stress value in MPA, positive values are traction.

3.3 The Kinematic Approach or How to Find the Worst Mechanism

The former static approach answers the question: what is the best equilibrium stress field according to some stress limitations? To find out the answer it extends numerically strut-and-tie rational and generalized it to any kind of structure. It must be pointed out that using a stress field as the only variable to find out a mechanical problem solution is rather uncommon. Indeed, discretization of the displacements, instead of stress, are known to give smaller size problems as explained in [12]. For instance, a much more classical method is a displacements formulation together with a Newton Raphson scheme that uses a given strain increment to find out the stress and the plastic strain. Thus, in displacements formulations, stress values are only retrieved "at the end" thanks to the constitutive law applied to a displacement field solution.

Regarding the presented method, a kinematic approach is also implemented. It consists of the search of a displacement field minimizing the potential energy plus the dissipation. Constrains on this minimization are only about boundary conditions, friction, and some slack variables.

The kinematic information completes advantageously the information already given by the equilibrium stress field, knowing that its computation cost is much less important. If the true solution is well bracketed, the retrieved displacement field gives access to a failure mechanism in adequation with the static approach stress field (Fig. 5).

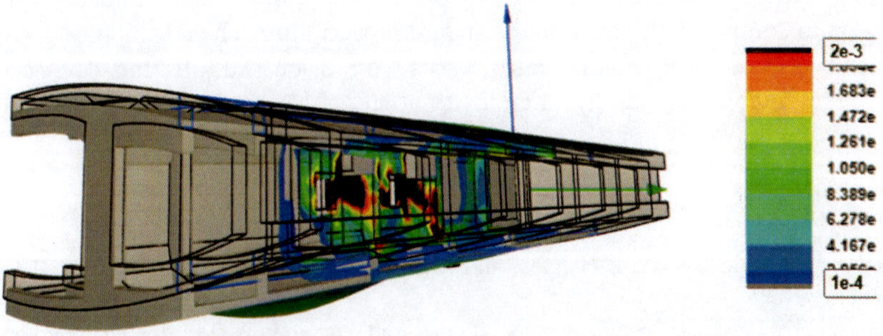

Fig. 5. Plastic strains on deformed geometry given by the kinematic approach (scaling factor 20).

3.4 At the End: Quantify the Error

Dual approach comparisons show that the choice to lead a static or a kinematic approach is far from having no impact. Indeed, with access to a kinematic result and a static result at the same time, it becomes possible to compute the distance between the two solutions. As the dual approach complies with all the mechanical equations except the constitutive law, this distance can be seen as an error in constitutive law (as shown in [19]).

It could seem expensive to lead two computations, meaning a static one and a kinematic one, with the only goal to assess the error. Indeed, numerous error estimators, post-treating a kinematic approach, give access to this error evaluation. This subject has been widely discussed in [12], and it appears that performing a dual approach is a very reliable way to assess the error and can be as expensive as some other error post-treatments. Moreover, it gives access to an equilibrium stress field which is of significant value regarding the understanding of the structure (as strut-an-ties are). Eventually, the lower bound theorem assures that the static approach under-estimates the resisting value, making the computation conservative (Table 1).

Table 1. Equations' verification (strong or weak) according to each approach

Equations verification	Static	Kinematic
Equilibrium	strong (co-diffusive)	weak
Compatibility	weak	strong
Constitutive Law	Only verified when true solution is found	Only verified when true solution is found

4 Conclusion

To quote the fib Model Code for Concrete Structures 2010 [20]: "Non-linear methods of analysis may be used for both ULS and SLS, provided that equilibrium and compatibility are satisfied and adequate non-linear behavior for materials is assumed". In spite of such recommendations, only the compatibility of displacements is widely used in modern software, whereas equilibrium seems much more difficult to achieve.

In this article we described an already operational numerical method, allowing civil engineers to comply with the quoted standard prescription. The advantages of this dual analysis based on strong mathematical bases were discussed, offering a powerful and reliable tool to cope with all the complexity of an existing or new 3D RC structure.

References

1. Ritter, W.: "Die Bauweise Hennebique," Schweizerische Bauzeitung Bd, XXXIII, No. 7, January 1899
2. Mörsch, E.: Der Eisenbetonbau, seine Theorie und Anwendung (Reinforced concrete, theory and application). Stuttgart, Verlag Konrad Wittwer. 2 (1912)
3. Schlaich, J. et Schäfer, K.: Design and detailing of structural concrete using strut-and-tie models. The Struct. Eng. 69(6), 3 (1991)
4. Bentz, E.C., Vecchio, F.J., Collins, M.P.: Simplified modified compression field theory for calculating shear strength of reinforced concrete elements (2007)
5. ACI Committee 318: Building Code Requirements for Structural Concrete (ACI 318–05) and Commentary (318R-05). American Concrete Institute, Farmington Hills, Mich., 2005, 430 p. 2
6. AASHTO, LRFD Bridge Design Specifications and Commentary, 3rd edn, American Association of State Highway Transportation Officials, Washington, D.C., 2004, 1264 p. 3
7. CEN: BS EN 1992–1–1:2004 Eurocode 2. Design of Concrete Structures. Part 1: General Rules and Rules for Buildings, 230 p. 4 (2004)
8. CSA Committee A23.3: Design of Concrete Structures (CSA A23.3–04). Canadian Standards Association, Mississauga, 214 p. 5 (2004)
9. JSCE: Specification for Design and Construction of Concrete Structures: Design. JSCE Standard, Part 1, Japan Society of Civil Engineers, Tokyo (1986)
10. Krabbenhøft, K., Lyamin, A.V., Sloan, S.W.: Formulation and solution of some plasticity problems as conic programs. Int. J. Solids Struct. **44**, 1533–1549 (2007)
11. Bleyer, J., de Buhan, P.: Lower bound static approach for the yield design of thick plates. Int. J. Numer. Methods Eng. **100**, 814–833 (2014)

12. Kempeneers, M.: Eléments finis statiquement admissibles et estimation d'erreur par analyse duale, thesis (2006)
13. Cervenka, V., Cervenka, J., Kadlec, L.: Model uncertainties in numerical simulations of reinforced concrete structures. Struct. Concr. **19**, 2004–2016 (2018)
14. Gerd-Jan Schreppers, Dr.ir.: Validation report Maekawa-Fukuura model and Cracked Concrete curves in Total Strain Crack model in DIANA (2017)
15. Fraeijs de Veubeke, B.: Displacement and equilibrium models in the finite element method. Stress analysis (1965): chapter-9
16. Pian, T.H.H.: Variational principles for incremental finite element methods. J. Franklin Inst. **302**(5–6), 473–488 (1976)
17. Almeida and Maunder: Equilibrium Finite Element Formulation (2017)
18. El Boustani, C.: Innovative optimization-based numerical methods for modeling the nonlinear behavior of steel structures, thesis (2020)
19. Ladevèze, P., Pelle, J.P.: Mastering calculation in linear and nonlinear mechanics (2006)
20. fib 2013 fib Model Code for Concrete Structures. Ernst und Sohn (2010)

The Role of Modelling in Structures Operation or Rehabilitation

Grimal Etienne[1](✉) and Chulliat Olivier[2]

[1] EDF Centre d'Ingénierie Hydraulique, La Motte Servolex, France
etienne.grimal@edf.fr
[2] EDF Hydro Alpes, Grenoble, France
olivier.chulliat@edf.fr

Abstract. In the context of regulatory changes, EDF CIH has been conducting research on concrete pathologies for more than 25 years. This research has led to the development of a numerical model for assessing the behaviour of affected structures. This model, which is now used industrially, is an integral part of the sustainable and safe operation of a structure with swelling. It enables the structure to be re-qualified, but also to efficiently design the maintenance programs for EDF Hydro's facilities.

Keywords: Modelling · Durability · Dam · Swelling · Rehabilitation

1 Introduction

1.1 Context

Almost thirty percent of hydraulic structures in the world suffer from internal concrete swelling pathologies. For dams, this pathology creates damages that can compromise the global stability of the structure at long term. For power plants, this can lead to various disorders: misalignment of turbines, ovalization of units, blocking of valves, etc.

However, it is not always essential to reinforce the structures affected. Numerical models are able to precisely describe the internal state of concrete as well as its probable evolution, which can be an effective tool in order to continue the safe operation of structures.

But, in some cases, reinforcements are necessary. The models then make it possible to design the most relevant restoration projects and optimize the cost of the works.

In this context, a methodology to manage the overall of structures, based on monitoring, laboratory tests and numerical modeling has been developed by EDF.

1.2 Global Methodology

This methodology makes it possible to ensure the operation of a structure affected by swelling in good safety conditions and is based on four successive phases.

P. Rossi and J.-L. Tailhan (Eds.): SSCS 2022, RILEM Bookseries 38, pp. 140–149, 2023.
https://doi.org/10.1007/978-3-031-07746-3_14

The first signs of swelling are often detected several years after the construction by cracking apparition in face or a side wall or by apparition of irreversible deformations (raising of the crest for example). The movements are then monitored in order to know their evolution and the swelling origin is studied in laboratory on samples drilled on the dam.

A second phase then begins with questions about the impact of swelling on affected structure. Will there be an evolution of the pressures? Will the spillway remain functional? Does the cracking that develops have an impact on the behavior of the structure?

In order to answer these questions, a third phase begins. A phase in which the structure is studied using numerical models able to reproduce the behavior of the dam. The models allow to identify failure mechanisms, to estimate a level of safety.

This step leads to a final step allowing long-term management of the structure, in which the need for reinforcement must be studied.

In the maintenance of a structure in operation, numerical modeling therefore takes a preponderant place.

2 From Modelling to Works: The Renovation of the Chambon Dam

Swelling modeling and in-situ reconnaissance proved to be essential for the design of the Chambon dam renovation works.

2.1 Historical Context

The Chambon dam is a 137 m high concrete gravity dam, located on the Romanche river in the french Alps at elevation of 1,000 m. It was built in 1934 and the reservoir was filled at the first time in 1935. The right bank and the central sections are straight, while the left bank section, where the spillway was located, is curved (Fig. 1). Its crest is near 300 m long and 5 m wide. The downstream face is H/V = 0.75/1 and the upstream face is vertical. It's made of cyclopean concrete, with a cement content varying from 150 to 250 kg/m^3.

Chambon dam suffers Alkali-Aggregate Reaction. Aggregates came from the local quarry of gneiss with numerous layers of black micaschists.

The first disorders, mainly cracks and construction joints opening, were detected since 1958 on both faces of the dam and in its galleries. At the same time, the first results of monitoring indicated an unusual behavior of the dam, whose half-left bank side was moving towards the upstream direction. Several series of tests were conducted from 1967 to 1996, confirming the presence of alkali-aggregate reaction and assigning it to the nature of the aggregates used in the concrete mix. It was also evident that the potential of the reaction and its consequences were very heterogeneous in the structure. The main swelling deformation are given in Fig. 2.

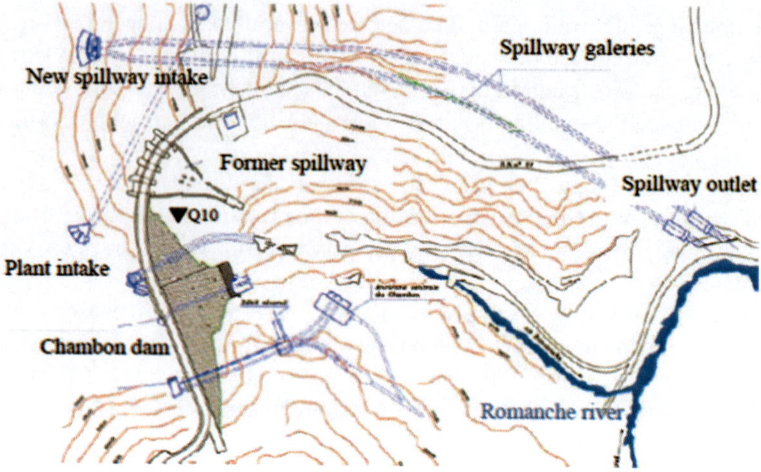

Fig. 1. Chambon dam plane view

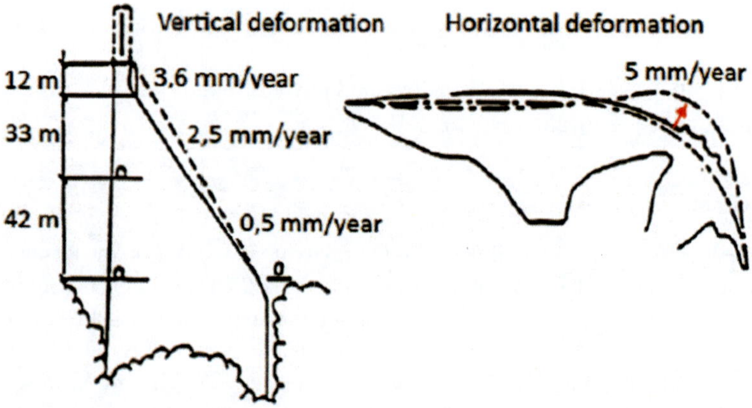

Fig. 2. Main swelling deformation rate at Chambon dam between 1984 and 1994

2.2 The First Reinforcement Works Campaign in 1992–1993

Significant work campaigns were conducted in the 1990's to treat the disorders created by AAR (Fig. 1), which main ones were:

a – Crack grouting: the more open construction joints were grouted in 1992 and 1993 in the upper 20 m of the downstream face to ensure the stability under earthquake and restore the integrity of the structure,

b – Closing of the old spillway openings: after construction of a new submerged spillway, the 4 openings of the former surface spillway were filled with concrete, to prevent a gate jamming or a spillway pier shear.

c – Upstream sealing: a waterproofing membrane was installed between 1991 and 1995 on the upper 40 m of the upstream face to prevent the creation of pore pressures in the open construction joints under earthquake.

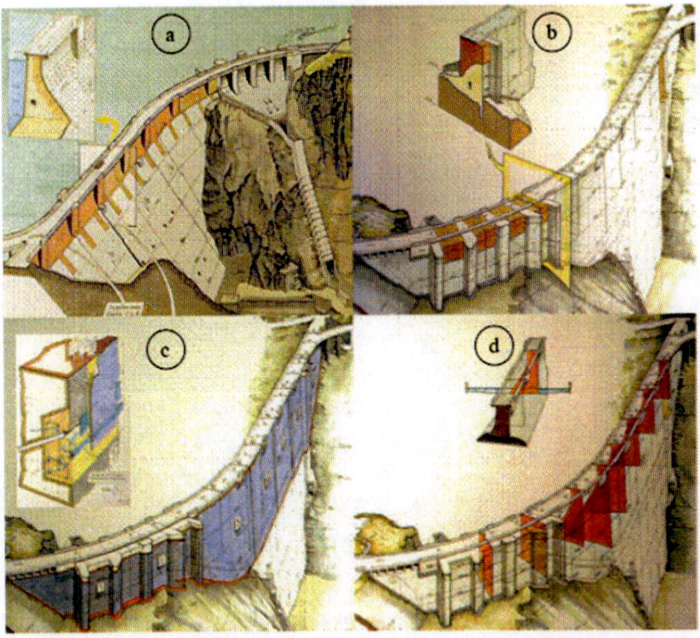

Fig. 3. Repair works conducted in the 1990's in relation with AAR

d – Slot cuts: 8 vertical diamond wire slot cuts were performed between 1995 and 1997 in the upper part of the dam. The upstream-downstream cuts were 32 m high at the maximum and 11 mm thick. Their purpose was to relieve internal stresses, reduce the risk of creating shear slides inside the left bank rock abutment and reduce the deformation in the curved zone.

Three campaigns of slot-cutting were required from 1995 to 1997 (Fig. 3d).

- 2 slots in 1995, 3 in 1996,
- 3 in 1997 were completed in depth below the crest from 18 m to 33 m with 11 mm diamond wire.

The main effects of slot cuts on the dam were:

- 1: the return of the curved part toward downstream,
- 2: the return of the bank side blocks toward the center of the dam.

During the last campaign, the 3 slots did not close and surrounding slots and joints opened, this was the sign of the effectiveness of the stress relief.

2.3 The New Behavior Analysis

Fifteen years after the first reinforcement works, the slots closure monitoring showed a slow crest compression, and the pendulums exhibited the restart of the curved left-wing movements toward upstream. Following this movement, the horizontal crack below the spillway reopened.

According to those observations, a structural FEM calculation including AAR in the constitutive law and an extensive investigation campaign was launched between 2007 and 2010. The aim was the diagnosis of the dam condition and the definition of the next works to continue the operation of the dam in safe conditions.

These investigations included drill holes in the dam body to identify the extent of crack networks, detailed inspection of the concrete-rock interface by digital borehole logging inside drill holes, identification and characterization of swelling laboratory tests on samples representative of the different areas of the dam, on-site stress measurement with flat jack devices inside the galleries and from the downstream face. Overcoring in-situ tests were also performed in order to get the three-dimensional stress tensors, leading to more relevant results of the compressive stresses. Foundation modulus measurements by dilatometer tests were also carried out.

This extensive campaign showed the almost systematic presence of a vertical cracking along the elevation drainage curtain in the upper part of the dam, which led to the conclusion that it was not possible to exclude some upstream concrete blocks instability in earthquake conditions. They might be precut by the elevation drainage curtain and extended horizontal or vertical discontinuities. The presence of these cracks was then explained by several factors:

- 1: vertical drains of diameter 800 mm (in red and green on Fig. 4) and 300 mm (in blue on Fig. 4) are separated by an average distance of 3.20 m; they could work as a "precut line" along the axis of the dam,
- 2: the differential expansion rate between the upstream concrete containing 250 kg/m^3 of cement and the downstream one containing 150 kg/m^3 could create shear stresses along the same surface (Fig. 4),
- 3: the geometry of the dam, curved in its left bank side.

Most of these cracks are observed in drilling holes through the curved zone and are several millimeters to more than 1 cm open. As the different cracks are most of time not linked inside the dam body, the presence of a continuous upstream slab can likely be excluded, but potentially unstable blocks under earthquake loading may exist. These virtual blocks are delimited: in the vertical longitudinal direction by the vertical longitudinal cracking and the upstream face (2.50 to 4.10 m), in the vertical transversal direction by dilatation joints and saw cuts (5.30 to 16.30 m) and by cracks potentially linking existing drains (singular points with an average distance of about 3 m) and in the horizontal direction by the construction joints (2.40 to 2.90 m).

Between structural cracks, the concrete displays good mechanical properties: compressive strength greater than 20 MPa, modulus of instantaneous deformation greater than 20 GPa. Moreover, the contact between limestone (Trias) and igneous rock (gneiss) in the left bank abutment is not disturbed by the thrust of the dam: it is closed and of good quality (with deformation moduli varying between 6 and 14 GPa) excluding any risk of shearing surface development inside the left bank abutment under the stresses developed by the dam swelling.

The swelling laboratory tests showed that the expansion of concrete should continue with a relatively constant rate for several decades, even if a gradual slowing is not excluded (Fig. 5).

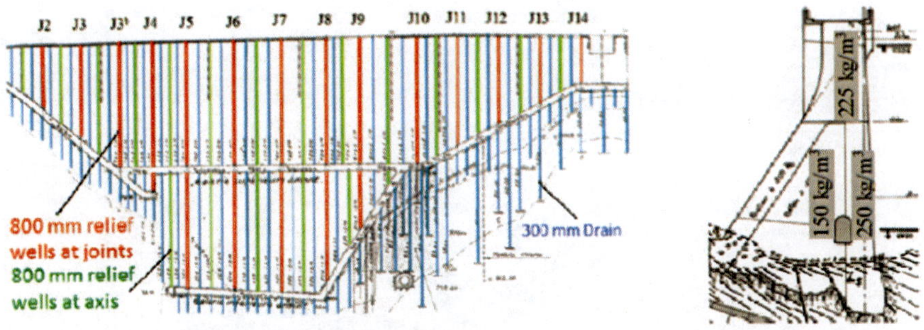

Fig. 4. Vertical drainage curtain seen from downstream and cross section

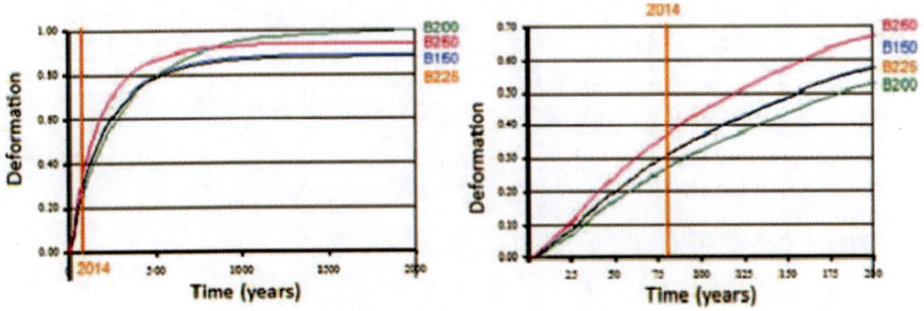

Fig. 5. Swelling potential at Chambon dam

A multi-scales swelling law for concrete developed by EDF CIH (Hydraulic Engineering Center) in collaboration of LMDC (Laboratoire Matériaux, Durabilité des Constructions) Toulouse and implemented in Code_Aster computer code developed by EDF was used at this occasion [1, 2]. It provided a very good fitting with the monitored dam behavior (Fig. 6). The FEM calculations showed that the benefits of the slot cuts done in the 1990's still remained in the upper part of the structure, confirmed by the monitoring of the deformations in the curved left wing.

The FEM calculations showed that the benefits of the slot cuts done in the 1990's still remain in the upper part of the structure, confirmed by the monitoring of the deformations in the curved right wing. They nevertheless displayed noticeable stresses parallel to the abutments. The swelling of concrete should continue with a constant rate, even if a gradual slowing is not excluded, for several decades. This swelling rate draw the attention on the shearing conditions of the upper concrete-rock interface in a medium term. The compressive stresses reach approximately 10 MPa in 2010 (Fig. 7).

Based on that diagnosis, a new reinforcement campaign was decided and received in 2010 an agreement from the Permanent Technical Committee on Dams and Hydraulic Structures (commissioned by the French Ministry of Industry).

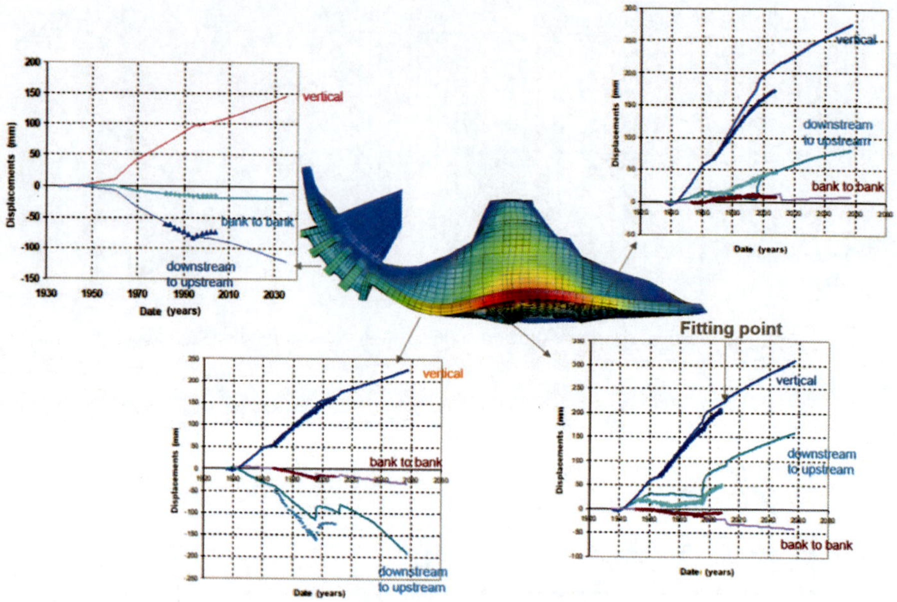

Fig. 6. Comparison between measured (points) and computed (line) displacements

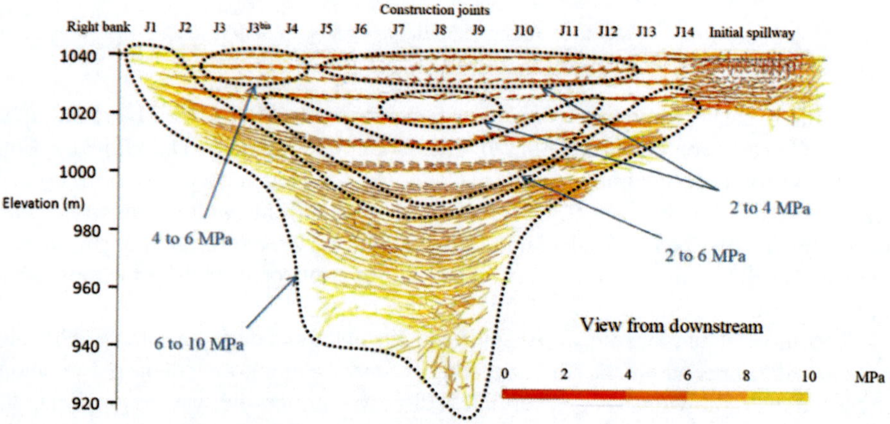

Fig. 7. Maximal principal stresses calculated in dan in 2010

3 The Second Reinforcement Works Campaign (2013–2014)

The main objectives of the second campaign were to reinforce the integrity of the upper part of the dam and to prevent any upstream block from falling that could lead to destabilization of the dam.

An upper part confinement was decided and obtained by the installation of 415 tendons (Fig. 8). Consisting of horizontal cables, with greased sheathed strands type T15, crossing the structure from upstream to downstream, they were pre-tensioned and non-grouted on the outer side of the general sheath. They are 3.70 m (horizontal) and 4 m (vertical) spaced, with about one tendon for 15 m^2. Upstream and downstream cable heads are fully embedded in reservations drilled in the dam, with diameter ranging between 500 and 700 mm.

As progressive tension increase due to concrete expansion remains possible, the downstream heads were designed to be adjustable and detensioning operations could be necessary in order to fit the required tension.

The 4180 m drill holes necessary for the tendons were cored from upstream with a systematically inspected by logging digital in order to map the cracks network and to represent them in a 3D digital model.

Fig. 8. The tendons pattern (red) and drainage curtain seen from downstream

In addition to tendons, a carbon fiber composite net has been set up on the up-stream face. It consisted in the sticking of 6,000 m carbon fiberstrips. The 20 to 30 cm wide strips link the tendon heads, along vertical, horizontal and diagonal lines (Fig. 9).

Fig. 9. The carbon fiber net seen from upstream

The carbon fiber net serves confinement of small blocks that could escape the tendons action. Strips composite material (carbon strips glued with epoxy resin) were designed to form a "chainstitch" and to resist tensile stress due to earthquake and further swelling.

While carbon fiber composite is a material commonly used in civil engineering, its implementation as an anti-seismic net is relatively innovative and requested qualification tests in laboratory.

The new campaign by diamond wire sawing was defined using the numerical model (i) to avoid the recompression of the upper part of the structure and (ii) to decompress the stress paths parallel to the abutments (Fig. 7). Slot cuts mostly followed the same tracks than in the 1990's, with the following exceptions:

- slot cut S3 in the central part of the dam was not sawn again, due the presence of a cable stuck during the previous campaign,
- both slot cuts flanking S3 (S2 and S4) were deepened to 42 m (650 m^2 each), in order to compensate the non-resawing of S3, but mainly to increase the stress reduction efficiency along the concrete/rock interface,
- the diamond wire was 16 mm wide (against 11 previously) to take advantage of technological progress made since the 1990's and avoid any risk of jamming in the event of non-fully closed former slot cuts,
- progresses in slot cutting technology allow nowadays far greater rates of cutting and made possible the 2,500 m^2 total cut in six months, with two equipments (Fig. 10).

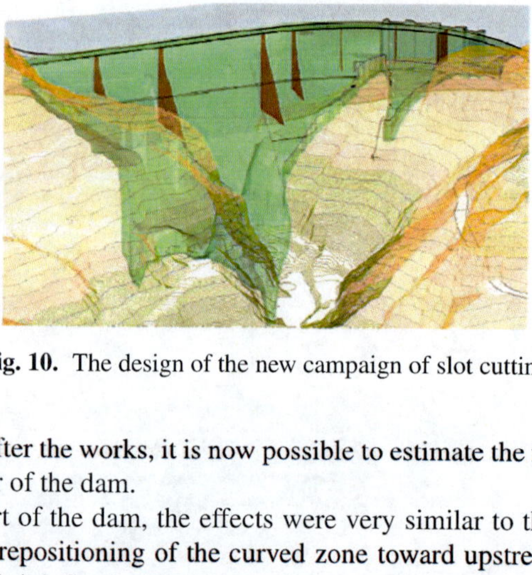

Fig. 10. The design of the new campaign of slot cutting

About 8 years after the works, it is now possible to estimate the influence of the slot cuts on the behavior of the dam.

In the upper part of the dam, the effects were very similar to those observed after the first campaign: repositioning of the curved zone toward upstream, acceleration of the central zone and right bank side deformation toward downstream, reopening of the former construction joints and the slot cuts, return of the side blocks toward the center of the dam. In the lower part of the dam, the repositioning of the side blocks toward the center of the dam was significantly greater following the new campaign, confirming the effectiveness of S2 and S4 deepening on the stresses along the abutments.

The new waterproofing membrane is conceptually identical to the one completed in 1995, with a few small modifications and improvements. It is tensioned by vertical fixings spaced 1.85 m apart, drained and divided into 12 independent compartments. The behavior of the previous geomembrane remained satisfactory after nearly twenty years of operation and gave the opportunity to re-use a large part of the stainless steel components.

4 Conclusion

This article presents the contribution of modeling in the management of structures affected by swelling.

The management methodology was presented through the case of Chambon dam. This methodology is based on three axes: monitoring, laboratory tests and behavior analysis.

The modeling of the structure then takes a prominent place in this methodology. In addition to justifying the behavior of the structure in its current state, the models also make it possible to predict the long-term deformations and stresses of concrete structures affected by swelling.

After an in-depth study of the mechanisms and consequences of swelling (AAR), the following renovation was proposed:

- installation of prestressed tie rods, in order to restore the monolithism of the upper part,
- new slot cutting works allowing the reduction of forces in concrete-rock contact,
- the installation of a waterproofing membrane in order to limit the pore pressures in the concrete.

The extended monitoring system enables to follow precisely the dam behavior and confirms that the effects of the slot cuttings are in line with the model predictions. Regular diagnoses will continue to be carried out in order to ensure that the dam operation continues under optimal safety conditions.

References

1. Code_Aster: Loi de comportement sur le fluage et l'endommagement du béton (RSI/RAG), R7.01.30
2. Morenon, P.: Modélisation des réactions de gonflement interne des bétons avec prise en compte des couplages poro-mécaniques et chimiques, Phd thesis, Université Paul Sabatier-Toulouse III (2017)

The Multi-scale Multi-technique Multi-physics Framework Required to Model the Thermal Properties of Cement-Based Materials

Tulio Honorio[1(✉)], Fatima Masara[1], Waleska Barbosa[2], and Farid Benboudjema[1]

[1] CentraleSupélec, ENS Paris-Saclay, CNRS, LMPS - Laboratoire de Mécanique Paris-Saclay, Université Paris-Saclay, 91190 Gif-sur-Yvette, France
thonorio@ens-paris-saclay.fr

[2] Universidade Federal do Paraná, Centro Politécnico, Engenharia e Ciência dos Materiais, 81531-990 Curitiba, PR, Brazil

Abstract. The physical origins of the properties controlling the thermal response of nanoporous materials are fundamentally related to atomic-scale processes. Therefore, the techniques that enable assessing the nanoscale are required to properly understand the thermal properties of cement-based materials. In this paper, we review the techniques of classical molecular simulations that can be used to link the structure of materials to their thermal properties. We focus on the heat capacity, thermal expansion, and thermal conductivity. Results on various phases present in cement systems (tricalcium silicate, C-(A)-S-H, Friedel's salt, and ettringite) are presented. The fundamental property data provided on the thermal properties are valuable input for multiscale modeling and prediction of the thermal behavior of cement-based materials.

Keywords: Molecular simulations · Heat capacity · Thermal expansion · Quantum corrections

1 Introduction

Thermal deformations are a major cause of cracking in cement-based materials (e.g.[19,27]). Thermo-chemo analysis are critical for massive concrete structures [2,18,27]. The physical origins of the properties controlling the thermal response of micro- and mesoporous materials are fundamentally related to atomic-scale processes. Therefore, techniques well-suited to assess the atomistic scale are required to properly understand the thermal behavior of cement-based materials or, in a more ambitious perspective, bottom-up engineer materials with a target thermal performance.

The financial support of the French National Research Agency (ANR) through the project THEDESCO (ANR-19-CE22-0004–01) is gratefully acknowledged.

P. Rossi and J.-L. Tailhan (Eds.): SSCS 2022, RILEM Bookseries 38, pp. 150–159, 2023.
https://doi.org/10.1007/978-3-031-07746-3_15

Fundamental component data regarding the properties of the main phases present in cement systems are critical input in multiscale modeling [20]. However, such data is missing for some key phases (e.g., AF-phases). Also, the effects of water content in the thermal properties of hydrated phases (especially C-S-H and AF-phases, which exhibits variable water content) are not fully understood [9].

In this paper, we review the techniques of classical molecular simulations that can be used to link the structure of materials to their thermal properties. We focus on the following thermal properties: the heat capacity, thermal expansion, and thermal conductivity. We show that the molecular simulations of the heat capacity of materials with crystalline domains, as it is the case of crystalline phases (e.g., tricalcium silicate) or microporous hydrated phases (e.g., C-(A)-S-H, ettringite, AFm, alkali-silica crystalline products), requires the incorporation of quantum corrections. As a result, the thermal properties of tricalcium silicate, ettringite, Friedel's salt, and C-A-S-H computed from classical molecular dynamics are presented and confronted with experimental data available from the literature. The fundamental property data provided in this work for the thermal properties of these phases are valuable input for multiscale modeling and prediction of the thermal behavior of cement-based materials.

2 Thermal Properties from Molecular Simulations

Strategies based on finite difference and fluctuations approaches are presented. In the case of the finite difference, the property definition is used to define the simulation protocol.

2.1 Coefficient of Thermal Expansion

Finite Difference. The tensor of coefficients of thermal expansion is defined in the anisotropic framework as (.e.g. [8,36]):

$$\alpha_{ij} = \left[\frac{\partial \epsilon_{ij}}{\partial T}\right]_{\sigma_{ij}} \tag{1}$$

where σ_{ij} is the stress, T is the temperature, and the strain ϵ_{ij} is the second-order Lagrangian deformation tensor (also given in its general anisotropic form):

$$\epsilon_{ij} = \frac{1}{2}\left[\mathbf{h}_0^{-T}\mathbf{h}^T\mathbf{h}\mathbf{h}_0^{-1} - \mathbf{I}\right] \tag{2}$$

with $\mathbf{I}$ being the unit 3×3 matrix, $[.]^T$ being the transpose operator; the $\mathbf{h}$ matrix being the 3×3 matrix formed by arranging the lattice vectors as $\{\mathbf{a}, \mathbf{b}, \mathbf{c}\}$, the reference state (at zero stress [36]) is denoted by the subscript 0.

In the finite difference approach, a number of simulations are run at the same stress σ_{ij} (of 1 atm) for various T in the vicinity of the target temperature. The derivative from Eq. 1 is finally evaluated assuming a linear behavior.

Fluctuation. In the fluctuation-dissipation approach, the thermal expansion tensor is obtained from the cross fluctuations of the strain ϵ_{ij} and the enthalpy H in an NσT simulation [8,25,36]:

$$\alpha_{ij} = \frac{\langle H\epsilon_{ij}\rangle_{N\sigma T} - \langle H\rangle_{N\sigma T}\langle\epsilon_{ij}\rangle_{N\sigma T}}{kT^2}. \tag{3}$$

where k is the Boltzmann constant, and $\langle . \rangle_{N\sigma T}$ is an NσT ensemble average.

2.2 Heat Capacity

Full Classical Estimates

Finite Difference. The heat capacity at constant volume C_v (respectively, the heat capacity at constant pressure C_p) is property quantifying how the internal energy E (respec., the enthalpy $H = E + PV$) varies with the temperature T:

$$C_V = \left[\frac{\partial E}{\partial T}\right]_V ; \quad C_P = \left[\frac{\partial H}{\partial T}\right]_\sigma \tag{4}$$

A number of simulations are run at the same volume (for C_V) or stress (for C_P) for various T in the vicinity of the target temperature. The derivatives in the definitions are finally evaluated assuming a linear behavior.

Fluctuations. In a full classical approach, the heat capacity at constant volume C_V is computed from the fluctuations of the internal energy E [1]:

$$C_V = \frac{\langle E^2\rangle_{NVT} - \langle E\rangle^2_{NVT}}{kT^2} \tag{5}$$

The heat capacity at constant pressure C_P is computed from the fluctuations of the enthalpy H [1]:

$$C_P = \frac{\langle H^2\rangle_{N\sigma T} - \langle H\rangle^2_{N\sigma T}}{kT^2} \tag{6}$$

Semi-Classical Estimates. The high-frequency lattice dynamics require that some quantum corrections are introduced in the estimation of the heat capacity of phases with crystalline domains. In other words, whenever the lattice frequency f happens to be on the order of $\frac{kT}{h}$ quantum effects are not negligible (or in terms of angular frequency ω, whenever $\hbar\omega$ is on the order of kT), with h being the Planck constant (and $\hbar = h/(2\pi)$).

Studies on clinker phases show indeed that full-classical approaches lead to overestimations of the heat capacity [5,10,34].

A semi-classical approach [13], in which classical simulations are used to compute input in quantum-corrected expressions, can be used to get heat capacity of crystal in agreement with the experimental evidence. In this semi-classical approach, the heat capacity at constant volume C_V is given by [32]:

$$C_V = \frac{dE_{vib}}{dT} = 3Nk \int_0^\infty \left(\frac{\hbar\omega \, n(\omega, T)}{kT} \right)^2 \exp\left[\frac{\hbar\omega}{kT} \right] g(\omega) d\omega \quad (7)$$

where $n(\omega, T) = \frac{1}{\exp\left[\frac{\hbar\omega}{kT}\right]-1}$ is the Bose-Einstein distribution, and $g(\omega)$ is the density of the states [13] (see [21] for a detailed demonstration):

$$g(\omega) = \frac{1}{\pi} \int_{-\infty}^{\infty} \sum_j \frac{\langle m_j \mathbf{v}_j(t).\mathbf{v}_j(0) \rangle}{\langle m_j \mathbf{v}_j^2(0) \rangle} \exp\left[i\omega t \right] dt \quad (8)$$

which is computed in the classical simulation using the mass-weighted velocity auto- correlation function $\langle m_j \mathbf{v}_j(t).\mathbf{v}_j(0) \rangle$, where m_j is the mass of particle j; $\mathbf{v}_j(t)$ is the velocity vector of particle j at time t, and "." is the scalar product. In the semi-classical approach, only the vibrational E_{vib} contribution for the internal energy is considered. The other contributions for the internal energy can be neglected: the translational and rotational contributions are negligible for temperatures below Debye temperature, and the lattice contribution is temperature-independent [13].

Hybrid Estimates. In hydrated minerals, it can be shown that the contribution of molecular water to the heat capacity is underestimated by semi-classical approaches [21]. An hybrid approach leads to experimentally sound estimates of the heat capacity of various hydrates minerals including clays [21], ettringite [17], C-A-S-H [29] and Friedel's salt [22]. In this hybrid approach, C_V of the hydrated mineral is computed using the semi-classical approach for the dry mineral $C_{v-min}^{semi-class}$ and the heat capacity of confined water C_{v-w}^{class} is computed using a full-classical approach [21]:

$$C_V = C_{v-w}^{class} + C_{v-min}^{semi-class} = C_{v-tot}^{class} - 3N_s k + C_{v-min}^{semi-class} \quad (9)$$

Since the full classical approach applied to the dry mineral returns Dulong-Petit law (i.e., $C_{v-min}^{class} = 3N_s k$, where N_s is the number of atoms in the dry mineral), we can write the following expression used in the equation above: $C_{v-w}^{class} = C_{v-tot}^{class} - 3N_s k$.

Mayer's Relation. The specific heat capacity at constant volume and pressure can be readily obtained from each other, once the tensor of thermal expansion α_{ij} and the elastic stiffness tensor C_{ijkl} are known, using Mayer's relation (e.g.[40]):

$$c_P - c_V = \frac{T}{\rho} C_{ijkl} \alpha_{ij} \alpha_{kl}, \quad (10)$$

where ρ is the density.

2.3 Thermal Conductivity

The full thermal conductivity tensor can be obtained from molecular simulations using equilibrium and non-equilibrium methods.

Non-Equilibrium Method. Non-equilibrium methods include direct thermostatting method [26], and reverse perturbation method [31].

Equilibrium Method. The main equilibrium method is the Green-Kubo formalism, detailed in the following. With Green-Kubo formalism, the thermal conductivity is the time integral of the ensemble average (denoted by $\langle . \rangle$ operator) of the auto-correlation of the heat flux J_i ($i = x, y, z$) (e.g. [1]):

$$\lambda_{ij} = \frac{V}{kT^2} \int_0^\infty \langle J_i(0) \otimes J_j(t) \rangle dt \tag{11}$$

where k is the Boltzmann constant, V is the volume, and $\otimes$ is the dyadic product. The heat flux vector is obtained from: $\mathbf{J}_p = \frac{1}{V} \left[\sum_p E_p \mathbf{v}_p - \sum_p \mathbf{S}_p \mathbf{v}_p \right]$ where E_p is the sum of the kinetics and potential energy, $\mathbf{v}$ is the velocity, and $\mathbf{S}_p$ is the stress tensor for atom p.

In some cases, the conductivity λ_{ij} computed from the standard Green-Kubo formalism might not converge. A solution to overcome this problem is to calculate the thermal conductivity tensor using the frequency-dependent response of the complex conductivity [39]:

$$\lambda_{ij}(\omega) = \frac{V}{kT^2} \int_0^\infty \langle J_i(0) \otimes J_j(t) \rangle \exp\left[i\omega t \right] dt \tag{12}$$

The (static) thermal conductivity λ_{ij}^0 is obtained from fitting at the low frequency range the modulus $|\lambda_{ij}(\omega)|$ of the complex conductivity with the expression $|\lambda_{ij}(\omega)| = \lambda_{ij}^0 + A\omega^\xi$, where λ_{ij}^0, A and ξ are free parameters in the fitting procedure [34].

2.4 Molecular Models and Methods

Simulation are performed with LAMMPS [33]. Ewald summation with a desired relative error in forces of 10^{-5}, and tail corrections for the Lennard-Jones potential are adopted for long-range interactions. The adopted atomic structures and force fields are detailed in the following.

C_3S. The triclinic C_3S (T1) obtained by Golovastikov [15] is adopted. Interactions are described using the Consistent Valence Force Field (CVFF) [12]. More details and structural validation are given in ref. [5].

C-A-S-H. The bulk C-A-S-H structure proposed by [28] is adopted. Interactions are described using ClayFF [11]. More details and structural validation are given in Ref. [29].

Ettringite and Metaettringite. The ettringite structure resolved by Moore and Taylor [30] is adopted. AFFF [23,24] is used to describe interactions between atoms. Ettringite dehydrates at low RH ($< 1\%$), forming a more disordered phase called metaettringite. Metaettringite structures are obtained from the molecular simulations described in [24]. More details and structural validation are given in Refs. [23,24].

Friedel's Salt. The monoclinic structure resolved by [35] of the associated with ambient temperatures (noted LT hereon) is adopted. AFFF [23,24] is used to describe interactions between atoms in Friedel's salt. Parameters for chlorides are taken as in the original version of ClayFF [11]. The high-temperature (HT, at 313 K) polymorph of Friedel's salt is obtained from simulations detailed in [22]. More details and structural validation are given in Ref. [22].

3 Results and Discussion

3.1 Thermal Expansion

The components of the tensor of coefficient of thermal expansion in the principal basis and the volumetric (i.e., $\alpha_v = Tr\ \alpha_{ij} = \alpha_{11} + \alpha_{22} + \alpha_{33}$) and linear (i.e., $\alpha_{lin} = \alpha_v/3$) coefficients of thermal expansion of phases studied here are gathered in Table 1. These results were obtained from molecular dynamics simulations using the fluctuation approach.

No experimental or simulation results are reported in the literature for the coefficient of thermal expansion of C-A-S-H and Friedel's salt. The results on C-A-S-H can be compared with the values reported for C-S-H obtained from molecular simulations of linear thermal expansion of $4.5 \pm 0.9 \times 10^{-5}$/K [34], and obtained from micromechanics inverse analysis 4.2×10^{-5}/K [14]. SImulations of the thermal expansion are provided only for monoclinic polymorphs of C_3S [10,34].

Experimental data on ettringite thermal expansion are reported by [16] following the crystallographic framework (in terms of lattice vector a, b, c): $\alpha_a = 4.2 \times 10^{-5}$K and $\alpha_c = 2.2 \times 10^{-5}$/K. In the hexagonal symmetry of the ettringite crystal, $\alpha_a = \alpha_b$ and a Cartesian frame $\alpha_{xx} = \alpha_{yy}$ is the vector a is aligned with x. Therefore, $\alpha_v = 10.6 \times 10^{-5}$/K. These values are fairly close to the one obtained from simulations when on take into account the standard deviations associated with the simulations of a few 10^{-5}/K. These results build confidence in using the simulation to get the properties of metaettringite, a phase for which almost no physical property has been experimentally assessed so far.

3.2 Specific Heat Capacity

Table 2 gathers the specific heat capacity obtained from molecular simulations. The semi-classical approach is adopted for C_3S (a non-hydrated phase). For the

Table 1. Coefficient of thermal expansion (in 10^{-5}/K) of phases relevant to cement systems obtained from molecular dynamics simulations using the fluctuation approach.

Phase	α_{11}	α_{22}	α_{33}	α_v	α_{lin}
C_3S	1.55	1.17	0.787	3.50	1.17
C-A-S-H	8.93	3.88	–0.52	12.33	4.11
Ettringite	7.61	5.03	2.83	15.45	5.15
Metaettringite	5.67-6.80	4.10–5.85	1.79-3.02	13.91–14.4	4.34–4.8
Friedel's salt LT	7.47	4.59	4.10	16.2	5.4
Friedel's salt HT	8.81	4.77	3.94	17.5	5.8

hydrated phases, a hybrid approach is adopted. Experimental or assessed data on c_p of hydrated phases are available in the literature: 1.006 to 0.849 kJ/(kg.K) for C-A-S-H [37]; 1.300 to 1.732 kJ/(kg.K) for ettringite [4,38] , and [4] 1.489 to 1.547 kJ/(kg.K) for metaettringite; 1.23 to 1.48 for Friedel's salt [3,7]. All these values compare well with the hybrid simulation results. In general, the heat capacity using full classical approaches leads to overestimations on the order of 50% (as is also the case for other hydrated phases such as clays [21]).

Table 2. Specific heat capacity (in kJ/(kg.K) of phases relevant to cement systems obtained from molecular dynamics simulations using methods incorporating quantum corrections: semi-classical for C_3S, hybrid for the hydrated phases.

Phase	c_p	c_v
C_3S	0.687	0.673
C-A-S-H	1.048	1.023
Ettringite	1.845	1.746
Metaettringite	1.422–1.879	1.366–1.805
Friedel's salt LT	1.47	1.38
Friedel's salt HT	1.44	1.39

3.3 Thermal Conductivity

The components of the tensor of thermal conductivity in the principal basis and the volumetric (i.e., $\lambda_v/3 = Tr\lambda_{ij}/3$) thermal conductivity of C_3S and (meta)ettringite are displayed in Table 1. These results were obtained from molecular dynamics simulations using Green-Kubo formalism. The C_3S value can be compared with the estimates for cement particles' thermal conductivity of 1.55 W/(m.K) [6]. No experimental results exist in the literature for (meta)ettringite, therefore, the molecular simulation results can be helpful as the first available data for this phase. The thermal conductivity of other phases relevant to cement systems considered here (C-A-S-H and Friedel's salt) are yet to be provided.

Table 3. Components of the thermal conductivity tensor in the princiapl basis (in W/(m.K) and volumetric thermal conductivity of phases relevant to cement systems obtained from molecular dynamics simulations using the Green-Kubo formalism.

Phase	λ_{11}	λ_{22}	λ_{33}	λ_v
C_3S	4.33	0.52	0.28	1.71
Ettringite	4.87	2.33	1.13	2.78
Metaettringite	3.12-8.05	1.61–4.12	1.18-1.45	2.06–4.45

4 Conclusion

In this work, the techniques of classical molecular simulations enabling the estimation of thermal properties from the composition and atomic structure of crystalline and hydrated phases in cement systems were reviewed. As results, the thermal properties of tricalcium silicate (meta)ettringite, Friedel's salt (LT and HT), and C-A-S-H computed from classical molecular dynamics are presented and confronted with experimental data available from the literature. A good comparison with the available experimental data on thermal properties builds confidence in using molecular simulation to provide thermal properties data on the various phases in cement systems. Such data complete the information on component properties in cement systems, being valuable input for multiscale modeling and prediction of the thermal behavior of cement-based materials.

Perspectives include computing the thermal properties of other phases relevant to cement systems.

References

1. Allen, M.P., Tildesley, D.J.: Computer Simulation of Liquids. Oxford University Press, New York (1989)
2. Azenha, M., et al.: Recommendations of RILEM TC 287-CCS: thermo-chemo-mechanical modelling of massive concrete structures towards cracking risk assessment. Mater. Struct. **54**(4), 1–13 (2021). https://doi.org/10.1617/s11527-021-01732-8
3. Balonis, M.: Thermodynamic modelling of temperature effects on the mineralogy of Portland cement systems containing chloride. Cem. Concr. Res. **120**, 66–76 (2019)
4. Baquerizo, L.G., Matschei, T., Scrivener, K.L.: Impact of water activity on the stability of ettringite. Cem. Concr. Res. **79**, 31–44 (2016)
5. Barbosa, W., Honorio, T.: Triclinic tricalcium silicate: structure and thermoelastic properties from molecular simulations. Cement and Concrete Research (accepted 2022)
6. Bentz, D.P.: Transient plane source measurements of the thermal properties of hydrating cement pastes. Materia. Struct. **40**(10), 1073 (2007)
7. Blanc, P., Bourbon, X., Lassin, A., Gaucher, E.C.: Chemical model for cement-based materials: thermodynamic data assessment for phases other than C-S-H. Cem. Concr. Res. **40**(9), 1360–1374 (2010)

8. Cagin, T., Karasawa, N., Dasgupta, S., Goddard, W.A.: Thermodynamic and elastic properties of polyethylene at elevated temperatures. In: MRS Online Proceedings Library Archive, vol. 278 (1992)
9. Chen, B., Kuznik, F., Horgnies, M., Johannes, K., Morin, V., Gengembre, E.: Physicochemical properties of ettringite/meta-ettringite for thermal energy storage: review. Sol. Energy Mater. Sol. Cells **193**, 320–334 (2019)
10. Claverie, J., Kamali-Bernard, S., Cordeiro, J.M.M., Bernard, F.: Assessment of mechanical, thermal properties and crystal shapes of monoclinic tricalcium silicate from atomistic simulations. Cem. Concr. Res. **140**, 106269 (2021)
11. Cygan, R.T., Liang, J.J., Kalinichev, A.G.: Molecular models of hydroxide, oxyhydroxide, and clay phases and the development of a general force field. J. Phys. Chem. B **108**(4), 1255–1266 (2004)
12. Dauber-Osguthorpe, P., Roberts, V.A., Osguthorpe, D.J., Wolff, J., Genest, M., Hagler, A.T.: Structure and energetics of ligand binding to proteins: Escherichia coli dihydrofolate reductase-trimethoprim, a drug-receptor system, vol. 4(1) (1988)
13. Dickey, J.M., Paskin, A.: Computer simulation of the lattice dynamics of solids. Phys. Rev. **188**(3), 1407–1418 (1969)
14. Ghabezloo, S.: Micromechanics analysis of thermal expansion and thermal pressurization of a hardened cement paste. Cem. Concr. Res. **41**(5), 520–532 (2011)
15. Golovastikov, N.I.: Crystal structure of tricalcium silicate, 3caosio_2 = c_3s. Sov. Phys. Crystallogr. **20**(4), 441–445 (1975)
16. Hall, C., Barnes, P., Billimore, A.D., Jupe, A.C., Turrillas, X.: Thermal decomposition of ettringite Ca6[Al(OH)6]2(SO4)3·26H2O. J. Chem. Soc. Faraday Trans. **92**(12), 2125–2129 (1996)
17. Honorio, T.: Thermal conductivity, heat capacity and thermal expansion of ettringite and metaettringite: effects of the relative humidity and temperature, Cem. Con. Res. (2022). (Accepted)
18. Honorio, T., Bary, B., Benboudjema, F.: Evaluation of the contribution of boundary and initial conditions in the chemo-thermal analysis of a massive concrete structure. Eng. Struct. **80**, 173–188 (2014)
19. Honorio, T., Bary, B., Benboudjema, F.: Factors affecting the thermo-chemo-mechanical behaviour of massive concrete structures at early-age. Mater. Struct. **49**(8), 3055–3073 (2015). https://doi.org/10.1617/s11527-015-0704-5
20. Honorio, T., Bary, B., Benboudjema, F.: Thermal properties of cement-based materials: multiscale estimations at early-age. Cement Concr. Compos. **87**, 205–219 (2018)
21. Honorio, T., Brochard, L.: Drained and undrained heat capacity of swelling clays, Physical Chemistry Chemical Physics (2022). https://doi.org/10.1039/D2CP01419J
22. Honorio, T., Carasek, H., Cascudo, O.: Friedel's salt: temperature dependence of thermoelastic propeties, p. 28, under review (2022)
23. Honorio, T., Guerra, P., Bourdot, A.: Molecular simulation of the structure and elastic properties of ettringite and monosulfoaluminate. Cem. Concr. Res. **135**, 106126 (2020)
24. Honorio, T., Maaroufi, M., Al Dandachli, S., Bourdot, A.: Ettringite hysteresis under sorption from molecular simulations. Cem. Concr. Res. **150**, 106587 (2021)
25. Honorio, T., Lemaire, T., Tommaso, D.D., Naili, S.: Molecular modelling of the heat capacity and anisotropic thermal expansion of nanoporous hydroxyapatite. Materialia **5**, 100251 (2019)

26. Ikeshoji, T., Hafskjold, B.: Non-equilibrium molecular dynamics calculation of heat conduction in liquid and through liquid-gas interface. Mol. Phys. **81**(2), 251–261 (1994)
27. Kanavaris, F., et al.: Enhanced massivity index based on evidence from case studies: towards a robust pre-design assessment of early-age thermal cracking risk and practical recommendations. Constr. Build. Mater. **271**, 121570 (2020)
28. Kunhi Mohamed, A., et al.: The atomic-level structure of cementitious calcium aluminate silicate hydrate. J. Am. Chem. Soc. **142**(25), 11060–11071 (2020)
29. Mahmoud Hawchar, B., Honorio, T.: C-A-S-H thermoelastic properties at the molecular and gel scales, J. Adv. Concr. Technol. **20**(6), 375–388 (2022). https://doi.org/10.3151/jact.20.375
30. Moore, A.E., Taylor, H.F.W.: Crystal structure of ettringite. Acta Crystallogr. Sect. B: Struct. Crystallogr. Cryst. Chem. **26**(4), 386–393 (1970)
31. Müller-Plathe, F.: A simple nonequilibrium molecular dynamics method for calculating the thermal conductivity. J. Chem. Phys. **106**(14), 6082–6085 (1997)
32. Oganov, A.R., Brodholt, J.P., David Price, G.: Comparative study of quasiharmonic lattice dynamics, molecular dynamics and debye model applied to MgSiO3 perovskite. Phys. Earth Planet. Inter. **122**(3), 277–288 (2000)
33. Plimpton, S.: Fast parallel algorithms for short-range molecular dynamics. J. Comput. Phys. **117**(1), 1–19 (1995)
34. Qomi, M.J.A., Ulm, F.J., Pellenq, R.J.M.: Physical origins of thermal properties of cement paste. Phys. Rev. Appl. **3**(6), 064010 (2015)
35. Rapin, J.P., Renaudin, G., Elkaim, E., Francois, M.: Structural transition of Friedel's salt 3CaO·Al2O3·CaCl2·10H2O studied by synchrotron powder diffraction. Cem. Concr. Res. **32**(4), 513–519 (2002)
36. Ray, J.R.: Molecular dynamics equations of motion for systems varying in shape and size. J. Chem. Phys. **79**(10), 5128–5130 (1983)
37. Thermodynamic properties of C-S-H, C-A-S-H and M-S-H phases: results from direct measurements and predictive modelling. Appl. Geochem. **92**, 140–156 (2018)
38. Struble, L.J., Brown, P.W.: Heats of dehydration and specific heats of compounds found in concrete and their potential for thermal energy storage. Solar Energy Mater. **14**(1), 1–12 (1986)
39. Termentzidis, K., Merabia, S., Chantrenne, P., Keblinski, P.: Cross-plane thermal conductivity of superlattices with rough interfaces using equilibrium and non-equilibrium molecular dynamics. Int. J. Heat Mass Transf. **54**(9-10), 2014–2020 (2011)
40. Wallace, D.C.: Thermodynamics of Crystals. Courier Corporation (1998)

Numerical Model for Explosive Spalling of High-Strength Concrete and Carbonation During and After Fire Exposure

Keitai Iwama[1], Koichi Maekawa[2(✉)], and Kazuaki Highuchi[3]

[1] Department of Civil and Environmental Engineering, Hong Kong Polytechnic University, Hong Kong, China
keitai.iwm.concrete@gmail.com
[2] Yokohama National University, Tokiwadai 79-5, Hodogaya, Yokohama 240-0044, Japan
maekawa-koichi-tn@ynu.ac.jp
[3] Maeda Corporation, Fujimi, Chiyoda-ku 102-0071, Tokyo, Japan

Abstract. This paper investigates the explosive spalling of high-strength concrete and the shear capacity of high-strength reinforced concrete (RC) beams, taking into consideration the impact of carbonation during and after a fire. First, the spalling model of the multi-scale platform was upgraded. In this study, the limit state of spalling was upgraded in consideration of the strain threshold of concrete in the strain softening range. This upgraded spalling model was validated by experiments on ultra-high-strength concrete during high-temperature heating. Based on the simulation results, explosive spalling was computationally predicted by the proposed model. Second, shear failure of high-strength RC beams subjected to high-temperature fire heating was analytically investigated and the proposed model was found to be able to roughly reproduce shear capacity and ductility after high-temperature heating. High concentration of carbon dioxide (CO_2) was found to have a large effect on the shear behavior of RC beams, increase their stiffness and shear capacity.

Keywords: Multi-scale modeling · Explosive spalling · High-strength concrete · Tension softening · Carbonation · Fire

1 Introduction

Explosive spalling during fire accidents causes severe damage to structural concrete. Thus, numerical [e.g. 1–3] and experimental [e.g. 4, 5] investigations have been conducted. Carbonation after high-temperature heating is also an important thermodynamic event in terms of the durability of RC [e.g. 6, 7]. Recently, carbonation during fire has become a focus of attention [8]. In this study, a numerical model is built for behavioral simulation of high-strength RC during and after a fire when spalling and carbonation are simultaneously combined.

The authors have sought to extend the applicability of current multi-scale modeling [9, 10] up to 1,000 °C [11–13] and carried out verification of the updated model [11–13], as

P. Rossi and J.-L. Tailhan (Eds.): SSCS 2022, RILEM Bookseries 38, pp. 160–169, 2023.
https://doi.org/10.1007/978-3-031-07746-3_16

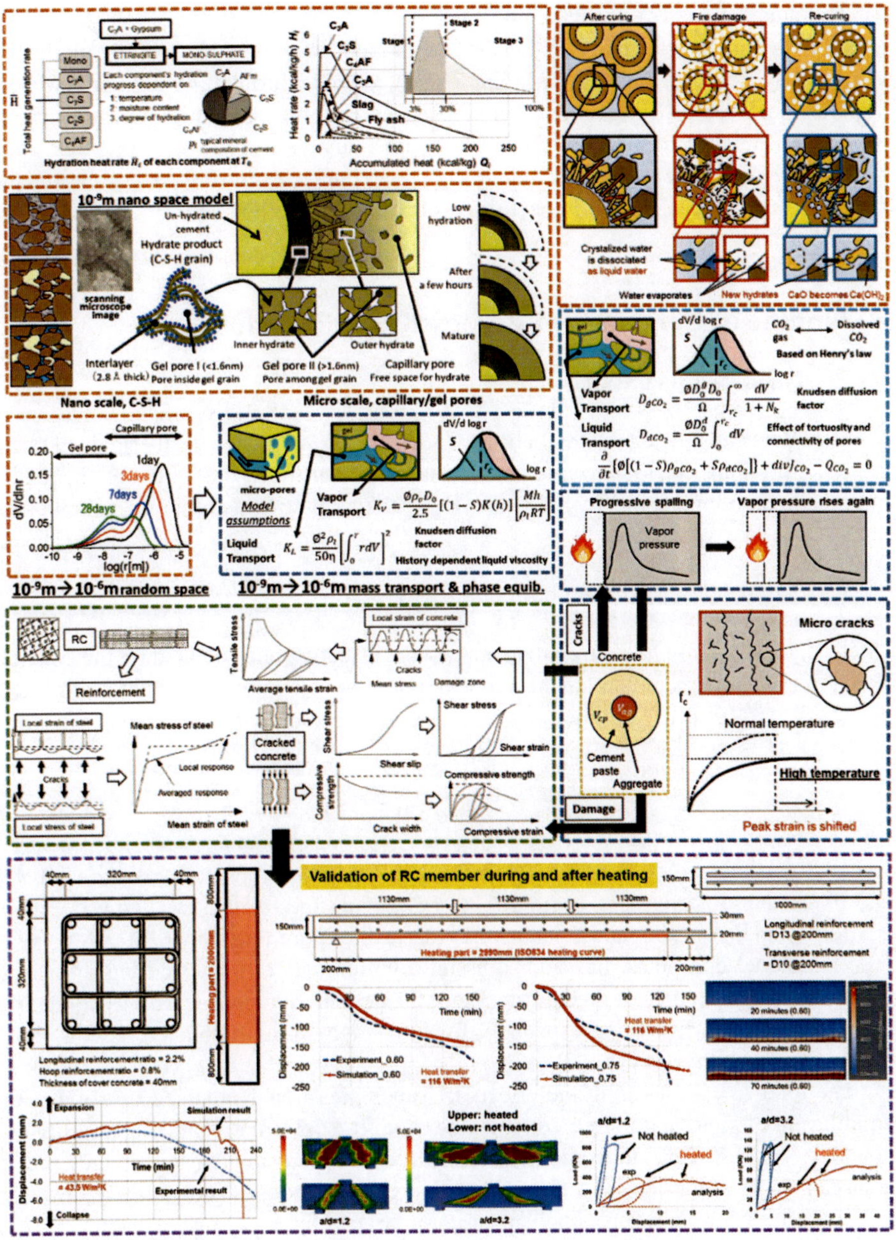

Fig. 1. Outline of current multi-scale platform and structural validation at a fire.

shown in Fig. 1. The previous updated model [11–13] can roughly capture the behavior of RC members during and after high-temperature heating (see the lower part of Fig. 1). The updated model has been also verified for carbonation in ambient environments [10, 14] and during fire with high concentration of CO_2.

This study is a part of strengthening the multi-scale platform [9–13]. First, the spalling model of the multi-scale platform [11] was upgraded. The upgraded spalling model was validated by experiments on ultra-high-strength concrete during high-temperature heating. Second, shear failure of high-strength RC beams subjected to high-temperature fire heating was analytically investigated. The coupled analysis of the upgraded thermodynamic model with the multi-directional crack model of structural concrete was validated by experimental results. The impact of carbonation on the behavior of RC beams was also confirmed.

2 Outline of Original and Proposed Spalling Model

2.1 Original Spalling Model

Major factors causing explosive spalling of structural concrete are the thermal stress and the vapor pressure due to the vaporization of moisture inside micro pores and the release of crystalized bound water [e.g. 15–17]. Here, the total stresses of concrete solids and pore media are expressed as

$$\sigma_{ij} = \sigma_{ij}' + \delta_{ij} p_{vap} \tag{1}$$

where σ_{ij} and σ_{ij}' are total stress tensors and the effective ones defined on the concrete skeleton (Pa), and p_{vap} denotes the vapor pressure developing inside both micro pores and crack gaps (Pa). The thermodynamic phase equilibrium of the idealized gas is shown as

$$p_{vap} = P_{sat} exp\left(\frac{P_l M_w}{\rho_l RT}\right) \tag{2}$$

where P_{sat} is the saturated vapor pressure (Pa), P_l is the pore water pressure (Pa), M_w and ρ_l are the molar volume of liquid water (kg/mol) and its density (kg/m^3), R is the gas constant (J/(mol·K)), and T is the absolute temperature (K).

This model computes the vapor pressure by automatically considering the vaporization of dehydrated crystalized water from the cement hydrates as well as condensed free water. This is important especially for ultra-high-strength concrete having much chemically bound water [15, 17].

The total stress including the effects of vapor pressure is obtained by solving the equilibrium of solid and pore media together with the deformational compatibility and boundary conditions. If the computed tensile stress exceeds the tensile strength, cracks are assumed to occur normal to the principal stress direction, as shown in Fig. 2.

The vapor pressure drops when cracking or large volume expansion occurs (Fig. 2). When the volumetric change is low at the time of crack occurrence with little strain increment, the authors assume that the cracked concrete still remains a part of the whole structure. However, when a large increase of volumetric change (strain) takes place, the authors define this as explosive spalling. As shown in Fig. 2, based on the strain softening range of concrete, the original model uses the expansion ratio of the micro- structure to analytically judge whether spalling occurs. The value α is

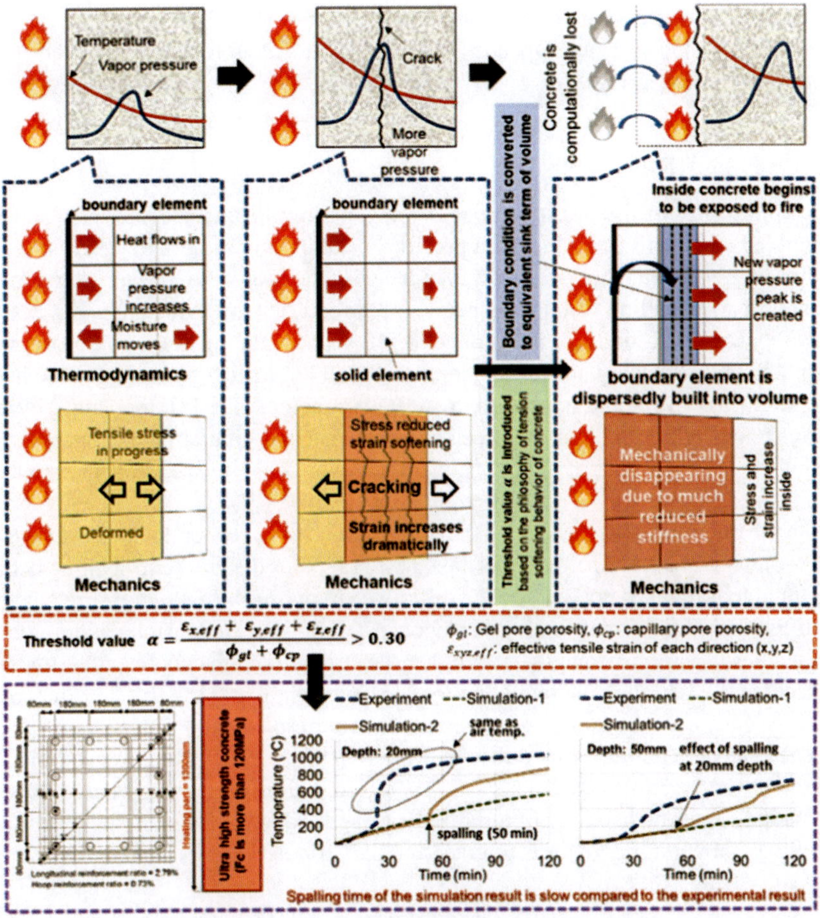

Fig. 2. Outline of original spalling model and its issue.

$$\alpha = \frac{\varepsilon_{x,eff} + \varepsilon_{y,eff} + \varepsilon_{z,eff}}{\phi_{gl} + \phi_{cp}} > 0.30 \tag{3}$$

where $\varepsilon_{x,eff}$, $\varepsilon_{y,eff}$ and $\varepsilon_{z,eff}$ are the effective tensile strain in each (x, y and z) direction, ϕ_{gl} and ϕ_{cp} are the porosity of gel and capillary pores, which are calculated by the micro-pore structure development model (see Fig. 1).

In the original model [11–13], if the expansion ratio of the micro-structure exceeds 0.30, explosive spalling is analytically deemed to occur. In the finite element scheme, it has been hard to reproduce geometrically vanishing elements of progressive spalling. To overcome this issue, the spalled-off elements are computationally replaced with the equivalent boundary transfer elements to link the ambient states and the elements so as to obtain the quick release of entropy and the mass of moisture after spalling, as shown in Fig. 2. Then, the inside element is computationally exposed to the environment of high temperature and vapor pressure. As the crack strain of spall-off elements is so large as to bear almost zero stresses in compression and tension, these elements are mechanically killed without load-carrying capacity if these elements are forced to leave their original position, as shown in Fig. 2.

2.2 Proposed Spalling Model

The applicability of the original model to the spalling behavior of normal- and high-strength concrete was confirmed in a previous validation [11–13]. However, in the case of ultra-high-strength concrete (with compressive strength greater than 120 MPa), significantly early spalling of cover concrete occurred after fire exposure compared with the case of normal-high-strength concrete [12]. Thus, the spalling computed by the original modeling took place later than in reality [11]. While conducting some heating experiments on ultra- and normal-high-strength concrete [12, 13], the authors found a difference in spalling behavior between ultra- and normal-high-strength concrete. In the case of the normal-high-strength concrete, the cover concrete spalls off in chunks of a certain size [13].

On the other hand, in the case of the ultra-high-strength concrete, the cover concrete spalls off in the form of fine debris as flaked off [12]. This difference in spalling behavior is thought to be caused not only by the difference in permeability to moisture but also to the difference of tension softening behavior. According to the tension softening curves of concretes of various strengths, the higher the strength of the concrete, the smaller the displacement when the concrete is no longer bearing a load [18–20]. This means that the cover concrete will soon leave the structure after cracking.

The change in permeability due to the variation of concrete strength is automatically considered in the multi-scale platform. Therefore, based on the knowledge related to the tension softening behavior of ultra-high-strength concrete [18–20], the limit state of spalling as shown in Eq. (3) was upgraded in the proposed model as

$$\alpha = \frac{\varepsilon_{x,eff} + \varepsilon_{y,eff} + \varepsilon_{z,eff}}{\phi_{gl} + \phi_{cp}} > 0.05 \tag{4}$$

where 0.05 is the upgraded threshold value for ultra-high-strength concrete (with compressive strength greater than 120 MPa). In the multi-scale platform [9], the local strength of meso-scale cement paste is computed based on the micro-pore structure and examined with a wide range of water to cement ratio, curing, and material age by Otabe and Kishi [21] as

$$f_c = f_\infty \{1 - exp(-\alpha D_{hyd.out}{}^{\beta})\} \tag{5}$$

where f_c is the compressive strength of concrete (MPa), f_∞ is the ultimate compressive strength (MPa), as determined in Otabe and Kishi [21], $D_{hyd.out}$ is the ratio of the volume of cement hydrates created outside of cement particles to the amount of capillary porosity when hydration is started [21], and α and β are constant values (3.0 and 4.0, respectively). Based on this strength model [21], the threshold value (0.30 or 0.05) is automatically determined in the proposed model.

3 Validation of Temperature Rise of Ultra-high-Strength Concrete After Explosive Spalling

The temperature rise of the plain concrete prism, whose compressive strength is about 150MPa, was examined using previous experimental results [22], as shown in Fig. 3. The plain concrete prism had a water to binder ratio of 0.22 and contained silica fume [22]. After casting, the specimens were covered with a plastic sheet at room temperature for the first day. Afterwards, they were demolded and stored in water at room temperature for 2 weeks. Then, the specimens were allowed to air for another 2.5 months [22]. The heating rate was 2 °C/min and the target temperature was 600 °C [22]. A simple outline of the heating test and comparison points of temperature rise are shown in the upper part of Fig. 3.

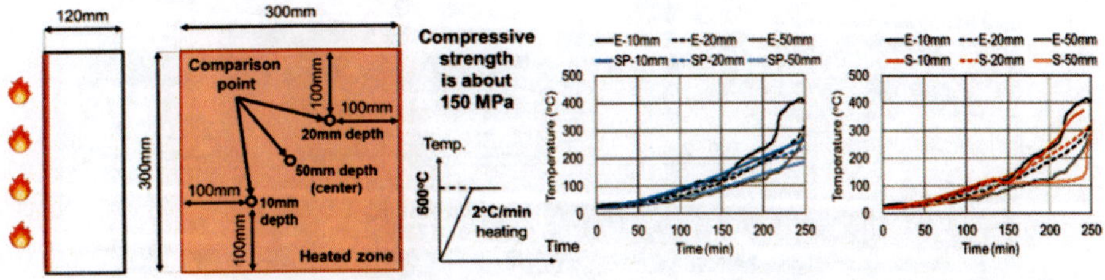

Fig. 3. Validation of temperature rise of plain concrete prism (Fc: 150MPa class): E = experiment, SP = simulation using original model, S = simulation using proposed model.

The validation results are shown in the lower part of Fig. 2. Although there is no sharp temperature rise after 200 min in the simulation results using the original spalling model (SP), the proposed model (S) can grasp the behavior of the temperature rise of the experiment. This is because heat was transmitted to the inside due to the occurrence of spalling near the heating surface in the proposed model. This means that spalling as if a thin layer of concrete at the surface peeled off occurs in the proposed model.

4 Shear Capacity and Deformation of High-Strength RC Beams After Post-fire-curing

4.1 Validation of Proposed Model Simultaneously Considering Spalling and Carbonation

Element models of RC beams to be subjected to heating are shown in Fig. 4. Three values for the shear span to depth ratio a/d, namely 1.2, 2.2, and 3.2, were used. The beams were subjected to a high temperature history using a large heating test furnace. Heating was performed for 2 h using the ISO 834 heating curve [13]. After heating, the RC beams were cured in laboratory condition as post-fire-curing [13]. Then, the specimens were subjected to shear loading by three-point bending using two simple supports and a load applied at the center of the beam [13].

Table 1. Shrinkage during cooling and expansion during post-fire-curing.

a/d	Shrinkage (mm)		Expansion (mm)	
	Upper	Lower	Upper	Lower
1.2	3.0	10.7	3.9	1.0
2.2	7.0	12.8	2.9	1.8
3.2	11.5	18.0	5.0	1.5

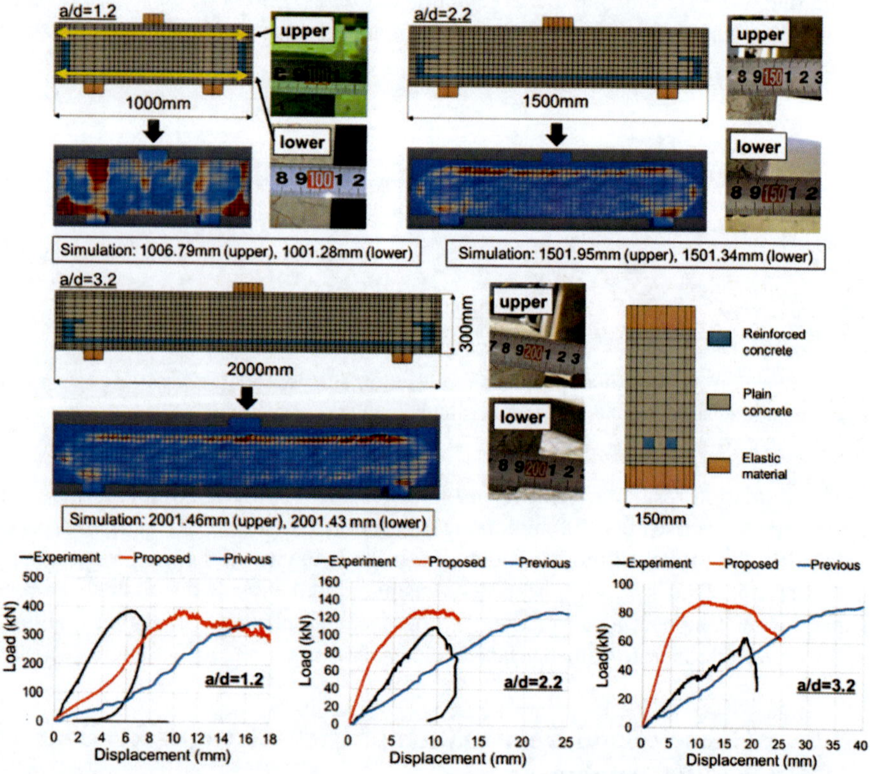

Fig. 4. Validation of shear capacity and deformation of high-strength RC beams after post-fire-curing.

The picture in Fig. 4 shows the length of RC beams immediately before the loading test. According to the measurement results, the length of the upper part of the RC beams is longer than their lower part. There are two possible reasons for this, at two different points: the point where the reinforcing bars shrink more than the concrete at cooling, and the point where the expansion of concrete during post-fire-curing is restrained by the reinforcements.

The proposed model can roughly grasp the trend of length change although the length calculated by the proposed model is shorter than the measured one. When examined separately for shrinkage during cooling and expansion during post-fire-curing, it was analytically indicated that the impact of the rebar was large, as shown in Table 1.

The analytical results of the proposed model and the previous model are shown in the lower part of Fig. 4. The proposed model can roughly grasp the experimental results, and the effect of carbonation can be seen in the increase of stiffness with respect to the shear behavior of RC beams subjected to high-temperature heating. The myriad cracks present in the specimen after heating cause carbonation in every part of the structure. As a result, the concrete that resists shear deformation becomes denser and stronger. Also, the shrinkage that occurs at the same time as carbonation alleviates the loosening after heating. It is considered that these factors greatly increase the stiffness of the RC beams in the proposed model.

In the cases of a/d = 2.2 and 3.2, the proposed model overestimates the stiffness of the RC beams. This is because the change in the length of the upper part of the beams in the analysis is smaller than the experimental value. The small change in length indicates that the specimen is less loose, and the stiffness is increased accordingly. In the future, the authors would like to continue to carefully study and upgrade the rehydration and carbonation model [13] during heating and post-fire-curing.

4.2 Investigation of High Concentration of CO_2 During Heating and Post-fire-curing

Using the validated proposed model, the effects of high concentrations of CO_2 during heating and post-fire-curing on the shear behavior of RC beams was investigated. The CO_2 concentration was set to 10% for both high CO_2 heating and high CO_2 post-fire-curing.

The comparison results are shown in Fig. 5. Based on the results of the parametric study, it can be seen that the stiffness and shear capacity of the RC beams increase in the case of exposure to a high concentration of CO_2 during post-fire-curing. Even if carbonation is promoted during heating, calcium carbonate ($CaCO_3$) generated by carbonation decomposes when the temperature of concrete exceeds 600 °C. On the other hand, carbonation during post-fire-curing densifies the micro-pore structure of concrete, and this can contribute to the increase in the shear capacity and stiffness of the RC members. In addition, during post-fire-curing, both expansion due to rehydration of calcium oxide (CaO) [13] and shrinkage due to carbonation [e.g. 23,24] occur at the same time. The proposed model simply models the structural behavior associated with said rehydration and carbonation. In the case where the shrinkage behavior is not taken into consideration, the stiffness and shear capacity of the RC beam are significantly reduced. In other words, it is shown that the expansion behavior associated with rehydration and greatly affects the mechanical response at the member level.

5 Conclusions

The authors aimed to extend the applicability of multi-scale modeling at spalling and carbonation during and after high temperature heating. The conclusions of this study are summarized as follows.

I. The limit state of spalling was upgraded in consideration of the strain threshold of concrete in the strain softening range. This upgraded spalling model was validated by experiments on ultra-high-strength concrete during high-temperature heating, and it could reproduce the trends of experimental results.

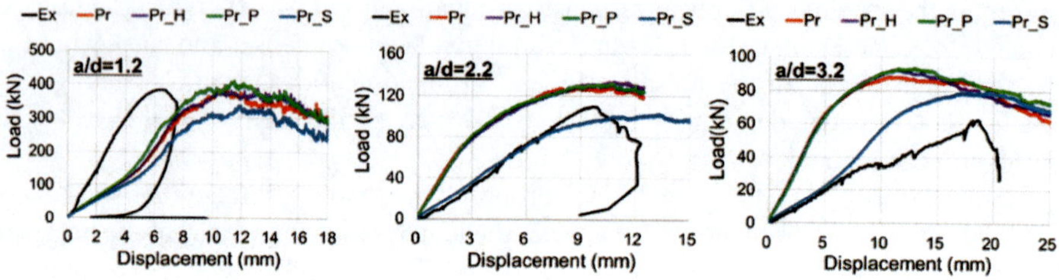

Fig. 5. Effects of high CO_2 concentration for shear capacity and deformation of high-strength RC beams: Ex = experiment, Pr = proposed model (same as Fig. 3), Pr_H = high CO_2 concentration during heating, Pr_P = high CO_2 concentration during post-fire-curing, Pr_S = without considering shrinkage at carbonation process.

II. The shear failure of high-strength RC beams subjected to high-temperature fire heating was analytically investigated and the proposed model was examined to be able to roughly reproduce the shear capacity and stiffness after high-temperature heating. High concentration of CO_2 after a fire was found to have a large effect on the shear behavior of RC beams, in the form of an increase of stiffness.

Acknowledgments. This study was financially supported by JSPS KAKENHI Grant Number 20H00260.

References

1. Gawin, D., Pesavento, F., Schrefler, B.A.: Towards prediction of the thermal spalling risk through a multi-phase porous media model of concrete. Comput. Methods Appl. Mech. Eng. **195**(41–43), 5707–5729 (2006)
2. Zhang, H.L., Davie, C.T.: A numerical investigation of the influence of pore pressures and thermally induced stresses for spalling of concrete exposed to elevated temperatures. Fire Saf. J. **59**, 102–110 (2013)
3. Kodur, V., Banerji, S.: Modeling the fire-induced spalling in concrete structures incorporating hydro-thermo-mechanical stresses. Cement Concrete Composit. **117**, Article ID: 103902 (2021)
4. Hertz, K.D.: Limits of spalling of fire-exposed concrete. Fire Saf. J. **38**(2), 103–116 (2003)
5. Banerji, S., Kodur, V., Solhmirzaei, R.: Experimental behavior of ultra high performance fiber reinforced concrete beams under fire conditions. Eng. Struct. **208**, Article ID: 110316 (2020)

6. Li, Q., Li, Z., Yuan, G., Shu, Q.: The effect of a proprietary inorganic coating on compressive strength and carbonation depth of simulated fire-damaged concrete. Mag. Concr. Res. **65**(11), 651–659 (2013)
7. Li, Q., Yuan, G., Shu, Q.: Effects of heating/cooling on recovery of strength and carbonation resistance of fire-damaged concrete. Mag. Concr. Res. **66**(18), 925–936 (2014)
8. Yatsushiro, D., Atarashi, D., Yoshida, N., Okumura, Y.: Study on carbonation mechanism of hardened cement paste due to fire damage (in Japanese). In: The 41th JCI Technical Conference, pp. 587–592, Japan Concrete Institute, Sapporo, Japan (2019)
9. Maekawa, K., Pimanmas, A., Okamura, H.: Nonlinear Mechanics of Reinforced Concrete. Spon Press, London and New York (2003)
10. Maekawa, K., Ishida, T., Kishi, T.: Multi-scale Modeling of Structural Concrete. Taylor & Francis, London (2009)
11. Iwama, K., Higuchi, K., Maekawa, K.: Thermo-mechanistic multi-scale modeling of structural concrete at high temperature. J. Adv. Concr. Technol. **18**(5), 272–293 (2020)
12. Iwama, K., Kato, Y., Baba, S., Higuchi, K., Maekawa, K.: Accelerated moisture transport through local weakness of high-strength concrete exposed to high temperature. J. Adv. Concr. Technol. **19**(2), 106–117 (2021)
13. Higuchi, K., Iwama, K., Maekawa, K.: Remaining shear capacity of fire-damaged high strength RC beams after moist curing. J. Adv. Concr. Technol. **19**(8), 897–912 (2021)
14. Ishida, T., Li, C.H.: Modeling of carbonation based on thermo-hygro physics with strong coupling of mass transport and equilibrium in micro-pore structure of concrete. J. Adv. Concr. Technol. **6**(2), 303–316 (2008)
15. Kalifa, P., Menneteau, F.D., Quenard, D.: Spalling and pore pressure in HPC at high temperatures. Cem. Concr. Res. **30**(12), 1915–1927 (2000)
16. Phan, L.T.: Pore pressure and explosive spalling in concrete. Mater. Struct. **41**(10), 1623–1632 (2008)
17. Liu, J.C., Tan, K.H., Yao, Y.: A new perspective on nature of fire-induced spalling in concrete. Constr. Build. Mater. **184**, 581–590 (2018)
18. Darwin, D., Barham, S., Kozul, R., Luan, S.: Fracture Energy of High-Strength Concrete. ACI Mater. J. **98**(5), 410–417 (2001)
19. Takagi, Y., Matsumoto, K., Fujita, M.: Study of fracture energy for high strength concrete (in Japanese). In: 2003 JSCE Annual Meeting, pp. 423–424. Japan Society of Civil Engineers, Tokushima, Japan (2003)
20. Niwa, J., Sumranwanich, T., Matsuo, T.: Experimental study to determine the tension softening curve of concrete (in Japanese). J. JSCE **41**(606), 75–88 (1998)
21. Otabe, Y., Kishi, T.: Development of hydration and strength model for quality evaluation of concrete (in Japanese). Indust. Sci. (Univ. Tokyo) **57**(2), 37–42 (2005)
22. Li, Y., Pimienta, P., Pinoteau, N., Tan, K.H.: Effect of aggregate size and inclusion of polypropylene and steel fibers on explosive spalling and pore pressure in ultra-high-performance concrete (UHPC) at elevated temperature. Cement Concr. Compos. **99**, 62–71 (2019)
23. Powers, T.C.: A hypothesis on carbonation shrinkage. J. PCA Res. Dev. Lab. **4**(2), 40–50 (1962)
24. Persson, B.: Experimental studies on shrinkage of high-performance concrete. Cem. Concr. Res. **28**(7), 1023–1036 (1998)

Modelling of Rock-Shotcrete Interfaces Using a Novel Bolted Cohesive Element

Ali Karrech(✉) and Xiangjian Dong

Department of Civil, Environmental and Mining Engineering, School of Engineering, The University of Western Australia, 35 Stirling Hwy, Crawley, WA 6009, Australia
ali.karrech@uwa.edu.au

Abstract. Rock-shotcrete interfaces are commonly encountered in mining and civil engineering infrastructures, which can trigger localised failure due to stress concentration. These interfaces are usually reinforced with support systems such as rock bolts and the behaviour of rock-shotcrete-bolt systems is often difficult to predict mechanically. In this study, we introduce a new technique to model rock-shotcrete interfaces embedding rock bolts using the finite element method. The proposed approach is implemented in the general purpose simulation package Abaqus via its user-defined element (UEL) subroutine. The proposed model takes into account the uneven interface roughness and the complex interaction between its components. The cohesive stiffness of the model degrades proportionally to the damage that occurs due to this interaction. The stiffness of the bolt connection and its location are also considered in the proposed mathematical formulation. The present bolted cohesive element has been validated experimentally; good agreement has been obtained between the measurement and numerical simulation under the conditions of direct shear test and bolt pull-out tests. Mesh independence has also been verified by examining the effect of mesh size on the overall force-displacement response of typical structures. With the model at hand, the effects of key installation parameters such as number of bolts, their inclinations and material properties have been investigated.

Keywords: Interface · Rock-shotcrete · Finite element method · User-defined element · Cohesive element

1 Introduction

Rock-shotcrete interfaces are common in mining and civil engineering infrastructures including underground tunnels and excavations. These interfaces are important to investigate since they represent irregularities that usually cause stress concentration. Factors such as rock roughness, shotcreting irregularities, properties of reinforcing elements (e.g. rock bolts and/or fibres), and the conditions of their installation influence the behaviour of these interfaces. These irregularities can be attributed to the heterogenous nature of rocks and also to

P. Rossi and J.-L. Tailhan (Eds.): SSCS 2022, RILEM Bookseries 38, pp. 170–177, 2023.
https://doi.org/10.1007/978-3-031-07746-3_17

the excavation process that is often based on drilling and blasting or mechanical digging, which makes it difficult to obtain smooth surfaces free of local defects. Shotcrete is usually sprayed on these surfaces to prevent detachment of loose rocks, especially in the presence of workers and/or heavy equipment. Naturally, shotcrete spraying results in non-uniform layers that slightly smoothen the original rock surface and strengthen it. The interface between the rock mass and the shotcrete layer relies essentially on the bounding that occur due to cement hydration reactions. Additional strength is gained by reinforcing the structure with rock bolts.

To the author's knowledge there are no cohesive element approaches in the literature to predict the behaviour of rock- shotcrete interfaces, especially when support systems such as rock bolts are used. Existing approaches to predict the behaviour of rock-shotcrete interfaces are essentially analytical as shown by the authors in previous contributions [1,2] and they ignore the effect of reinforcing systems. The present formulation builds on a similar element that the authors suggested for fiber-reinforced concrete [2]. However, the different scales between fibers and bolts as well as the absence of interface in plain concrete make this approach fairly different. It is worthwhile noting that the proposed formulation does not apply only to mining and civil engineering infrastructures embedding such interfaces but also to other applications where detachment between different materials can occur in a localised manner.

In this study, we introduce a new technique to model rock-shotcrete interfaces embedding rock bolts using the finite element method. We developed a so called "bolted cohesive element" through the user-defined element (UEL) subroutine of Abaqus to describe this complex interface. The proposed model takes into account the uneven interface roughness and the complex interaction between its components.

2 Bolted Cohesive Element Model

Cohesive modelling was introduced by Dugdale and Barenblatt [3,4] in 1960s s and it has been applied to various engineering problems [5–7]. It describes joints or interfaces that differ from the bulk materials that surround them in terms of mechanical behaviour. The governing equations are derived from the principle of virtual work which reads

$$\int_{\Omega} \boldsymbol{\sigma}^T \delta\boldsymbol{\varepsilon} d\Omega + \int_{\Gamma_c} \boldsymbol{T_c}^T \delta\boldsymbol{\Delta} dS = \int_{\Gamma_e} \boldsymbol{T_e}^T \delta\boldsymbol{u} dS \tag{1}$$

where $\boldsymbol{\sigma}$, $\delta\boldsymbol{\varepsilon}$, $\delta\boldsymbol{\Delta}$ and $\delta\boldsymbol{u}$ represent Cauchy stress, virtual strain, virtual separation and virtual displacement field, respectively. The principle is applied within a 3D domain Ω embedding a cohesive region Γ_c as shown in Fig. 1. External forces $\boldsymbol{T_e}$ and/or essential boundary conditions $\boldsymbol{u}$ are applied through the surface Γ_e. The contribution of body forces is neglected in this formulation.

At the crack interface (Γ_c), the cohesive traction $\boldsymbol{T_c}$ is related to the relative separation between the interface boundaries $\boldsymbol{\Delta}$ by:

$$\boldsymbol{T_c} = \boldsymbol{D_c}\ \boldsymbol{\Delta} \tag{2}$$

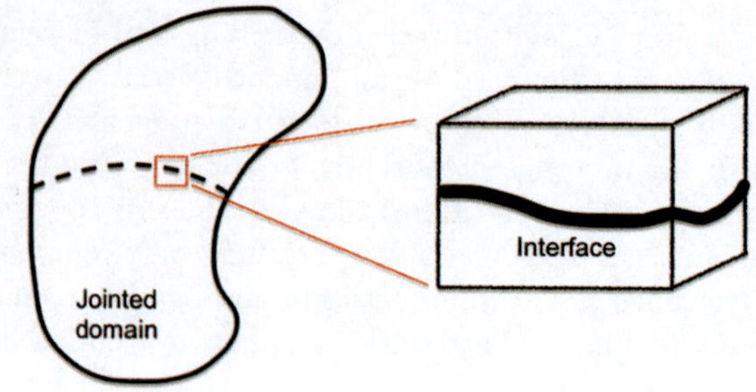

Fig. 1. Schematic representation of a 3D cohesive element.

where the constitutive matrix $\boldsymbol{D_c}$ is determined experimentally. Embedding a bolt into the interface at a location (x_b, y_b, z_b) and applying a pretension force $\boldsymbol{P_0}$ (see in Fig. 2), results in the following expression

$$\int_{\Omega} \boldsymbol{\sigma}^T \delta\boldsymbol{\varepsilon} dV + \int_{\Gamma_c} \boldsymbol{T_c}^T \delta\boldsymbol{\Delta} dS + \int_{\Gamma_c} \boldsymbol{T_b}^T \delta\boldsymbol{\Delta} \delta(x - x_b, y - y_b, z - z_b) dS \\ = \int_{\Gamma_e} \boldsymbol{T_e}^T \delta\boldsymbol{u} dS + \int_{\Gamma_c} \boldsymbol{P_0}^T \delta\boldsymbol{u} dS \tag{3}$$

where $\boldsymbol{T_b}$ is an internal point load that equilibrates the bolt pretension and $\delta(x - x_b, y - y_b, z - z_b)$ is the Dirac delta function.

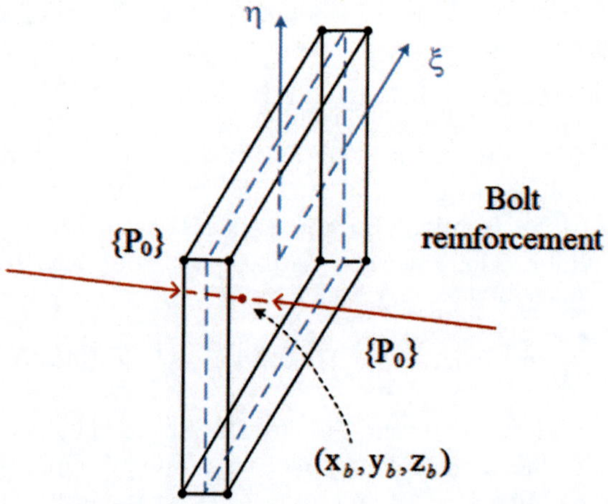

Fig. 2. Cohesive element with pre-stressed bolt reinforcement.

Based on Eq. (3) it can be shown that each cohesive element is governed by $\boldsymbol{K}\boldsymbol{d} = \boldsymbol{F}$, where $\boldsymbol{K}$ is the stiffness matrix, $\boldsymbol{d}$ is the vector of degrees of freedom, and $\boldsymbol{F}$ is the vector of nodal forces. It can be shown that the stiffness matrix is

$$\boldsymbol{K} = \int_{-1}^{1}\int_{-1}^{1} \tilde{\boldsymbol{B}}^T \boldsymbol{D}_{local}\tilde{\boldsymbol{B}}|\boldsymbol{J}|d\xi d\eta + \int_{-1}^{1}\int_{-1}^{1} \delta(\boldsymbol{\xi} - \boldsymbol{\xi}_b)\tilde{\boldsymbol{B}}^T \boldsymbol{D}_s\tilde{\boldsymbol{B}}d\xi d\eta \tag{4}$$

where $\tilde{\boldsymbol{B}} = \boldsymbol{R}\boldsymbol{B}$ and $\boldsymbol{R}$ is a rotation matrix. The first term of Eq. (4) represents the cohesive stiffness and the second denotes the bolt stiffness. Similarly, the vector of nodal forces reads

$$\begin{aligned}\boldsymbol{F} = &\int_{-1}^{1}\int_{-1}^{1} \boldsymbol{B}^T\boldsymbol{R}^T\boldsymbol{T}_c|\boldsymbol{J}|d\xi d\eta + \int_{-1}^{1}\int_{-1}^{1} \boldsymbol{B}^T\boldsymbol{R}^T\boldsymbol{T}_b\delta(\xi - \xi_b, \eta - \eta_b)d\xi d\eta \\ &+ \int_{-1}^{1}\int_{-1}^{1} \delta(\boldsymbol{\xi} - \boldsymbol{\xi}_b)\tilde{\boldsymbol{B}}^T \boldsymbol{P}_0 d\xi d\eta\end{aligned} \tag{5}$$

The vector of nodal forces includes the effects of cohesive traction, bolt force, and initial pretension applied on the bolt which are the three terms of Eq. (5), respectively. The last term can be expressed by

$$\boldsymbol{F}^e{}_p = \boldsymbol{B}_{(\xi_b,\eta_b)}\boldsymbol{R}\boldsymbol{P}_0 \tag{6}$$

The rock-shotcrete interface has a complex structure with uneven interface behaviour as illustrated in Fig. 2-a. The response of this interface is governed by a multi-dimensional stiffness having normal (K_n) and shear (K_t) components. Given the uneven interface, K_t is further decomposed into cohesion and friction components. The interface behaviour can be measured by using normal loading tests [8], static Brazilian disc split tests [9] and direct shear tests [10], respectively. A bilinear traction-separation law is adopted for the normal traction component in this simulation. In addition, a mixed 'cohesive' and 'Coulomb-friction' model is used to approximate the non-linear behaviour of rock-shotcrete interfaces. This results in a three-stages behaviour that includes elastic, bond failure and friction sliding regions, as suggested by [10]. As shown in Fig. 2-b, the shear traction increases linearly with displacement until it reaches the onset of failure (T'_{max}) at a displacement δ_0. Beyond this threshold, the shear traction reduces non linearly until T'_{max} is reached at a displacement δ_c. A residual shear stress is then maintained as displacement increases.

The degradation of the cohesive stiffness is described with the damage variable D, which varies from zero to one and can be expressed as a function of displacement as follows

$$D = \begin{cases} 0, & for \quad \delta < \delta_0 \\ f(\delta), & for \quad \delta_0 \leqslant \delta < \delta_c \\ 1, & for \quad \delta \geqslant \delta_c \end{cases} \tag{7}$$

where the function $f(\delta)$ is fitted experimentally to reflect the effects of material strength and joint roughness. A quintic polynomial function is used in the current

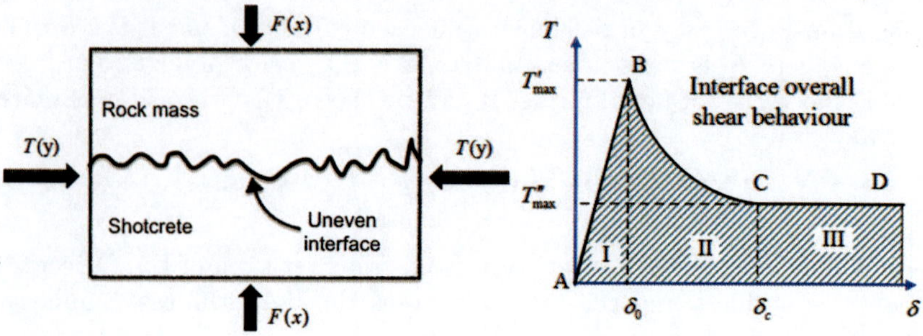

Fig. 3. Illustrations of (a) uneven rock-liner interface and (b) interface shear constitutive model.

case. In addition, we considered that the shear stiffnesses in the two tangential directions are equal which leads to the following constitutive relation for the shear component:

$$T_c = \begin{cases} k_s^c \delta, & for \;\; \delta < \delta_0 \\ k_s^c(1 - f(\delta))\delta + k_{fs}^c \delta, & for \;\; \delta_0 \leqslant \delta < \delta_c \\ \mu_n T_n, & for \;\; \delta \geqslant \delta_c \end{cases} \tag{8}$$

where $k_s^c = T'_{max}/\delta_0$ and $k_{fs}^c = T''_{max}/(\delta_c - \delta_0))$.

3 Results and Discussion

Before the model can be used for field applications, simple cases of loading are considered for validation. The first case simulates a simple shear test and the results are compared to published experimental data [10]. Cubic specimens of size 150 mm with artificial rock-concrete interfaces are considered. The specimens are subjected to shear displacement at a rate of 0.005 mm/s under a normal compression stress of 4 MPa. The bulk material is assumed to have an elastic modulus of 21 GPa and a Poisson ratio of 0.3. The friction coefficient at the interface is set to $\mu_n = 0.8$ and the joint roughness coefficient (JRC) of the interface is less than 2. The measured normal compression and residual stresses are 4 MPa and 3.2 MPa, respectively. Figure 3-a shows the reaction force over the area of shear interface versus applied shear displacement. It can be seen that a good agreement is obtained between the proposed model and the experimental test. As a second validation case, a single bolt pull-out test is simulated and the results are compared to published experimental data [11]. The pull-out tests were conducted using an MTS Criterion 60 testing machine; a bolt of 20 mm in diameter is inserted into a steel tube and bonded with resin. Resin-based bolting is widely used by the mining industry to ensure fast curing (the mixture solidifies in minutes unlike cement bolting). Again, excellent agreement was obtained between the numerical and experimental data, as shown in Fig. 3-b.

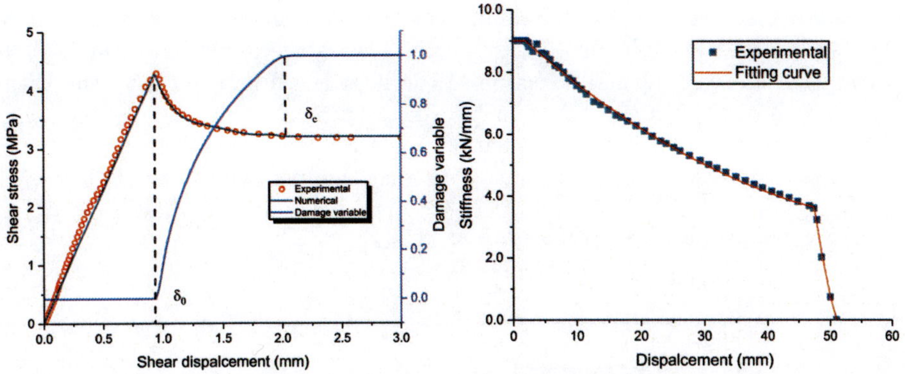

Fig. 4. Model validation for (a) direct shear test and (b) normal stiffness of the resin bonded bolt system.

After validation using the direct shear and single bolt pull-out tests, the model was used to conduct a parametric study. The first parameter of interest is the number of bolts within a given interface, as the common practice in engineering infrastructures is to apply more than one bolt to enhance the overall behaviour of support systems. Figure 5-a shows the response of the structure with respect to displacement for various numbers of bolts. It can be seen that the total force increases with the number of bolts, which suggests that the denser the bolts, the higher the shear resistance of the system. However, Fig. 5-b indicates that the average force does not change much with the number of bolts. This means that the overall force increases linearly with the bolt numbers.

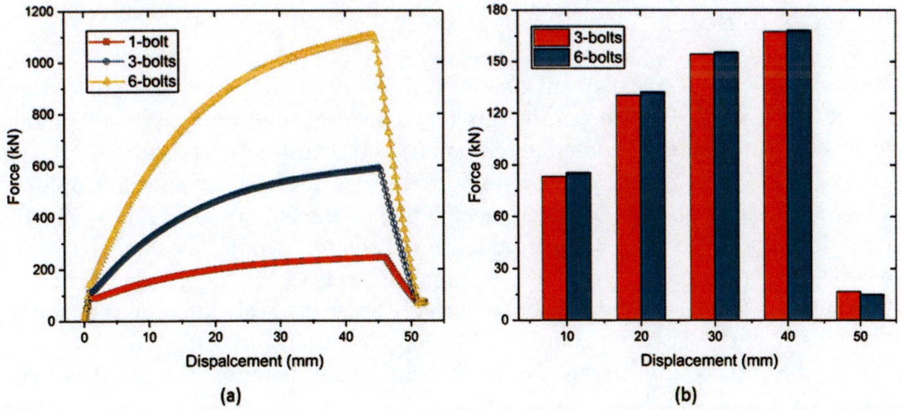

Fig. 5. The effect of bolt number (a) overall force variation against displacement (b) average force increase under 3 and 6 bolts situations.

Another key parameter that influences the performance of bolt-based support systems is the bolt installation angle. This is the angle between the bolt orientation and the normal direction of the interface. Figure 6 shows the shear force versus displacement at various bolt installation angles. It can be seen that the force increases with the installation angle and reduces abruptly when the bolt fails. The results indicate that the higher the installation angle the stiffer the bolted system. Similar results were obtained experimentally and reported in previous studies by Dight [12] and Li et al. [13].

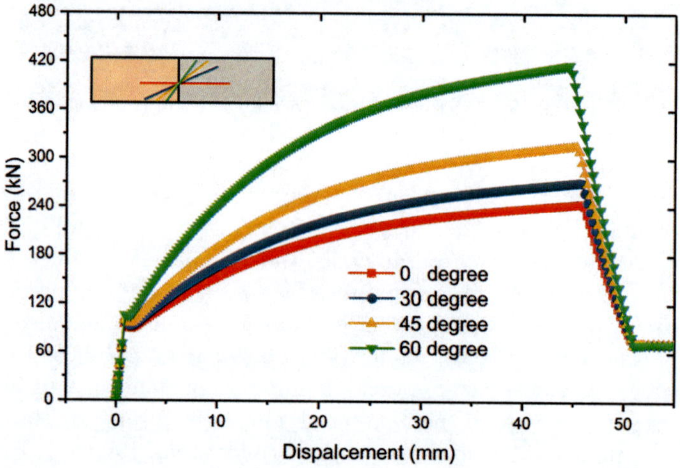

Fig. 6. The shear force variation under the effect of bolt installation angle.

4 Conclusions

A novel bolted cohesive element has been introduced to model support systems of rock bolts reinforcing shotcrete applied to underground excavations. The proposed element has been implemented using the Fortran user element subroutine UEL of ABAQUS. The model was validated using direct shear and bolt pull-out tests and it showed excellent agreement between the experimental data and the numerical results. Based on this model, a parametric study has shown that the overall behaviour of the system is strongly influences by the geometry of bolt installation. The model has indicated that the shear resistance of the system increases linearly with the number of reinforcing bolts. In addition, the model has als shown that increasing the installation angle increases the effective stiffness of the system and its overall strength.

References

1. Dong, X., Karrech, A., Elchalakani, M., Qi, C., Manka, M.: 3D bolted cohesive element for the modelling of bolt-reinforced rough rock-shotcrete interfaces. Comput. Geotech. **125**, 103659 (2020)
2. Manka, M., Karrech, A., Dight, P., Ciancio, D.: Dual cohesive elements for 3D modelling of synthetic fibre-reinforced concrete. Eng. Struct. **174**, 851–860 (2018)
3. Dugdale, D.S.: Yielding of steel sheets containing slits. J. Mech. Phys. Solids **8**(2), 100–104 (1960)
4. Barenblatt, G.I.: The mathematical theory of equilibrium cracks in brittle fracture. In: Advances in Applied Mechanics, Vol. 7, pp. 55–129. Elsevier (1962)
5. Park, K., Paulino, G.H.: Cohesive zone models: a critical review of traction-separation relationships across fracture surfaces. Appl. Mech. Rev. **64**(6), 060802 (2011)
6. Park, K., Paulino, G.H.: Computational implementation of the ppr potential-based cohesive model in abaqus: educational perspective. Eng. Fract. Mech. **93**, 239–262 (2012)
7. Spring, D.W., Paulino, G.H.: A growing library of three-dimensional cohesive elements for use in abaqus. Eng. Fract. Mech. **126**, 190–216 (2014)
8. Bae, G., Chang, S., Lee, S., Park, H.: Evaluation of interfacial properties between rock mass and shotcrete. Int. J. Rock Mech. Min. Sci. **41**, 106–112 (2004)
9. Luo, L., Li, X., Tao, M., Dong, L.: Mechanical behavior of rock-shotcrete interface under static and dynamic tensile loads. Tunn. Undergr. Space Technol. **65**, 215–224 (2017)
10. Tian, H., Chen, W., Yang, D., Yang, J.: Experimental and numerical analysis of the shear behaviour of cemented concrete-rock joints. Rock Mech. Rock Eng. **48**(1), 213–222 (2015)
11. Feng, X., Zhang, N., Yang, S., He, F.: Mechanical response of fully bonded bolts under cyclic load. Int. J. Rock Mech. Min. Sci. **109**, 138–154 (2018)
12. P. Dight, A case study of the behaviour of rock slope reinforced with fully grouted rock bolts. In: Proceedings International Symposium on Rock Bolting, Abisko, Sweden, pp. 523–538 (1983)
13. Li, L., Hagan, P., Saydam, S., Hebblewhite, B., Li, Y.: Parametric study of rockbolt shear behaviour by double shear test. Rock Mech. Rock Eng. **49**(12), 4787–4797 (2016)

Numerical Simulations of CNT/CNF Reinforced Concrete Using ANSYS

Sofija Kekez(✉)

Silesian University of Technology, 44100 Gliwice, Poland
sofija.kekez@polsl.pl

Abstract. Concrete mix design is an important stage in the design of concrete structures, especially in the case of specific structures such as high-rise buildings, long bridges, and other massive structures, or when a specific type of concrete is used, such as green, recycled, high-strength, lightweight, etc. The proper design of the mix proportions assures that the concrete is as economical as possible, and that the successful construction and exploitation stages are provided. Concrete mix design methods include analytical, experimental, and statistical methods. Determination of the optimal mix proportions of concrete, according to the relevant concrete property is commonly obtained by using a hybrid of these methods, called the semi-experimental methods which usually include some statistical method combined with experimental testing. This paper determines the feasibility of application of ANSYS for the mix design of concrete reinforced with carbon nanotubes or carbon nanofibers. The work here explores the possibilities given by the program, examines the feasibility of numerical modeling in concrete mix design, and compares the results of the numerical simulations with the results of previous experimental testing of CNT/CNF reinforced concrete.

Keywords: Concrete mix design · Nano-reinforced concrete · Numerical simulations · Mechanical properties

1 Introduction

Concrete mix design can be defined as the process of selecting suitable ingredients for concrete mixtures and determining their relative quantities with the purpose of producing an economical concrete, which has certain minimum properties, notably, workability, strength, and durability [1]. The precedence of concrete mix design is especially important in the construction of large voluminous structures such as dams, or of megastructures, which require a specific type of concrete such as high performance concrete. Concrete mix design assures that the required properties are achieved and keeps the use of costly ingredients at the necessary minimum, while making the construction stage and the structure itself as economically feasible as possible. Classification of concrete mix design procedures generally includes analytical, semi-experimental, experimental, and statistical methods. Semi-experimental methods present the combination of two types of methods, usually, some analytical or statistical prediction is combined with experimental testing of the prediction [2]. One of those combinations is presented through the

P. Rossi and J.-L. Tailhan (Eds.): SSCS 2022, RILEM Bookseries 38, pp. 178–186, 2023.
https://doi.org/10.1007/978-3-031-07746-3_18

predictions made by the numerical simulation models. Cai et al. [3] present the investigation of coarse aggregate settlement under vibration in fresh concrete by developing a 3-D multiphase numerical model. The prediction model shows that the vibration time has the greatest effect on the settlement, and furthermore, experimental testing verifies the validity of those predictions. The properties of hardened concrete are also investigated in the works such as Gan et al. [4], where the numerical models of heterogeneous microstructure of hardened cement pastes are implemented from the images obtained by X-ray computed microtomography. The creep and fracture represent the investigated properties, and the results are verified through the comparison with the experimental testing results.

Micromechanical models are the most commonly used for numerical predictions of the properties of any type of concrete, providing the direct link between the microstructure and the mechanical properties [5]. Although it is the most promising approach for predictions of concrete properties, it includes very advanced models and implements techniques, which are not widely available. Most of the techniques include equipment such as Scannin Electron Microscopy (SEM) and Transimission Electron Microscopy (TEM), as well as specific programming such as CEMHYD3D and HydraticCA. Therefore, the approach of micromechanical models is not readily usable for practical purposes and situations where the designers' skills and equipment are lacking. This work employs numerical models developed in ANSYS to establish the possibilities of predicting the mechanical properties of concrete reinforced with carbon nanotubes (CNTs) and carbon nanofibers (CNFs). The novel features of the software, namely homogenization, is used for modeling the materials, which were experimentally investigated in [7–17].

2 Numerical Modeling

Cementitious composite materials reinforced with CNTs or CNFs are fabricated and investigated as described in the literature [7–17]. The works are chosen based on the materials, procedure of dispersion and fabrication, and results of SEM analysis and strength tests. These materials have been collected in terms of the ingredients, mix proportions, and the weight ratio of the nano-reinforcement and then used for numerical simulations in ANSYS. Table 1 summarizes the ingredients of the collected cementitious composite materials. The total of 161 cement, mortar, and concrete samples are simulated, from that fifty samples are used for the validation of the models.

Table 1. Summary of materials.

Material	Cement [kg/m^3]	Water [kg/m^3]	Fine aggregate [kg/m^3]	Coarse aggregate [kg/m^3]	Superplasticizer [kg/m^3]	CNT [wt.%]	CNF [wt.%]
Min	317.61	121.6	0	0	0	0	0
Max	1875	789.48	1994.4	1284	27.27	2	2.5

Material models are developed within the new Material Designer feature of ANSYS, which may provide RVEs (representation volume elements) of several structural types and give results in the form of a tensor of elasticity. The initial idea was to develop the model by incorporating each material using different RVEs for each scale of the concrete structure. However, the main issue of the Material Designer is that it cannot support multiple models within a single project. Another issue is that the Material Designer cannot recognize aggregate states other than solid. Hence, the cementitious matrix part of the composite is introduced based on the experimental results, and further simulations are referred only to the applicability of the nano-reinforcement. To this end, Random UD Composite RVE is used to model the cementitious matrix with CNTs or CNFs as randomly distributed fibrous additions. The RVE is defined by the matrix and the fiber material, geometry of the fibers, and their orientation. Meshing of the RVE is provided only as the conformal meshing without limiting the size of the finite element due to the irregular geometrical distribution of the two materials. Figure 1 shows an example of the Random UD Composite RVE of the concrete reinforced with CNTs.

Fig. 1. Random UD Composite RVE of concrete matrix with CNTs.

After the homogenization is finished and the elasticity properties of the CNT/CNF concrete composite materials are determined, further steps include reproducing the experimental tests in order to compare the results and establish the applicability of ANSYS in practice. Figure 2 shows the flow of the ANSYS project used for each sample of the composite material. As it shows, Material Designer is building the data library, which is further used for two models; one for the axial compression test, and the other for the three-point bending test. The geometry and loading of the samples is modeled as realistically as possible. Following the laboratory testing, the models are provided with the geometry of 40 × 40 × 160 mm for the bending test and 40 × 40 mm of exposed surface for the compression test, according to the procedure described in EN 196-1.

Meshing is provided with 3-D 8 node cube elements with the edge length of 1 mm. During the analysis, smaller mesh elements were tested, however, the results did not show any change. Therefore, the 1 mm elements are used to shorten the analysis duration.

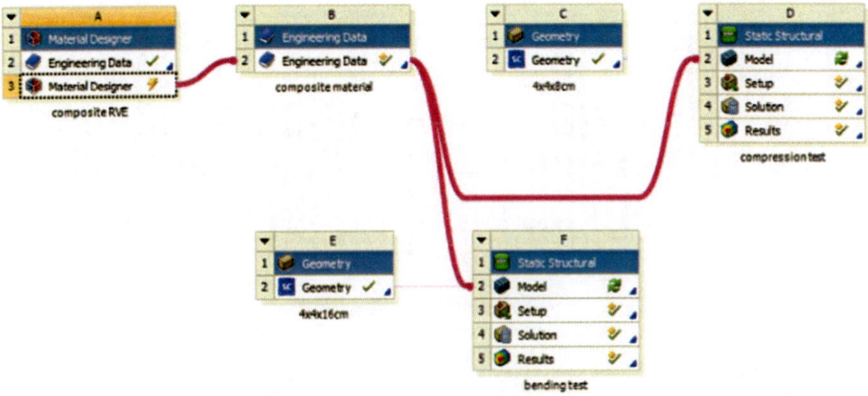

Fig. 2. ANSYS project flow schematic.

3 Results and Discussion

Validation of the models is done using fifty samples including plain and nano-reinforced cement, mortar, and concrete samples with or without the superplasticizer. Validation is provided to establish the behavior of the homogenized composite materials in every available mixture scenario, including the cement classes of 42.5 and 52.5 MPa, and the presence or the absence of the superplasticizer and the nanomaterials i.e., carbon nanotubes or carbon nanofibers. The fifty validation samples included two or three samples from each laboratory testing. Figure 3a and 3b show the comparison between the results obtained from the experimental testing and the numerical simulations for compressive strength and flexural strength of the validated composite materials, respectively. Compressive strength shows higher values for the experimental testing, while keeping a somewhat constant difference in values. Since the results of the simulations give lower strength, it may be concluded that the simulation is on the safe side, and mostly falls within the same strength class as the laboratory samples. However, the bending test shows an entirely different situation. The simulation results are higher than the experimental and the change appears to be significant for several samples. In some cases, the flexural strength exceeds 20 MPa, which may be considered as unrealistic, even for the heavily reinforced samples with the weight ratio over two percent. It may be observed that with the increase of the weight ratio of the nanofiller, the flexural strength shows significant increase compared to the experimental results. Due to the completely different results for compressive strength and for flexural strength, further analysis primarily implies testing of the rest of the cementitious composite material samples.

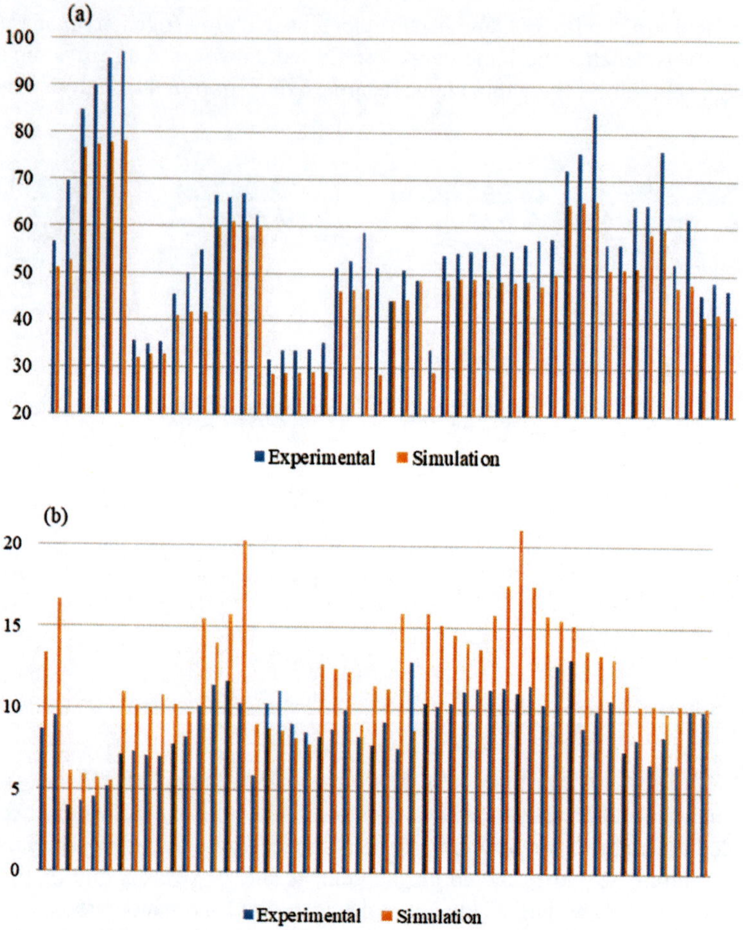

Fig. 3. Validation samples compared between experimental and simulated testing: (a) compressive strength; (b) flexural strength.

Even though the validation of the samples proved successful only partially, further investigation is made for both compressive and flexural strength of the samples to observe their behavior on a wider range of the nano-reinforcement weight ratio. To this end, other 111 samples are modeled in ANSYS in the same manner. Figures 4a, 4b, 4c, and 4d show resulting maps for maximum principal bending stress, maximum principal compressive stress, minimum principal compressive stress, and normal stress under compression, respectively, for a single mortar sample with the mix ratio of 0.5/1/0.33 for water/cement/sand, with cement class of 52.5, the addition of superplasticizer, and 1.85 wt.% of carbon nanofibers, labeled in the data library as mor52505033splNF185.

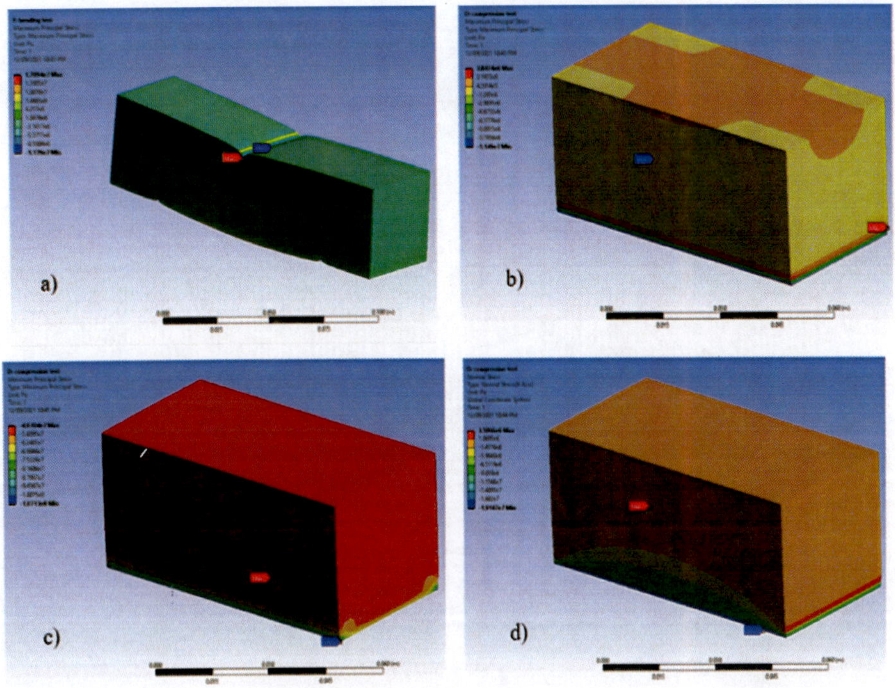

Fig. 4. Results for mor52505033splNF185: (a) maximum principal tensile stress; (b) maximum principal compressive stress; (c) minimum principal compressive stress; (d) normal stress under compression.

Figures 5a and 5b present the comparison between the experimental results and the results obtained through numerical simulations of compressive strength and flexural strength, respectively. The specific occurrence shown for the validation samples continues. The compressive strength simulation results show a certain decrease compared to the experimental results, however, considering the uncertainties of the model, this change may be assumed as relatively small and rather acceptable. The difference is relatively constant, it is smaller for the lower strength composites and greater for the higher strength composites, indicating that ANSYS is making an overestimation of the influence of the nanofiller. Disparate occurrence is regarding the group of samples where the CNT/mortar is reinforced with 0.025, 0.05, and 0.075 wt.% of CNTs. The simulation results show a decrease in compressive strength of around 44% compared to the laboratory results. Other 108 samples show results with up to 10% decrease from the experimental results for the lower weight ratios (up to 0.05 wt.% of CNTs or CNFs). Some results show deviations even up to 20% for the weight ratios over 0.05%, and all samples with weight ratios higher than 0.1% show this kind of decrease.

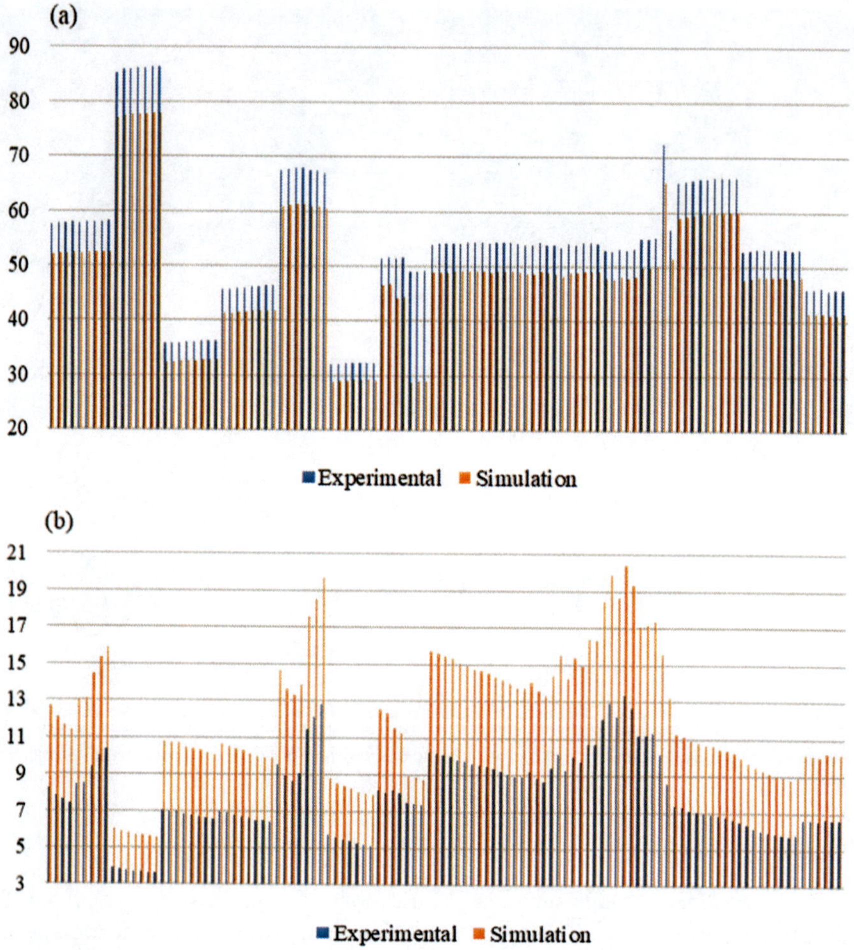

Fig. 5. Comparison between experimental and simulation results for: (a) compressive strength; and (b) flexural strength.

Furthermore, the results of the bending tests show an even higher overestimation of the effect that the nanofillers have on the cementitious matrix by the software. Unlike for the compression, modeled materials have significantly higher flexural strength then the laboratory samples. In some cases, the results report change even higher than 200 percent. It may be assumed that due to the extreme difference in the flexural strength between the matrix and the fiber component of the model, the homogenization process allowed an unrealistic equalization of the final flexural capacity of the composite material. It is surely a direct consequence of the matrix material idealization, and the lack of consideration of the complex heterogenic microstructure of concrete. Interestingly, it may be observed in Fig. 5b that the difference between the experimental and simulation results is rather constant, which may imply the possibility that some changes to the model could resolve the issue. Additional possibility may be that the RVE does not describe the randomization of the nanomaterials in a completely realistic manner. During fabrication of these composite materials, it is impossible to control the orientation of the individual fibers, hence, only part of them can positively affect the flexural strength of the hardened sample.

4 Conclusions

This paper investigates the possibility of using ANSYS software for predictions of the mechanical properties of the CNT/CNF reinforced concrete. To this end, the composite materials are modeled using Material Designer feature, and the laboratory 3-point bending and compression tests are reproduced. Considering that the micromechanics of the cementitious matrix was not included in the analysis, it may be concluded that the compression models have been relatively successful, especially for the lower weight ratios of the nano-reinforcement. Vis-à-vis, the bending tests showed extreme increase in flexural strength compared to the experimental tests, as a direct consequence of the matrix idealization. It is a possibility that adjusting the material model would improve the results of the bending test. However, it may be concluded that Ansys software is still not a preferred option for the concrete mix design. Material Designer is capable of recognizing only solids, cannot combine more shapes and sizes, and is incapable of incorporating more than a single RVE. The only possibility would be to manually model the microstructure and make the analysis on the smallest possible geometry. However, it would include an extreme amount of time, rendering such undertaking as completely impractical. For now, the application of carbon materials using Ansys would be most practical for carbon threads and strands, which would work on the macroscale.

References

1. Neville, A.M., Brooks, J.J.: Concrete Technology. 2nd edn. Pearson Education Limited, Harlow, England (2010). ISBN 978-0-273-73219-8
2. Ahmad, S., Alghamdi, S.A.: A statistical approach to optimizing concrete mixture design. Sci. World J. **2014**, e561539 (2014). https://doi.org/10.1155/2014/561539
3. Cai, Y., et al.: An experimental and numerical investigation of coarse aggregate settlement in fresh concrete under vibration. Cement Concr. Compos. **122**, 104153 (2021). https://doi.org/10.1016/j.cemconcomp.2021.104153
4. Gan, Y., Romero Rodriguez, C., Zhang, H., Schlangen, E., van Breugel, K., Šavija, B.: Modeling of microstructural effects on the creep of hardened cement paste using an experimentally informed lattice model. Comput. Aided Civil Infrast. Eng. **2021**, 1–17 (2021). https://doi.org/10.1111/mice.12659
5. Zhang, H., Xu, Y., Gan, Y., Chang, Z., Schlangen, E., Šavija, B.: Microstructure informed micromechanical modelling of hydrated cement paste: techniques and challenges. Constr. Build. Mater. **251**, 118983 (2020). https://doi.org/10.1016/j.conbuildmat.2020.118983
6. Gan, Y., Zhang, H., Liang, M., Schlangen, E., van Breugel, K., Šavija, B.: A numerical study of fatigue of hardened cement paste at the microscale. Int. J. Fatigue **151**, 106401 (2021). https://doi.org/10.1016/j.ijfatigue.2021.106401
7. Kahidan, A., Shirmohammadian, M.: Properties of Carbon Nanotube (CNT) reinforced cement. Int. J Eng. Res. **5**(6), 497–503 (2016). https://doi.org/10.17950/ijer/v5s6/616

8. Li, G.L., Wang, P.M., Zhao, X.: Pressure-sensitive properties and microstructure of carbon nanotube reinforced cement composites. Cement Concr. Compos. **29**, 377–382 (2007). https://doi.org/10.1016/j.cemconcomp.2006.12.011
9. Liew, K.M., Kai, M.F., Zhang, L.W.: Mechanical and damping properties of CNT-reinforced cementitious composites. Compos. Struct. **160**, 81–88 (2017). https://doi.org/10.1016/j.compstruct.2016.10.043
10. Zhang, J., Kea, Y., Zhang, J., Han, Q., Dong, B.: Cement paste with well-dispersed multi-walled carbon nanotubes: mechanism and performance. Constr. Build. Mater. **262**, 120746 (2020). https://doi.org/10.1016/j.conbuildmat.2020.120746
11. Konsta-Gdoutos, M.S., et al.: Effect of CNT and CNF loading and count on the corrosion resistance, conductivity and mechanical properties of nanomodified OPC mortars. Constr. Build. Mater. **147**, 48–57 (2017). https://doi.org/10.1016/j.conbuildmat.2017.04.112
12. Konsta-Gdoutos, M.S., Danoglidis, P.A., Falara, M.G., Nitodas, S.F.: Fresh and mechanical properties, and strain sensing of nanomodified cement mortars: the effects of MWCNT aspect ratio, density and functionalization. Cement Concr. Compos. **82**, 137–151 (2017). https://doi.org/10.1016/j.cemconcomp.2017.05.004
13. Parveen, S., Rana, S., Fangueiro, R., Paiva, M.C.: Microstructure and mechanical properties of carbon nanotube reinforced cementitious composites developed using a novel dispersion technique. Cem. Concr. Res. **73**, 215–227 (2015). https://doi.org/10.1016/j.cemconres.2015.03.006
14. Hawreen, A., Bogas, J.A., Dias, A.P.S.: On the mechanical and shrinkage behavior of cement mortars reinforced with carbon nanotubes. Constr. Build. Mater. **168**, 459–470 (2018). https://doi.org/10.1016/j.conbuildmat.2018.02.146
15. Galao, O., Zornoza, E., Baeza, F.J., Bernabeu, A., Garcés, P.: Effect of carbon nanofiber addition in the mechanical properties and durability of cementitious materials. Mater. Constr. **62**(307), 343–357 (2012). https://doi.org/10.3989/mc.2012.01211
16. Li, G.Y., Wang, P.M., Zhao, X.: Mechanical behavior and microstructure of cement composites incorporating surface-treated multi-walled carbon nanotubes. Carbon **43**, 1239–1245 (2005). https://doi.org/10.1016/j.carbon.2004.12.017
17. Konsta-Gdoutos, M.S., Danoglidis, P.A., Shah, S.P.: High modulus concrete: effects of low carbon nanotube and nanofiber additions. Theoret. Appl. Fract. Mech. **103**, 102295 (2019). https://doi.org/10.1016/j.tafmec.2019.102295

Computational Performance Assessment and Failure Analysis of Reinforced Concrete Wall Buildings Under Seismic Loads

Ioannis Koutromanos[1(✉)], Marios Mavros[2], Marios Panagiotou[3], Jose I. Restrepo[4], and Rodolfo Alvarez[5]

[1] Virginia Tech, Blacksburg, VA 24061, USA
ikoutrom@vt.edu
[2] University of Cyprus, Nicosia, Cyprus
[3] Nabih Youssef Structural Engineers, Los Angeles, CA, USA
[4] University of California, San Diego, La Jolla, CA, USA
[5] MM Engineers SC, Mexicalli, Mexico

Abstract. This paper employs computational simulation to investigate the seismic performance of a prototype 14-story building with core-walls, located in Downtown Los Angeles and designed in accordance with the pertinent minimum code requirements. The computational representation of the building employs the beam-truss model (BTM) for the walls and floor slabs. The BTM allows capturing the cyclic inelastic behavior of RC walls and slabs, while being conceptually simpler than, e.g., continuum finite element models. Nonlinear static and dynamic analyses are conducted. The latter consider a set of triaxial ground motions scaled at two intensity levels, namely, those of the design earthquake and the risk-targeted maximum considered earthquake (MCE_r).

The analyses provide insights into the evolution of structural system damage and lateral strength degradation, while also elucidating the complex interaction between the webs and flanges of the core wall and the system effects associated with coupling between the walls, beams, slabs, and columns. The potentially detrimental effect of the multidirectional nature of the seismic motion on the inelastic deformability of the wall components is also demonstrated. The implications of the obtained results, in the context of the currently employed performance-based seismic design procedures in the United States, are also discussed.

Keywords: Reinforced concrete · Structural walls · Seismic design

1 Introduction

Reinforced concrete walls constitute a very popular lateral-load resisting system for multi-story buildings. A very popular configuration for such buildings employs nonplanar flanged core-walls, such as C-shape walls, to provide strength and stiffness against lateral loading, while also serving as enclosures for elevator shafts. Understanding the inelastic behavior, damage patterns and collapse modes for such structures is of uttermost

P. Rossi and J.-L. Tailhan (Eds.): SSCS 2022, RILEM Bookseries 38, pp. 187–196, 2023.
https://doi.org/10.1007/978-3-031-07746-3_19

significance for evaluating the performance of newly designed and existing buildings in seismically active regions.

Core-walls in multi-story buildings are typically detailed in accordance with design standards as so-called special walls, wherein adequate seismic performance and safety are ensured through the development of inelastic flexural deformations in well-detailed, plastic hinge regions, located at the lowest levels of the wall section. Specific design documents which are used in the United States includes the ASCE 7-16 standard [1] and the ACI 318-14 code [2]. For RC wall buildings up to 73 m (240 ft) high that are located on a Seismic Design Category D or E, ASCE 7-16 [1] allows the use of response spectrum analysis on a building model which uses effective stiffness modifiers for the various members. For buildings taller than 73 m (240 ft), RCSW are permitted as an *Alternative Structural System* after an independent review and approval of the structural design by the local jurisdiction. The design verification of such buildings must rely on nonlinear dynamic analysis in accordance with performance-based seismic design (PBSD) guidelines such as those proposed by the Los Angeles Tall Buildings Structural Design Council (LATBSDC [3]). The use of accurate computational models is indispensable for the reliable PBSD of RCSW buildings.

A number of approaches have been proposed and used for the simulation of RC walls under cyclic loading, including beam-based models or similar approaches enforcing a plane-section hypothesis (e.g., [4]). Finite element models, typically based on shell elements (e.g., [5]), have also provided some level of accuracy for capturing flexure-dominated walls, but they may be problematic for cases involving shear failures associated with, e.g., web crushing or formation of strongly localized inclined cracks.

A powerful method for the simulation of structural walls under earthquake loading is the beam-truss model (BTM), originally proposed by Lu et al. [6, 7], which allows the simulation of nonplanar walls and other components such as coupling beams and slabs. The BTM has been shown to capture all common shear-flexure failure modes, such as diagonal tension and compression as well as horizontal and vertical web crushing of rectangular and flanged walls. A detailed review of all developments pertaining to the use of the BTM for seismic analysis of RC walls and other components is provided in [8].

The present study employs the BTM for the nonlinear seismic analysis of a hypothetical, 14-story RC wall building system located in Downtown Los Angeles, California. The building has been designed in accordance with minimum requirements of pertinent code standards. Static and dynamic time history analyses are conducted for the building. The latter focus on simultaneous application of three ground motion components. A set of 11 triaxial records from previous earthquakes, scaled to the design earthquake and the risk-targeted maximum considered earthquake (MCE_r) for the hypothetical location of the building, is considered. The most significant results from the static and dynamic analyses, together with their implications for the performance-based seismic design of RC wall buildings based on computational simulation, are provided.

2 Description of Building

The present study is focused on a 14-story, RC core-wall building, located in Downtown Los Angeles, and designed in accordance with the minimum requirements of code standards. An elevation view of the building is provided in Fig. 1a, while the plan configuration is depicted in Fig. 1b. The specific figures also provide a set of coordinate axes X and Y, which are aligned with the edges of the floor plan, and an axis Z along the height of the building. The floors consist of 203 mm (8 in.) thick post-tensioned RC flat slabs, having a square plan configuration with a dimension of 28 m (92 ft). All stories have a height of 3.66 m (12 ft), and the total building height H is 51.2 m (168 ft). The core wall consists of two C-shaped walls, such that the wall webs (for X-direction loading) are configured as coupled piers along the X-direction as shown in Fig. 1a and as solid walls along the Y-direction. The plan view of the building is shown in Fig. 1b. Coupling in the X-direction is provided by 1016 mm (40 in.) deep beams. The floor gravity loads are carried by the core-wall and a total of 12 columns which are distributed along the perimeter of the plan of the building.

The building has been designed in accordance with the California Building Code, which is in turn based on ASCE 7-16 [1] and ACI 318-14 [2]. The wall shear stress design demand values in the X- and Y-direction were less than 50% of the shear stress limits in the ACI 318-14 building code [2]. Each coupling beam was provided with diagonal reinforcement and was fully confined along its length. The columns were designed according to ACI 318-14 [2] resulting to a longitudinal reinforcement ratio of 1.8% and a confining reinforcement volumetric ratio of 1.74%. Finally, each floor slab has a longitudinal steel ratio of 0.66% in both directions (0.33% at the top and 0.33% at the bottom) and is post-tensioned with 12 mm (0.5 in.) unbonded strands spaced at 304 mm in both directions, with an effective post-tensioning force of 120 kN. A more detailed discussion of the design is provided in Mavros et al. [9] and in Alvarez [10].

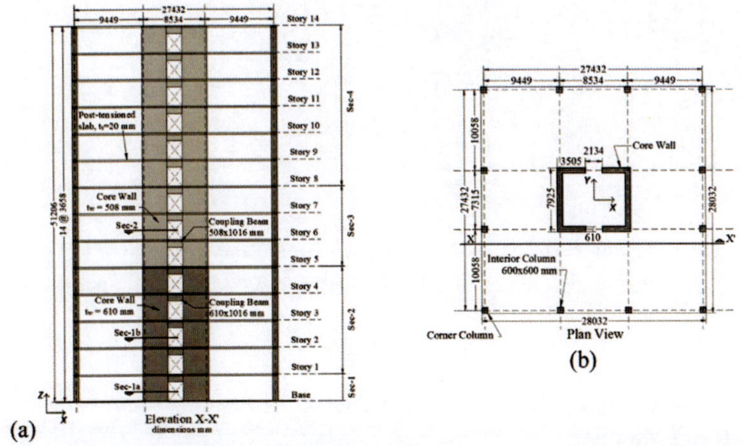

Fig. 1. Prototype building considered in analytical study: a) Elevation, b) Plan configuration.

3 Analysis Method

3.1 Overview

The computational simulation of the building relies on the beam-truss-model (BTM) methodology, originally described in detail in Lu and Panagiotou [6] and Lu et al. [7]. As shown in Fig. 2a, the BTM uses an assemblage of horizontal and vertical beam elements and inclined truss elements to represent a wall, beam, or slab component. The diagonal truss elements are aimed to represent the inclined in-plane compression field of RC components. The beam elements include a fiber sectional model, wherein the cross-sectional dimension along the thickness of the member is subdivided into multiple fibers, each having its own stress-strain (constitutive) law.

The analyses presented herein employ a user-friendly implementation of the BTM described in Mavros et al. [9], wherein each BTM cell, consisting of two horizontal beam elements, two vertical beam elements and two inclined truss elements, is represented as a rectangular shell macro-element. This approach, schematically summarized in Fig. 2b, facilitates the model definition and interpretation of analysis results, while maintaining the predictive capability of the BTM.

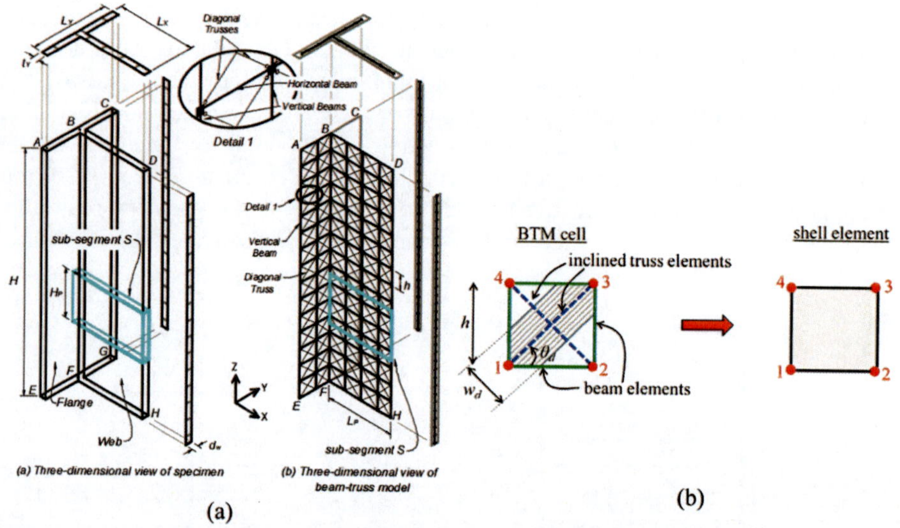

Fig. 2. Beam-Truss Model (BTM) and implementation: a) Representation of RC wall as assemblage of line elements (figure from [6]), b) Formulation of a BTM standard panel assemblage as a rectangular, four-node macro-element.

3.2 Material Models

The concrete material in the BTM is modeled using the uniaxial stress-strain law schematically summarized in Fig. 3a. The specific law is capable of accounting for strength and stiffness degradation due to tensile cracking and compressive crushing, while also including appropriate hysteretic laws. For the diagonal truss elements, the concrete constitutive law accounts for the effect of normal tensile strain, ε_n, on the compressive resistance, in accordance with the work of Vecchio and Collins [11]. Specifically, compressive axial stresses for the truss elements are multiplied by a reduction coefficient, β, which expresses the effect of normal tension on the compressive stress of concrete. The value of β depends on the normal tensile strain of each truss element following the law presented in Fig. 3b. The material model for concrete involves softening, which is bound to lead to spurious mesh-size effects in an analysis due to strain localization.

The behavior of the reinforcing steel is described through the constitutive model by Kim and Koutromanos [12]. In the absence of buckling or rupture, the specific material model can be fully calibrated if the stress-strain curve in monotonic tension, schematically shown in Fig. 4a, is known. The constitutive model includes a hysteretic law to capture the cyclic response of reinforcing steel, and can also adjust the material resistance to account for the impact of inelastic buckling, as depicted in Fig. 4b.

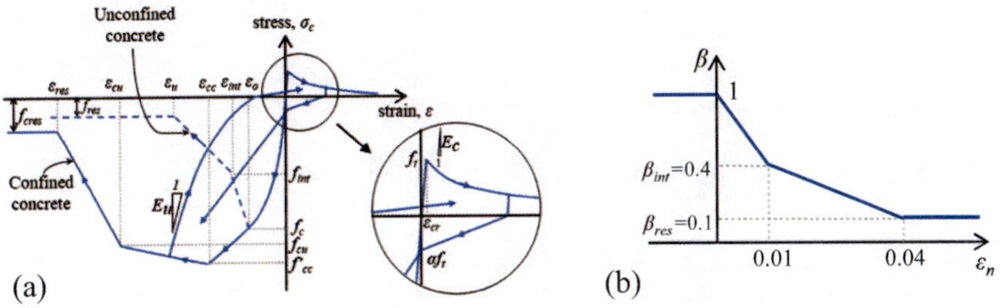

Fig. 3. Material model for concrete: a) Hysteretic law for tension and compression, b) Reduction factor for compression in inclined elements, accounting for the effect of transverse tensile strains.

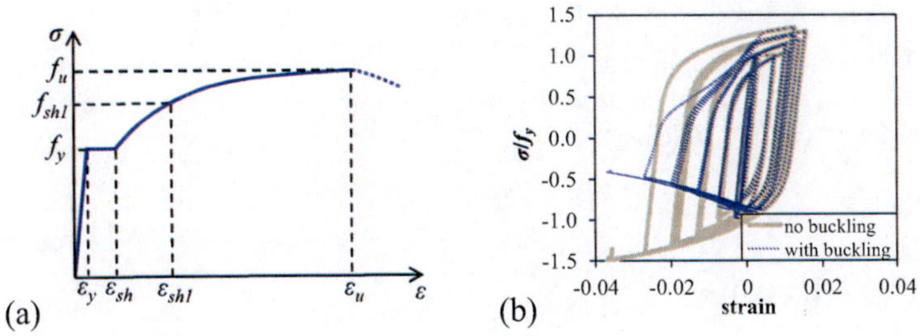

Fig. 4. Material model for reinforcing steel: a) Behavior for monotonic tension, b) Sample hysteretic response, with and without buckling.

3.3 Setup for Static and Dynamic Analysis

The static analyses are conducted for a combination of vertical (gravity) loads and lateral loads representing the impact of earthquake forces. The vertical loads are applied first, and their magnitude remains constant throughout the analysis. The equilibrium equations are enforced through a Newton-Raphson iterative procedure. Two lateral load patterns, i.e. distributions of forces with height, are considered in the analysis, as shown in Fig. 5a. These two patterns are meant to provide an upper- and lower-bound estimate of the effective height of the building (i.e., the height at which the resultant seismic lateral force is applied).

The dynamic analyses consider a set of 11 ground motions, appropriately scaled at two distinct levels of seismic intensity. The first level corresponds to a design-level earthquake for downtown Los Angeles, while the second level corresponds to the risk-targeted maximum considered earthquake (MCE_r) intensity for the same location. The pseudo-acceleration and displacement spectra for the ground motions scaled to MCE_r intensity are compared to the MCE spectrum for downtown Los Angeles per ASCE 7-16 in Fig. 5b. The equations of motion are solved in time through a Hilber-Hughes-Taylor (HHT) time-stepping scheme, combined with a Newton-Raphson iterative strategy. A Rayleigh proportional viscous damping matrix, based on the initial stiffness of the structure, is also used in the analysis. The values of the mass- and stiffness-proportional coefficients are determined so that a viscous damping ratio of 0.5% is obtained for period values of 0.5 s and 5 s.

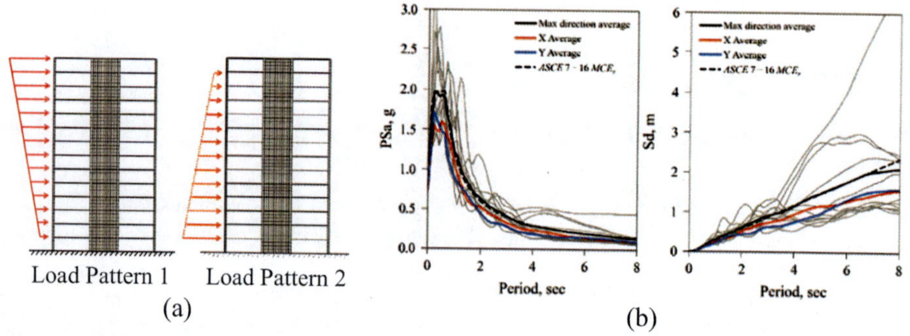

Fig. 5. Analysis setup: a) Load patterns considered for static analysis, b) Response spectra for 11 ground motions used in dynamic analysis, compared to the MCE level spectrum for Downtown Los Angeles per ASCE 7-16 [1].

4 Analysis Results

4.1 Static Analysis Results

The hysteretic curves obtained from the static analysis of the building, for loading with each of the two load patterns along directions X and Y, are presented in Fig. 6. The deformation in these figures is quantified by the roof (average) drift ratio, θ_r, while the resistance is expressed through the ratio of the base shear force, V_b, over the building weight, W. Each of the plots in Fig. 6 includes the curves corresponding to monotonic loading applied to the building. The point in each analysis corresponding to the peak lateral resistance is also marked as a solid circle in each plot.

It can be observed that the load pattern has a profound impact on the drift ratio that the structure can develop before the occurrence of strength degradation. Load pattern 2 entails a much lower deformability of the structure. This is because this pattern entails the development of larger shear stresses on the wall web segments, and it is well-established that the flexural ductility of RC walls is negatively impacted by increased shear stresses. The impact of shear-flexure interaction on the behavior of the building under static loading is also evident by observing the deformed patterns and maximum principal strain contours. These are presented in Fig. 7a for loading in the X-direction with Pattern 1 and Fig. 7b for loading in the Y-direction with Pattern 2. Before the occurrence of strength degradation, the behavior of the walls is flexure-dominated. At the instant of peak load, both cases provide fairly distributed strains along the bottom stories of the building. Strength degradation is associated with localized, shear-induced failure of wall piers at the bottom 1–2 stories, as deduced from the same figures. These observations stress the necessity of simulation tools such as the BTM, which can capture the inelastic shear-flexure interaction, for the seismic analysis of RC wall systems. Another important finding, deduced from Fig. 12, is the need to accurately describe the nonlinear behavior of the coupling beams in a model especially within the plastic hinge regions of coupled walls.

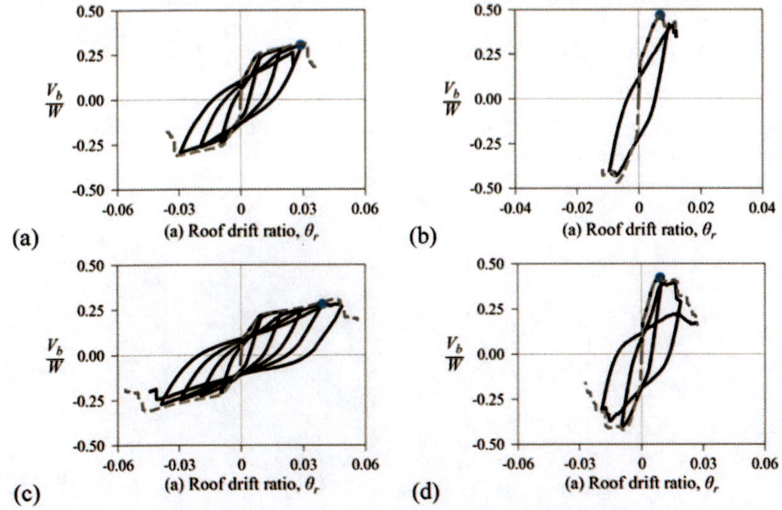

Fig. 6. Hysteretic force-versus roof drift ratio plots obtained from static analyses: a) Loading with Pattern 1 along X direction, b) Loading with Pattern 2 along X direction, c) Loading with Pattern 1 along Y direction, d) Loading with Pattern 2 along Y direction.

Figure 8a compares the level of shear stresses for application of Load Pattern 2 along the X-direction of the building. The figure also shows the upper and lower limit values for the shear stresses in the webs per the Los Angeles Tall Building Seismic Design Council (LATBSDC) performance-based seismic design guidelines [3]. The significant amount of shear stresses leads to damage patterns which include extensive shear damage in wall webs, as deduced from the maximum principal strain contour plots shown in Fig. 7a.

Figure 8b shows the deformed shape of the building model corresponding to a roof story drift ratio of 3%, for cyclic static loading in the Y-direction with Load Pattern 1. The deformed shapes in the figure also include vertical displacement contours. The axial elongation and rotation of the core-wall imposes significant flexural deformations in the slab. The vertical displacement contours of Fig. 8b indicate the formation of yield lines in the slabs associated with downwards vertical displacement. It is worth mentioning that the slab coupling also affects the demands in the columns. It becomes apparent that system effects such as the interaction between the floor slabs and the other structural components have profound impact on the seismic resistance mechanism. It is worth noting that such system effects are typically not considered in nonlinear seismic analysis practice for slab span lengths longer than 6 m according to LATBSDC [3].

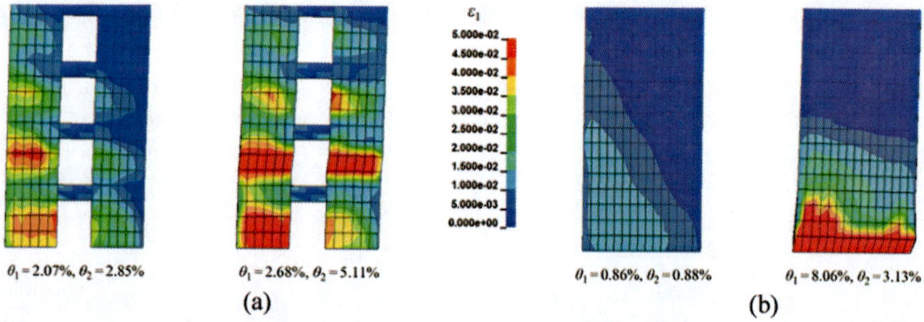

Fig. 7. Hysteretic force-versus roof drift ratio plots obtained from static analyses, at the instant of peak load and at the end of the analysis: a) Loading with Pattern 1 along X direction, b) Loading with Pattern 2 along Y direction.

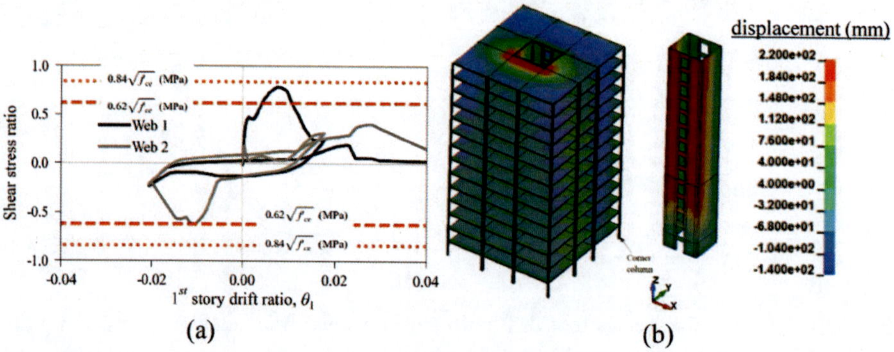

Fig. 8. Results obtained from nonlinear static analysis: a) Comparison of shear stresses on web piers to upper-bound values allowed in performance-based guidelines by the LATBSDC [3], b) Deformed shape and vertical displacement contours for building and core wall.

4.2 Dynamic Analysis Results

Figure 9a presents a summary of key response quantities obtained for the dynamic analyses, namely, the peak interstory drift ratio (IDR) and peak floor accelerations along the height of the building, obtained for the MCE_r-level ground motions. An important remark stemming from the specific plots is the significant impact of higher modes on the response, reflected on the fact that the upper stories have significant IDR values and large floor accelerations.

The potentially detrimental effect of the multidirectional nature of the seismic motion on the inelastic deformability of the wall components is also demonstrated. For instance, five of the MCE_r ground motions led to vertical concrete crushing of the confined corners of the core wall. This is deduced from Fig. 9b which provides the maximum vertical compressive strains recorded for each motion, for the design-level and MCE_r-level intensities. The level of compressive strains recorded for several motions may imply that repair of the building after an MCE_r earthquake may not be possible. This finding is critical for the accurate calculation of residual deformations (lateral and vertical) that determine the functional recovery performance of earthquake-prone communities. It also indicates that superior detailing schemes may be necessary to ensure that the building will ensure repairability and quick return to post-earthquake service after a major seismic event.

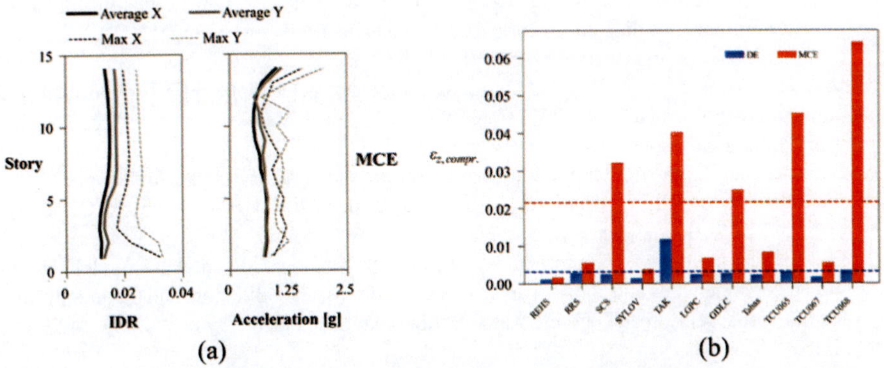

Fig. 9. Dynamic Analysis results: a) Peak interstory drift ratio (IDR) and floor accelerations along building height, b) Maximum vertical strains recorded during the design-level earthquake (DE) and the MCE_r-level earthquake (MCE).

5 Conclusions

This study has used the nonlinear beam-truss model (BTM) for the performance evaluation of a prototype 14-story building located in Downtown Los Angeles, California. The building was designed in accordance with the minimum requirements of the California Building Code. Static analyses for two different load patterns provided insights on the damage patterns, inelastic deformability and response mechanisms of the building system. The significant impact of shear-flexure interaction, coupling slabs, and the deformation patterns of the core-wall which deviate from a plane-section hypothesis, indicate that commonly adopted beam-based models may not be accurate enough for the nonlinear analysis of RC core-wall systems.

Results from the dynamic analyses demonstrated the impact of multi-directional excitations, which may introduce excessive compressive strains on the core-walls and severe damage, which will require demolition and replacement of the building. Ensuring enhanced performance (with, e.g., limited damage) for multi-story buildings in seismically active regions may require improved detailing strategies to ensure adequate confinement in regions where compressive crushing is probable.

References

1. American Society of Civil Engineers (ASCE): Minimum design loads and associated criteria for buildings and other structures: ASCE/SEI 7-16. ASCE, Reston, VA (2017)
2. American Concrete Institute (ACI): Building code requirements for reinforced concrete. ACI 318-14. ACI, Farmington Hills, MI (2014)
3. Los Angeles Tall Buildings Structural Design Council (LATBSDC): LATBSDC-20, An Alternative Procedure for Seismic Analysis and Design of Tall Buildings Located in the Los Angeles Region (2020). https://drive.google.com/file/d/1cBv7S4Eh_c8IoPtrch7uxIHP9FbQELO_/view
4. Kolozvari, K., Kalbasi, K., Orakcal, K., Massone, L.M., Wallace, J.: Shear–flexure-interaction models for planar and flanged reinforced concrete walls. Bull. Earthq. Eng. **17**(12), 6391–6417 (2019). https://doi.org/10.1007/s10518-019-00658-5
5. Vásquez, J.A., Jünemann, R., de la Llera, J.C., Hube, M.A., Chacón, M.F.: Three-dimensional nonlinear response history analyses for earthquake damage assessment: a reinforced concrete wall building case study. Earthq. Spectra **37**(1), 235–261 (2021)
6. Lu, Y., Panagiotou, M.: Three-Dimensional Cyclic Beam-Truss model for nonplanar reinforced concrete walls. J. Struct. Eng. **140**(3), 04013071 (2014). https://doi.org/10.1061/(asce)st.1943-541x.0000852
7. Lu, Y., Panagiotou, M., Koutromanos, I.: Three-dimensional Beam-Truss model for reinforced concrete walls and slabs – part 1: modeling approach, validation, and parametric study for individual reinforced concrete walls. Earthquake Eng. Struct. Dynam. **45**(9), 1495–1513 (2016)
8. Panagiotou, M., Koutromanos, I., et al.: Nonlinear Beam-Truss Model (BTM) for seismic performance evaluation of reinforced concrete wall buildings. In: Proceedings of the 2021 SEAOC Convention, San Diego, CA (2021)
9. Mavros, M., Panagiotou, M., Koutromanos, I., Alvarez, R., Restrepo, J.I.: Seismic analysis of a modern 14-story reinforced concrete core wall building system using the BTM-shell methodology. Earthquake Eng. Struct. Dynam. **51**(6), 1540–1562 (2022). https://doi.org/10.1002/eqe.3627
10. Alvarez, R.: Seismic response verification of reinforced concrete structural wall systems. Ph.D. dissertation, University of California, San Diego (2020)
11. Vecchio, F., Collins, M.: The modified compression-field theory for reinforced concrete elements subjected to shear. ACI J. Proc. **83**(2), 219–231 (1986)
12. Kim, S., Koutromanos, I.: Constitutive model for reinforcing steel under cyclic loading. ASCE J. Struct. Eng. **142**(12), 1–14 (2016)

Fuzzy Logic-Based Approach for the Uncertainty Modelling in Cementitious Materials

Philipp Kunz[1], Antonio Cibelli[2(✉)], Giovanni Di Luzio[2], Liberato Ferrara[2], and Viktor Mechtcherine[1]

[1] Institute of Construction Materials, Technische Universität Dresden, Georg-Schumann-Str. 7, 01187 Dresden, Germany
{philipp.kunz,viktor.mechtcherine}@tu-dresden.de

[2] Department of Civil and Environmental Engineering, P.za Leonardo da Vinci 32, 20133 Milano, Italy
antonio.cibelli@polimi.it

https://tu-dresden.de/bu/bauingenieurwesen/ifb

https://www.dica.polimi.it

Abstract. Refined physics-based models generally present a relevant number of parameters to calibrate against experimental data, which might be unavailable for the mixture or the service scenario of interest. This represents one of the most relevant issues in material modelling, especially when descriptive models are adapted to serve as predictive ones. Additionally, accurate small-scale models are particularly suitable for simulating laboratory tests. The up-scaling to structural members, or even entire structures, requires the identification of bridging parameters, responsible for bringing the small-scale models' accuracy into the engineering models adopted for the long-term prediction at the structural level. This paper presents a methodological approach to deal with the uncertainty featuring the models calibration in case of limited experimental data. In addition, a strategy of up-scaling, relying on the fuzzy logical approach is presented. The activity performed is framed into the Horizon 2020 project ReSHEALience [1].

Keywords: Fuzzy logic · Uncertainty modelling · Material modelling · Cementitious materials · Fibre reinforced concrete · Lattice discrete particle model · Hygro-thermo-chemical model · ONIX

1 Introduction

Material numerical modelling is steered by (i) the scale at which the models are developed, and (ii) how much affordable the calibration of the governing parameters results. Whichever resolution is chosen, the phenomena occurring at either that or smaller observation scales must be explicitly implemented into the governing equations. As an example, a microscale mechanical model is expected

P. Rossi and J.-L. Tailhan (Eds.): SSCS 2022, RILEM Bookseries 38, pp. 197–206, 2023.
https://doi.org/10.1007/978-3-031-07746-3_20

to rely on constitutive equations whose formulation refers to all the mechanisms detectable at the microscopic observation scale. Then, if well-formulated, when applied to larger domains, the modelling framework should allow for catching the main response patterns at the macroscale, without any further adjustment, net of specific issues which could arise at larger scales (e.g. size effect).

Focusing on cementitious materials, the models conceived at small scales (nano-, micro-, mesoscale) present no theoretical limitations in being applied to larger elements, as structural members or even entire structures. These models generally take advantage of sound chemo-physical backgrounds. Furthermore, to be formulated at a small scale permits to simulate the material behaviour very accurately, enabling, on turn, an effective calibration and validation of the governing parameters against laboratory measurements. A number of trends in the material response might be more easily associated with cause-effect relationships to either mechanical or environmental (controlled) boundary conditions. However, besides the higher computational cost, the use of small scale models for the simulation of large structural components, eventually produced in real sites, yields the rise of the uncertainty discussed above. In fact, on one hand, the performance of structures built on site is affected by aleatoric factors like the quality of the production process, and the not controlled environmental conditions and loading regimes. In addition, the elements size, likely significantly larger than that of laboratory specimens, introduces a size effect which cannot be neglected.

Therefore, two types of uncertainty feature the material modelling: (i) data uncertainty at the material scale, due to the large number of parameters to calibrate, some of which even hardly experimentally detectable, and (ii) model uncertainty after up-scaling, mainly due to the lack of information for whole building structures. In order to deal with these aleatoric aspects the feasibility of numerical approaches as predictive tools should be investigated.

In the context of the Horizon 2020 project ReSHEALience [1,2], the concept of Ultra High Durability Concrete (UHDC) is introduced to refer to those materials belonging to the wide family of cementitious materials, whose mixture recipe is tailored to enhance the performance in terms of durability compared to ordinary concretes. The UHDC long-term performance is modelled through a discrete multi-physics model, M-LDPM [3–10], resulting from the full coupling between the Lattice Discrete Particle Model (LDPM) [11,12] and the Hygro-Thermo-Chemical model (HTC) [13–15]. The fuzzy logic and fuzzy set theory have been adopted to turn the M-LDPM model into an actual predictive tool. In this paper this methodology to cope with uncertainty in cementitious materials modelling is presented.

2 State-of-the-Art

Uncertainty can be considered as the gradual assessment of the truth content of a statement with relation to an event. It is useful to differentiate between stochastic uncertainties, respectively randomness and epistemic uncertainty. Epistemic uncertainty is resulting from insufficient information or lack of (experimental) data, while randomness will always occur [16]. Furthermore, it is required to separate between model uncertainty and data uncertainty.

Among other methods for mathematically describing and quantifying uncertainty the probability theory and fuzzy set theory could be used [17]. The probability theory is a mathematical theory of chance, stating the likelihood that an event will occur by quantifying it with a number between zero and one. The logic of fuzzy set theory, instead, allows the quantification of degrees of truth or membership to a set.

Fuzzy logic itself is a concept which can be obtained by using mathematical expressions of fuzzy sets. Unlike the basic principle of bivalence logic, the fuzzy one allows not just statements as "true" or "false" - as it is used in binary systems - but provides the possibility of using linguistic terms such as "much smaller" or "a little higher". Therefore, the fuzzy logic can be used for numerical modelling approaches to simulate the human process of drawing conclusions, quantifying non-determined information and deriving decisions based on limited, or even lacking, data [18]. Then, it could be derived that fuzzy logic is a principle that "everything is a matter of degree" [19]. As it is derived from the fuzzy logic, the fuzzy-set theory can be used to describe epistemic uncertainty, allowing a flexible assessment of the possibility, particularly useful to deal with complex governing parameter systems and limited experimental databases.

Fuzzy logical modelling approaches were already presented in civil engineering research, especially with regards to the structural analysis and uncertain loading regimes applied to structural elements [20,21]. Furthermore the use of fuzzy logic towards the prediction of compressive strength of cement was investigated in [22]. In the aforementioned works, the fuzzy logic was found suitable to deal with uncertainty due to either lack of information in parameters or models. Within the framework of the ReSHEALience project, this approach is applied to address the uncertainty modelling with respect to the durability modelling of UHDC.

Because of this, there is an increasing interest in developing consistent numerical strategies which might turn this phase into a physics-based estimation, rather than an experimental-driven best fitting between direct measurements and simulation results. In addition, this approach is worthwhile being explored because many parameters governing M-LDPM models are not actually measurable, being them empirical coefficients.

3 Uncertainty Modelling

The aim is to address two relevant issues featuring the development of predictive models for long-term performance of cementitious materials: (i) the uncertainty in calibrating physics-based models at small scales of observation, and (ii) the residual uncertainty in modelling upscaling. The first point is addressed by means of a fuzzy set controller, namely a fuzzy logic-based approach in five steps, for the calibration in case of limited experimental data. Then, a strategy to deal with the residual uncertainty at structural level is presented. Since the latter refers to those uncertainty sources spawning at the macroscale, the label residual is adopted to emphasise that they cannot be treated through the fuzzy set controller at the material level. The residual uncertainty is mainly statistical, due to a size effect-induced randomness. The strategy to cope with it relies on the definition of a fuzzy-probabilistic concept to improve the modelling of degradation mechanisms in cracked state at the structural scale. Such improvement derives from the identification of bridge parameters, which are in charge of transferring information out of the mesoscale models to the macroscale modelling in building members.

3.1 Material Level

The M-LDPM model allows to simulate a wide range of phenomena. However, since its predictive potential is enabled only if the governing parameters are properly calibrated against experimental data, it can be referred to as a descriptive model. The fuzzy set controller approach is firstly conceived as a surrogate for calibration process, though it would permit also to turn M-LDPM in an actual predictive model when limited information are available for the material of interest. Even without a proper calibration against laboratory evidence, by using the fuzzy logic-based idea the governing parameters can be identified, accounting for both scenario-related variables and inherent cross-dependencies among the parameters themselves.

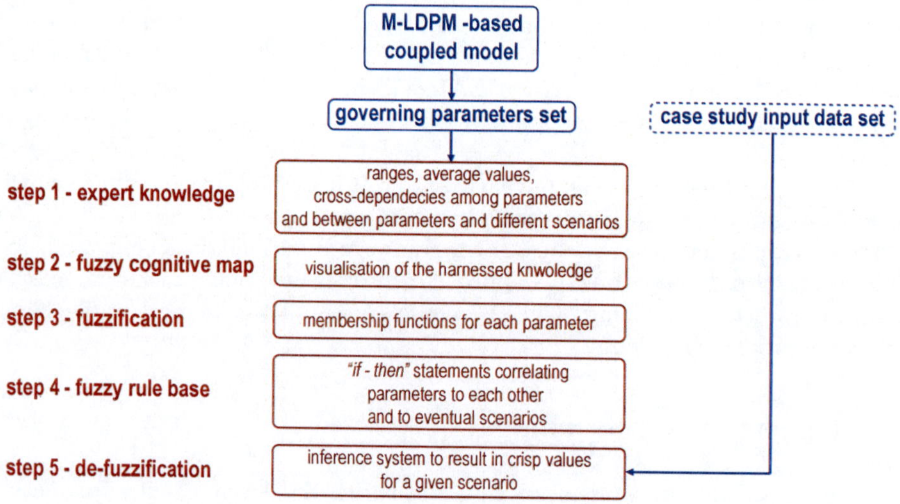

Fig. 1. Fuzzy logic and set theory-based approach for the calibration of the governing parameters in five steps.

The proposed approach consists of five steps (see Fig. 1).

1. Collection of the expert knowledge on the material itself and on the relationships between peculiar aspects, such as material composition, environmental conditions, loading regimes, and the models governing parameters. The latter are previously organised in order to figure out which of them are actually affected by the uncertainty.
2. Definition of a fuzzy cognitive map, namely a fuzzy graph for representing causal reasoning [23], which allows the extraction, analysis, and visualisation of the knowledge based on hard evidence as well as subjective expert judgement.
3. Fuzzification of the parameters, which govern the models adopted, by associating a membership function to each of them. This operation is carried-out for those parameters whose value is affected by a scenario-related uncertainty.
4. Implementation of a fuzzy rule base with if-then conditions. This fuzzy rule base is serving as the main combination between the membership functions, the causal relations, as it is stated in the cognitive map, and the expert knowledge provided by surveys or literature of similar materials.
5. De-fuzzification process: the membership functions, interval limitations and expert knowledge are combined into an inference system to identify a crisp value for each governing parameter.

Expert Knowledge Gathering. Since the experimental investigation and the uncertainty treatment should consider all the unknown parameters, it is important to firstly visualise all the parameters governing the numerical models. The latter are listed in Table 1, separated for the HTC and LDPM models, respectively.

In order to mitigate as much as possible the controller complexity, an investigation on the uncertainty of the governing parameters is performed. In [24] the authors presented the ONIX model (ONly (the) mIX design as input), which extracts the HTC parameters governing the self-desiccation process by taking advantage of simulations at microscale of the hydration process conducted through the CEMHYD3D model [25]. The database behind CEMHYD3D model is deemed to be adequate for the identification of those parameters governing the hydration process modelling. Due to the comprehensive CEMHYD3D database, it is not necessary to implement any additional uncertainty modelling. Using ONIX, the only parameters that need to be identified are the ones related to the moisture transport phenomena.

A literature survey is conducted in order to populate a database gathering the values adopted for the governing parameters for different cementitious materials. Furthermore, for each HTC-, LDPM-, and LDPM-F-based simulation available in the literature, the boundary conditions are collected. This permits to gain a consistent knowledge about the qualitative relationships between, on one side, mixture compositions and testing conditions, and, on the other side, the values calibrated for each parameter.

3.2 Structural Level

The uncertainty model implemented for the material model can be modified and used for all levels of consideration. Nonetheless, the computational performance for highly comprehensive numerical simulations are always limited to a certain specimen size. Therefore, it is required to develop an idea to scale-up the results. Since cracks always have a crucial role in the transport and deterioration processes of reinforced concrete, it is considered as the starting point for the up-scaling approach.

Table 1. Models parameters

Model	Governing parameters
HTC	η_c, $A_{c.1}$, $A_{c.2}$, $E_{a.c}$, $\tilde{\alpha}_c^{\infty}$, ζ_c for the cement hydration
	η_{sl}, $A_{sl.1}$, $A_{sl.2}$, $E_{a.sl}$, $\tilde{\alpha}_{sl}^{\infty}$, ζ_{sl} for the slag reaction
	D_0, D_1, n, E_{ad}, n_c, w_{c0}, w_{c1} for the moisture transport phenomena
	g_1, g_2 for the evaporable water content
	κ_c, κ_{sl} for the non-evaporable water content
	κ_{sh}, $\Delta\tilde{w}_c$, $\Delta\tilde{w}_{sl}$ for the chemical shrinkage
	λ_T, c_T, $\tilde{Q}_c^{\infty}$, $\tilde{Q}_{sl}^{\infty}$ for the heat transport phenomena
	A_λ, n_λ, α_0, T_{ref}, T_{max} for the ageing law
	$\tilde{A}_{sh.1}^{(m)}$, $\tilde{A}_{sh.2}^{(m)}$, $E_{sh}^{(m)}$, $w_{c0.sh}^{(m)}$, T_{ref} for the self-healing of the matrix cracks
	$\tilde{A}_{sh.1}^{(f)}$, $\tilde{A}_{sh.2}^{(f)}$, $E_{sh}^{(f)}$, $w_{c0.sh}^{(f)}$, T_{ref} for the self-healing of the tunnel cracks
	a_w, b_w, a_h, b_h **for the effect of** h and w_c on the self-healing
LDPM	E_0^{∞}, α for the elastic behaviour
	σ_t^{∞}, l_t^{∞} for the softening tensile fracturing behavior
	n_t for the shear-tension interaction during softening
	r_{st}^{∞} for the shear-to-tensile strength ratio
	μ_0, μ_∞, σ_{N0}^{∞} for the frictional behaviour
	σ_{c0}^{∞}, κ_{c0}, κ_{c1}, κ_{c2}, E_d/E_0 for the compression behaviour
	k_t, k_s, k_c for the unloading-reloading in tension, shear and compression
	n_a, m_a, k_a for the parameters ageing
	c_{sh}, γ_{sh} for the mechanical impact of matrix and tunnel cracks self-healing
LDPM-F	G_d, τ_0, β, γ_p for the bond law
	k_{sn}, k_{sp}, k_{rup} for the micromechanical matrix-fibre interaction

In order to transfer information from material scale modelling to the transport model of chlorides into cracked and non-cracked structural elements scale-up/bridge parameters are defined. To the purpose, the fuzzy probabilistic design approach for chloride ingress presented by Altmann et al. in [26] is used as basis for the crack related transport mechanism. This adapted concept provides the possibility to take advantage of the information gained through calibrated simulations at the material level whilst considering the aforementioned residual uncertainty at the structural level: e.g. average crack density or evenness of crack distribution, which might vary due to different size effects.

The proposed bridging parameters for chloride ingress are:

- the crack width distribution;
- the crack space distribution;
- the normalised self-healing reaction degree;
- and the mechanical healing impact factor.

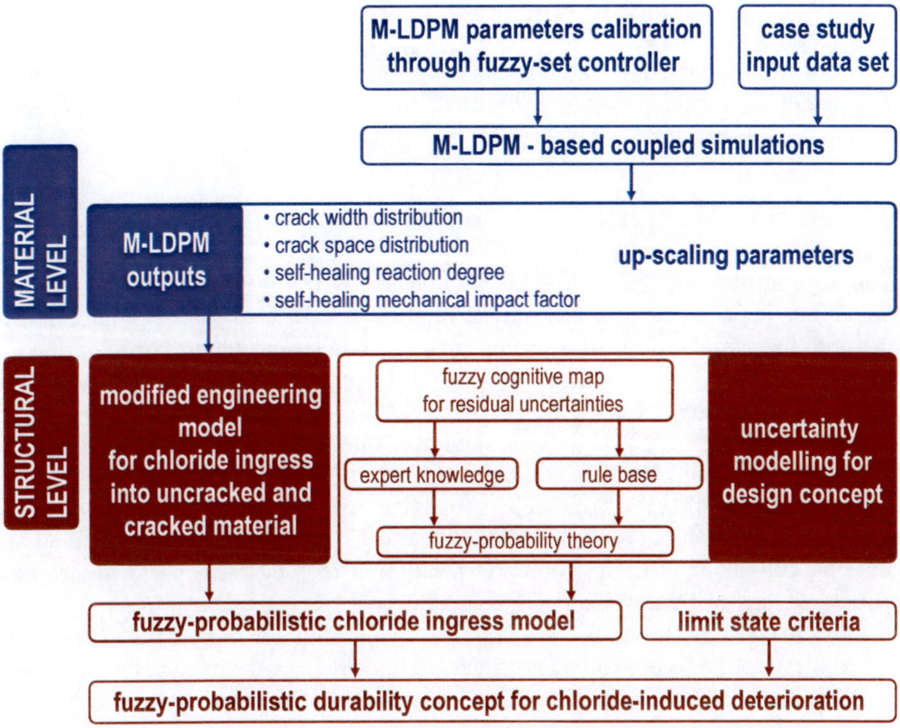

Fig. 2. Schematic approach for implementing up-scaling parameters from material to structural level for chloride ingress.

Crack Width and Crack Spacing Distribution. Cracks have a large influence on the chloride penetration into an UHDC component by acting as "transport highway" for chlorides into structural elements up to the the reinforcement layer. Even in very narrow cracks, the chloride ingress can be much higher in comparison to non-cracked areas, and thus the depassivation of steel reinforcement could be reached faster. Since cracks can trigger accelerated deterioration of a structural element, the crack width distribution was set as one of the four important bridging parameter. The effect of crack spacing is even more difficult to assess, compared to that of crack width: accelerated transport mechanisms are favored in case of closely spaced narrow cracks. In general, hence, it is difficult to provide a valid statement about the influence of proximity of cracks. Nonetheless since it might lead to accelerated deterioration it is considered as one of the up-scaling parameter [27].

Mechanical Regain and Self-Healing Degree. The self-healing degree, in the mesoscale model conceived as measure of cracks closure, might limit the transport mechanism, whereas the mechanical healing impact factor governs the crack healing-induced regain in terms of stiffness and load-bearing capacity. Because of this, both parameters are identified as bridging parameters between meso- and macroscale models. Even though the modelling approach on structural level only considers the chloride transport process, the mechanical regain is considered a bridge parameter because it could be related to the crack closure degree in further investigations.

4 Closing Remarks

The approach presented addresses two relevant issues in cementitious materials modelling, representing an innovation in the field. Due to its fuzzy nature, the methodology is a self-optimising and self-learning tool. Then it might be easily improved as soon as new set of data are published for the models and the materials of interest. At the same time, thanks to the variation ranges based on the expert knowledge collected, the governing parameters resulting from the calibration-free algorithm should never lead to unrealistic numerical simulation results. Furthermore, the approach at the structural level could help in improving the engineering models for practitioners and professionals, especially for considering the effect of damage on the prediction of long-term performance and structures design in aggressive environments.

The fuzzy logic-based approach will be validated for one of the Ultra High Durability Concrete mixtures identified within the activity of the H2020 project ReSHEALience. The accuracy of the fuzzy logic-based predictions will be then compared to the values of the models' parameters, gained through the calibration against experimental data.

Acknowledgements. The work described in this paper has been performed in the framework of the project ReSHEALience - Rethinking coastal defence and green-energy Service infrastructures through enHancEd-durAbiLity high-performance cement-based materials, whose funding the authors gratefully acknowledge. This project has received funding from the European Union Horizon 2020 research and innovation programme under grant agreement No 760824. The information and views set out in this publication do not necessarily reflect the official opinion of the European Commission. Neither the European Union institutions and bodies nor any person acting on their behalf, may be held responsible for the use which may be made of the information contained therein.

References

1. Di Luzio, G., Ferrara, L., Alonso y Alonso, M.C., Kunz, P., Mechtcherine, V., Schröfl, C.: Predicting the long-term performance of structures made with advanced cement-based materials in extremely aggressive environments: current state of practice and research needs - the approach of H2020 project ReSHEALience. In: Symposium on Concrete Modelling, CONMOD 2018, vol. 2018, pp. 467–477. RILEM Publications SARL (2018)
2. Cuenca, E., Mezzena, A., Ferrara, L.: Synergy between crystalline admixtures and nano-constituents in enhancing autogenous healing capacity of cementitious composites under cracking and healing cycles in aggressive waters. Constr. Build. Mater. **266**(B):121447 (2021)
3. Wan, L., Wendner, R., Liang, B., Cusatis, G.: Analysis of the behavior of ultra high performance concrete at early age. Cement Concr. Compos. **74**, 120–135 (2016)
4. Pathirage, M., Bousikhane, F., D'Ambrosia, M., Alnaggar, M., Cusatis, G.: Effect of alkali silica reaction on the mechanical properties of aging mortar bars: experiments and numerical modeling. Int. J. Damage Mech. **28**(2), 291–322 (2019)
5. Alnaggar, M., Di Luzio, G., Cusatis, G.: Modeling time-dependent behavior of concrete affected by alkali silica reaction in variable environmental conditions. Materials **10**(5), 291–322 (2017)
6. Abdellatef, M., Alnaggar, M., Boumakis, G., Cusatis, G., Di Luzio, G., Wendner, R.: Lattice discrete particle modeling for coupled concrete creep and shrinkage using the solidification microprestress theory. In: CONCREEP, vol. 10, pp. 184–193 (2015)
7. Angiolilli, M., Pathirage, M., Gregori, A., Cusatis, G.: Lattice discrete particle model for the simulation of irregular stone masonry. J. Struct. Eng. **147**(9), 04021123 (2021)
8. Li, W., Bousikhane, F., Carey, J.W., Cusatis, G.: Discrete modeling of the fracture-permeability behavior of shale. In: 51st US Rock Mechanics/Geomechanics Symposium, OnePetro (2017)
9. Cibelli, A., Pathirage, M., Cusatis, G., Ferrara, L., Di Luzio, G.: A discrete numerical model for the effects of crack healing on the behaviour of ordinary plain concrete: Implementation, calibration, and validation. Eng. Fract. Mech. **263**, 108266 (2022)
10. Cibelli, A., Ferrara, L., Di Luzio, G.: A discrete numerical model for the effects of crack healing on the behaviour of fibre reinforced concrete: Implementation, calibration, and validation. Unpublished
11. Cusatis, G., Pelessone, D., Mencarelli, A.: Lattice Discrete Particle Model (LDPM) for failure behavior of concrete. I: theory. Cement Concr. Compos. **33**(9), 881–890 (2011)
12. Schauffert, E.A., Cusatis, G.: Lattice discrete particle model for fiber-reinforced concrete. I: theory. J. Eng. Mech. 138(7), 826–833 (2012)
13. Di Luzio, G., Cusatis, G.: Hygro-thermo-chemical modeling of high performance concrete. I: theory. Cement Concr. Compos. **31**(5), 301–308 (2009)
14. Di Luzio, G., Cusatis, G.: Hygro-thermo-chemical modeling of high performance concrete. II: numerical implementation, calibration, and validation. Cement Concr. Compos. **31**(5), 309–324 (2009)
15. Di Luzio, Cusatis, G.: Solidification-microprestress-microplane (SMM) theory for concrete at early age: theory, validation and application. Int. J. Solids Struct. **50**(6), 957–975 (2013)

16. Möller, B., Beer, M.: Fuzzy Randomness. Uncertainty in Civil Engineering and Computational Mechanics, 2nd Edition. Springer Science & Business Media, Berlin, Germany (2004)
17. Altmann, F., Mechtcherine, V.: Modelling chloride ingress into cracked and crack-free Strain-Hardening Cement-based Composites (SHCC). In: Ferreira, R.M., Gulikers, J., Andrade, C. (eds.) 3rd International PhD Workshop on Modelling the Durability of Reinforced Concrete, pp. 75–83. RILEM Publications SARL Guimarães, Portugal (2009)
18. U.S. Department of commerce: Fuzzy logic: A key technology for future competitiveness (1991)
19. Kosko, B.: Fuzzy thinking: The new science of fuzzy logic. Hyperion (1993)
20. Götz, M., Graf, W., Kaliske, M.: Enhanced uncertain structural analysis with time- and spatial-dependent (functional) fuzzy results. Mech. Syst. Signal Process. **119**, 23–38 (2019)
21. Weber, W.E., Reuter, U.: Fuzzy modeling of wave-shielding under consideration of cost-effectiveness for an efficient reduction of uncertainty. Adv. Eng. Softw. **109**, 53–61 (2017)
22. Akkurta, S., Tayfur, G., Can, S.: Fuzzy logic model for the prediction of cement compressive strength. Cem. Concr. Res. **34**(8), 1429–1433 (2004)
23. Altmann, F.: A Durability Concept for Strain-Hardening Cement-Based Composites, PhD Thesis, Institute of Construction Materials, Faculty of Civil Engineering, Technische Universität Dresden, Desden, Germany (2012)
24. Pathirage, M., Bentz, D.P., Di Luzio, G., Masoero, E., Cusatis, G.: The ONIX model: a parameter-free multiscale framework for the prediction of self-desiccation in concrete. Cement Concr. Compos. **103**, 36–48 (2019)
25. Bentz, D.P.: CEMHYD3D: A Three-Dimensional Cement Hydration and Microstructure Development Modeling Package. Version 3.0. NIST Interagency/Internal Report (NISTIR), National Institute of Standards and Technology, Gaithersburg, Maryland, United States (2005)
26. Altmann, F., Sickert, J.-U., Mechtcherine, V., Kaliske, M.: A fuzzy-probabilistic durability concept for strain-hardening cement-based composites (SHCCs) exposed to chlorides Part 1: Concept development. Cement Concr. Compos. **34**, 754–762 (2012)
27. Kunz, P., Mechtcherine, V.: Diffusion of chlorides in cracked Strain-Hardening-Cement-Based-Composites (SHCC), Proceedings of the 4th International RILEM Conference on Microstructure Related Durability of Cementitious Composites (2020)

Bayesian Inverse Modelling of Early-Age Stress Evolution in GGBFS Concrete Due to Autogenous Deformation and Aging Creep

Minfei Liang(✉), Erik Schlangen, and Branko Šavija

Microlab, Faculty of Civil Engineering and Geosciences, Delft University of Technology, 2628 Delft, The Netherlands
{M.Liang-1,Erik.Schlangen,B.Savija}@tudelft.nl

Abstract. Stress evolution of restrained concrete is directly related to early-age cracking (EAC) potential of concrete, which is a tricky problem that often happens in engineering practice. Due to the global objective of carbon reduction, Ground granulated blast furnace slag (GGBFS) concrete has become a more promising binder comparing with Ordinary Port-land Cement (OPC). Although GGBFS concrete produces less hydration heat which further prevents thermal shrinkage, the addition of GGBFS highly increases the autogenous shrinkage and thus increases EAC risk. This study presents experiments and numerical modelling of the early-age stress evolution of GGBFS concrete, considering the development of autogenous deformation and creep. Temperature Stress Testing Machine (TSTM) tests were conducted to obtain the autogenous deformation and stress evolution of restrained GGBFS concrete. By a self-defined material sub-routine based on the Rate-type creep law, the FEM model for simulating the stress evolution in TSTM tests was established. By characterizing the creep compliance function with a 13-units continuous Kelvin chain, forward modelling was firstly conducted to predict the stress development. Then inverse modelling was conducted by Bayesian Optimization to efficiently modify the arbitrary assumption of the codes on the aging creep. The major findings of this study are as follows: 1) the high autogenous expansion of GGBFS induces compressive stress at first hours, but its value is low because of high relaxation and low elastic modulus; 2) The codes highly underestimated the early-age creep of GGBFS concrete. They performed well in prediction of stress after 200 h, but showed significant gaps in predictions of early-age stress evolution; 3) The proposed inverse modelling method with Bayesian Optimization can efficiently adjust the aging terms which produced best modelling results. The adjusted creep compliance function of GGBFS showed a much faster aging speed at early ages than the one proposed by original codes.

Keywords: Concrete · Bayesian Optimization · Early age cracking · Creep · Relaxation · Autogenous shrinkage

1 Introduction

Early-age cracking (EAC) is one of the trickiest problems happened to early-age concrete structures. The hydration-induced volume shrinkage (i.e., autogenous shrinkage) and

P. Rossi and J.-L. Tailhan (Eds.): SSCS 2022, RILEM Bookseries 38, pp. 207–217, 2023.
https://doi.org/10.1007/978-3-031-07746-3_21

external restraint result in accumulation of tensile stress, which may exceed the tensile strength of concrete and cause EAC. During this process, the evolution of elastic modulus, shrinkage, creep and environmental conditions (e.g., temperature) together determines the buildup of tensile stress.

Ground granulated blast furnace slag (GGBFS) has been used as a popular supplementary cementitious material (SCMs), which can highly decrease the hydration heat but increase autogenous shrinkage [1, 2]. Thereby, for EAC analysis of GGBFS concrete, the autogenous shrinkage appears to be a more influential factor than thermal shrinkage. Another factor that strongly influences the EAC risk is the early-age creep/relaxation, which directly influences the accumulation of tensile stress and therefore should not be neglected in all cases of EAC analysis [3]. Comparing to elastic modulus, early-age creep is much more difficult to measure, as the effects of hydration process and creep/relaxation cannot be fully decoupled, especially in early-age. A potential solution is minutes-repeated short-term creep test, because the hydration effects can be neglected in such a short-term [4]. However, minutes-long short-term creep tests are not likely to fully reflect the creep behaviour of concrete in a longer time-range. EAC analysis requires the input of creep compliance function *J (t, t')* (or relaxation modulus *R (t, t')*) that is continuous at time of loading *t'* and longing duration *t*.

To evaluate the EAC risk, a number of restraint tests have been performed, such as rigid cracking frame test, internal restraint test, ring test, and temperature-stress testing machine test (TSTM) [5–9]. Among these tests, TSTM stands out due to its advantages of tunable temperature control, loading scheme and restraint degree. In this paper, we aim to establish a modelling framework for early-age stress evolution due to autogenous shrinkage and aging creep. Based on Rate-Type Creep Law and exponential algorithm, the creep/relaxation of early-age concrete is quantified by a 13-Unit Kelvin Chain. Then, based on Bayesian Optimization and stress results of TSTM tests, the aging creep is quantified efficiently within the framework of EURO code and ACI code.

2 Methods

The stress evolution of restrained concrete is the basic and direct index for evaluating the EAC risk. Based on Boltzmann superposition, the stress evolution initiated from different time of loading t_0 by shrinkage and creep is quantified. However, integrating the whole stress history poses to be a computational dilemma because it requires to save all stress results at every step and FEM elements. Thereby, this paper incorporates the Rate-type Creep Law, which avoids the complex stress integration and only needs to solve an incremental quasi-elastic constitutive equation at every step [10, 11]. Furthermore, in view of the unattainable experimental input for aging creep, this paper conducts Bayesian Optimization to achieve fast and efficient inverse modelling for the continuous creep compliance function *J (t, t')*.

2.1 Viscoelastic Material Subroutine

Based on the Boltzmann superposition, the creep strain can be expressed by the following integration:

$$\varepsilon(t) = \int_0^t J\left(t, t'\right)\mathrm{d}\sigma\left(t'\right) \tag{1}$$

where J is the creep compliance function. Writing Eq. (1) in an incremental form and approximate a linear stress variation in each time interval, one gets the aforementioned quasi-elastic constitutive equation that needs to be solved at every time step:

$$\Delta\sigma = E^* \Delta\varepsilon - \sigma^* \tag{2a}$$

$$E^* = \frac{\Delta t}{\int_{t_i}^{t_{i+1}} J(t_{i+1}, t')dt'} \tag{2b}$$

$$\sigma^* = E^* \int_0^{t_i} \left[J\left(t_{i+1}, t'\right) - J\left(t_i, t'\right)\right]\dot{\sigma}\, dt' \tag{2c}$$

where $\Delta\sigma$ and $\Delta\varepsilon$ are difference of stress and strain between two consecutive time steps t_i and t_{i+1}. The creep compliance curve can be expressed as a Dirichlet series, which can be the governing equations of Kelvin chain rheological model with sets of spring and dashpot:

$$J\left(t, t'\right) = \frac{1}{E_0(t')} + \sum_{j=1}^{N} \frac{1}{E_j(t')}\left(1 - e^{-\frac{t-t'}{\mu_j}}\right) \tag{3}$$

where N is the number of Kelvin chain units; E_j and μ_j is the elastic modulus and retardation time of j-th Kelvin chain units. Substituting Eq. (3) in Eq. (2) and calculate the integral by mid-point rule, one can obtain the E^* and σ^* in Eq. (2) as follows [10, 11]:

$$E^*\left(t^*\right) = \frac{1}{\frac{1}{E_0(t^*)} + \sum_{j=1}^{N} \frac{1}{E_j(t^*)}\left(1 - \left(1 - e^{-\frac{\Delta t}{\mu_j}}\right)\frac{\mu_j}{\Delta t}\right)} \tag{4a}$$

$$\sigma^*(t_i) = E^*\left(t^*\right)\sum_{j=1}^{N}\left(1 - e^{-\frac{\Delta t}{\mu_j}}\right)\varepsilon_j^*(t_i) \tag{4b}$$

$$\varepsilon_j^*(t_i) = \int_0^{t_i} \frac{1}{E_j(t')} e^{-\frac{t_i - t'}{\mu_j}}\, \dot{\sigma}\, dt' \tag{4c}$$

$$\varepsilon_j^*(t_{i+1}) = e^{-\frac{\Delta t}{\mu_j}}\varepsilon_j^*(t_i) + \frac{1}{E^*(t_i)}\left(1 - e^{-\frac{\Delta t}{\mu_j}}\right)\frac{\mu_j}{\Delta t}\Delta\sigma \tag{4d}$$

where t^* is the average of two consecutive time steps t_i and t_{i+1}. Equation (2a) and Eq. (4) form the incremental quasi-elastic constitutive equation of this paper.

2.2 FEM Configuration

Incorporating the influence of autogenous shrinkage, the aforementioned quasi-elastic constitutive equation Eq. (2a) becomes:

$$\Delta\sigma = E^*(\Delta\varepsilon - \Delta\varepsilon_{ad}) - \sigma^* \tag{5}$$

where $\Delta\varepsilon_{ad}$ is the autogenous strain happening in each time interval. As shown in Eqs. (4) and (5), there are mainly 3 input parameters: 1) Elastic modulus, 2) autogenous shrinkage and 3) Creep compliance curve characterized by Kelvin chain parameters. The elastic modulus is computed by the empirical formulas of Model Code 2010 [12] based on the compressive strength test, and the autogenous shrinkage is directly extracted from ADTM tests, which are presented in our dual paper submitted to this conference and previous works [13, 14]. The creep compliance formula of ACI and EURO codes are tentatively used as input of this paper [15, 16]. Their creep compliance formula can be expressed as:

$$J(t, t') = \frac{1}{E(t')} + C_0 C_1(t') C_2(\xi) \tag{6}$$

where C_0 is a coefficient related to general concrete properties and environmental conditions; C_1 is a power function to describe the aging of creep compliance; C_2 is a power function representing the non-aging term, which mainly depends on the time length of loading ξ ($\xi = t - t'$). To fit the Eq. (10) into a Dirichlet series as in Eq. (3), the procedure proposed by Bazant is adopted, which guarantees a unique and stable solution when fitting a continuous spectrum of Kelvin chain parameters [17–19]. Firstly, the retardation time μ_j is chosen as a priori to prevent ill-conditioned equation system as follows:

$$\mu_j = 10^{-6+j}, \ j = 1:13 \tag{7}$$

The continuous form of the non-aging term is expressed as follows:

$$C_2(\xi) = \int_0^\infty \frac{1}{E_j}\left(1 - e^{-\frac{\xi}{\mu_j}}\right) d(ln\mu_j) \tag{8}$$

Using the Laplace transform and Widder's formula, the solutions of E_j can be derived as:

$$\frac{1}{E_j} = -ln10 * \lim_{k\to\infty} \frac{(-k\mu)^k}{(k-1)!} C_2^{(k)}(k\mu) \tag{9}$$

In this paper, the spectrum of third order (k = 3) is used. The fitting results of the mixes with w/c = 0.35 based on EURO code is shown as an example in Fig. 1, which shows that the fitted Kelvin chain can mimic the codes with good precision.

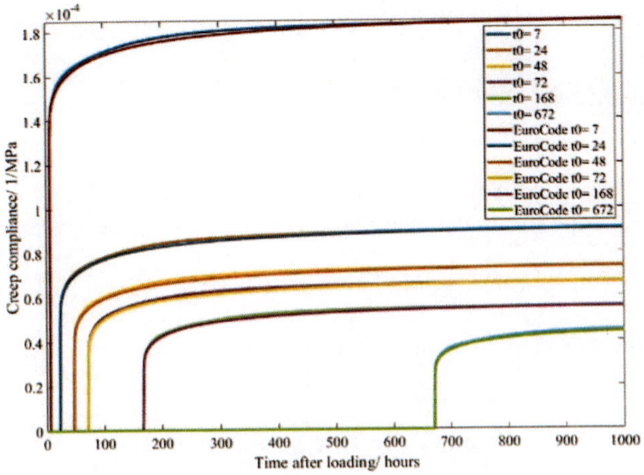

Fig. 1. Fitting results of Kelvin chain with codes (Euro Code w/c = 0.35)

Having set up the constitutive equation and input parameters, the TSTM tests can be simulated properly. The dog-bone specimen used in TSTM, and corresponding mesh and boundary settings are shown in Fig. 2. Roller boundaries are attached to the highlighted purple area in Fig. 2(b) (i.e., bottom sides and the lateral sides of the two ends), which ensues zero displacement in the normal direction of the boundary surface.

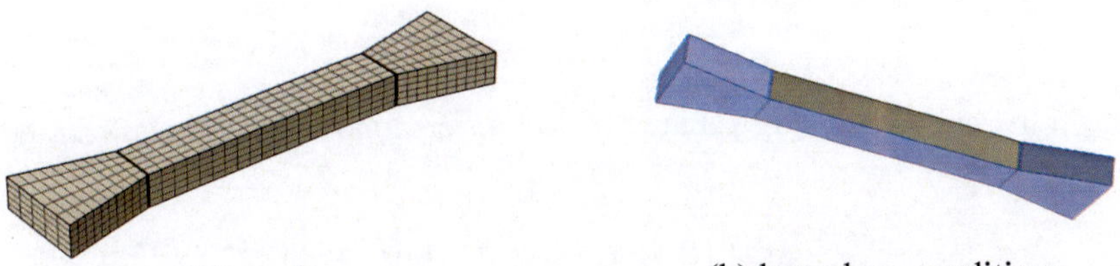

(a) mesh (b) boundary conditions

Fig. 2. Numerical dog-bone specimen

The modelling result of stress evolution σ_M is calculated at the middle section of the specimen and expressed as follows:

$$\sigma_M(t) = \frac{\iint \sigma_{xx} dydz}{A} \tag{10}$$

where σ_{xx} is the component of stress tensor in xx direction (i.e., axial direction of dog-bone specimen); A is the cross-section area. The metric Root Mean Squared Error (RMSE) is adopted to quantify the modelling accuracy by averaging the residual error at each time step:

$$RMSE = \sqrt{\frac{\sum_t (\sigma_M(t) - \sigma_T(t))^2}{t_{total}}} \tag{11}$$

where $\sigma_M(t)$ and $\sigma_T(t)$ is modelling and testing results of stress at time step t.

2.3 Bayesian Inverse Modelling

To cope with the testing dilemma of aging creep tests, this study implements Bayesian Inverse Modelling to derive a continuous aging creep compliance surface. Based on Sects. 2.1 and 2.2, the following black-box function can be obtained:

$$RMSE = f\left[\varepsilon_{ad}(t), E(t), J\left(t, t'\right), \sigma_T(t)\right] \tag{12}$$

Then the inverse modelling process can be defined as the following optimization process:

$$\arg\min f\left\{J\left(t, t'\right)|\varepsilon_{ad}(t), E(t), \sigma_T(t)\right\} \tag{13}$$

Recalling the creep compliance formulas given by the codes in Eq. (7), one can find that the constant term C_0 and non-aging term C_2 both depends on specific mi parameters and environmental conditions, while the aging term C_1 remains unchanged for any mixes and environmental conditions. The codes cannot reflect the difference of aging creep of different mixes [20, 21]. Thereby, the inverse modelling process focuses on the aging term C_1, which can be formulated as follows for both codes:

$$\text{Euro code}: \ C_1\left(t'\right) = \frac{c}{b + t'^a}; \tag{14a}$$

$$\text{ACI code}: \ C_1\left(t'\right) = \frac{c}{t'^a} \tag{14b}$$

where a, b, and c are fitting parameters. The inverse modelling process can then be described as follows:

$$\arg\min f\left\{C_1\left(t', a, b, c\right)|\varepsilon_{ad}(t), E(t), \sigma_T(t), C_0, C_2\left(t - t'\right)\right\} \tag{15}$$

Implementing the optimization process described by Eq. (15), one has to reiterate the FEM model described in Eq. (12) to find the lowest RMSE optima. This process can be inefficient when coping with complex FEM models which requires considerable amount of time to run. Thereby, we propose to implement Bayesian Optimization to achieve fast solution of Eq. (15). Bayesian Optimization is composed by two major parts: a Gaussian Process predictor and an acquisition function [22, 23], as shown in Fig. 3.

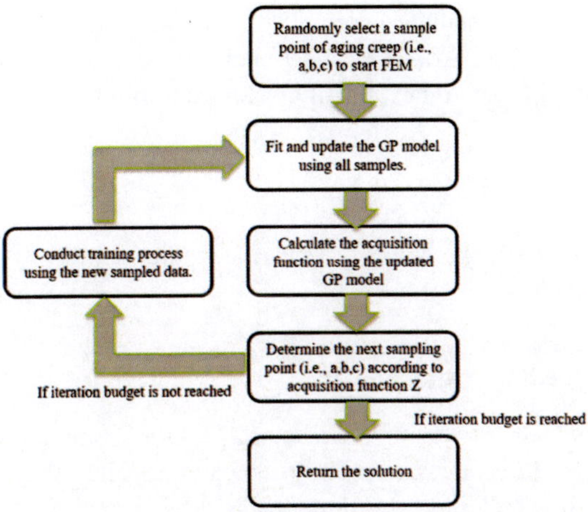

Fig. 3. Workflow of Bayesian optimization

Assuming the sampling points *X (a,b,c)* and corresponding *RMSE* of each FEM trial follows the multivariate Gaussian distribution, denoted as $D_{1:n} = (X_{1:n}, Y_{1:n})$, the Bayesian inference of next sampling points can be expressed as:

$$Y_{n+1}|\mathrm{D}_{1:n} \sim \mathrm{N}\left(\mu(\mathrm{X}_{n+1}), \sigma^2(\mathrm{X}_{n+1}) + \sigma^2_{noise}\right) \tag{16}$$

in which

$$\mu(\mathrm{X}_{n+1}) = \boldsymbol{k}^T K^{-1} Y_{1:n} \tag{17a}$$

$$\sigma^2(\mathrm{X}_{n+1}) = k(\mathrm{X}_{n+1}, \mathrm{X}_{n+1}) - \boldsymbol{k}^T\left(K + \sigma^2_{noise}\boldsymbol{I}\right)^{-1}\boldsymbol{k} \tag{17b}$$

$$\boldsymbol{k} = \left[k(X_{n+1}, X_1)k(X_{n+1}, X_2)\cdots k(X_{n+1}, X_n)\right] \tag{17c}$$

$$K = \begin{bmatrix} k(X_1, X_1) & \cdots & k(X_1, X_n) \\ \vdots & \ddots & \vdots \\ k(X_n, X_1) & \cdots & k(X_n, X_n) \end{bmatrix} + \sigma^2_{noise}\boldsymbol{I} \tag{17d}$$

where k is the covariance kernel between any two sampling points X_i and X_j, calculated by an exponential function of the second norm of the difference value between two samples; $\boldsymbol{k}$ is a n-by-1 covariance matrix of the new sampling point X_{n+1} and n sampling points $X_1 \sim X_n$; K is the n-by-n covariance matrix of any two sampling points assembled by k. With the GP predictor (Eqs. 16 and 17), the inference of RMSE based on any combinations of (a,b,c) can be made. Then, the acquisition function Expected Improvement is adopted to infer the improving potential of every possible sample point, expressed as below:

$$I(X_n) = \left(y_{best} - \mu(\mathrm{X}_n)\right)\Phi\left(\frac{y_{best} - \mu(\mathrm{X}_n)}{\sigma(\mathrm{X}_n)}\right) + \sigma(\mathrm{X}_n)\phi\left(\frac{y_{best} - \mu(\mathrm{X}_n)}{\sigma(\mathrm{X}_n)}\right) \tag{18}$$

where $\Phi(\cdot)$ and $\phi(\cdot)$ are the standard normal density and distribution function; y_{best} is the tentative optimal value (i.e., lowest RMSE) in current sample space. At evert iteration, the sampling point with the highest I will be chosen to run the next FEM simulation.

3 Results and Discussion

In this section, the modelling results of the testing results of w/c = 0.35 in our dual paper [13] will be presented.

3.1 Forward Modelling with Codes

The modelling results directly using the creep formulas of codes are shown in Fig. 4. An obvious gap between the code results and tests can be seen at very early-age, approximately before 150 h, which indicates that the codes' assumptions of the constant aging pattern for concrete creep are not applicable, as expected. However, after 150 h, the curves of modelling stress tend to be more parallel with the testing results, which indicates that both codes show good conformity with TSTM testing results at a latter age. Therefore, most modelling error lie in the early-age creep inference made by codes. If the time-zero of the models and tests are fixed at a time point after 150 h, good conformity between the modelling and testing results can be expected. But in the meantime, this will also cause significant errors since the stress before 200 h is overlooked.

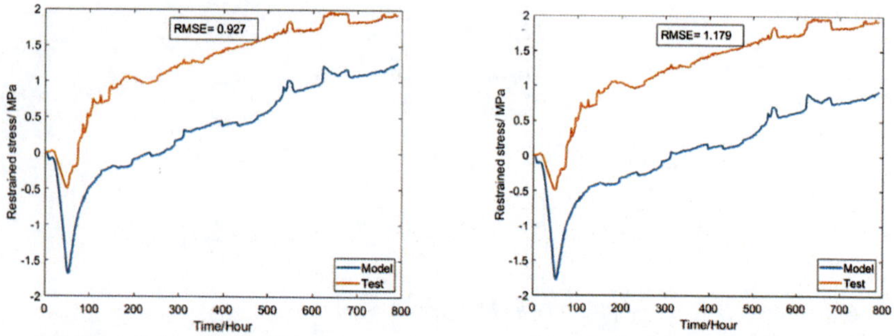

(a) model with ACI creep (b) model with EURO creep

Fig. 4. Simulated stress evolution with different codes for creep compliance

3.2 Inverse Modelling with Adjusted Codes

Following the procedure described in Sect. 2.3, the adjusted terms for aging creep can be obtained. The optimization history is shown in Fig. 5. The optimization history shows a fast convergence of this method, within around 5 steps for ACI and 12 steps for EURO code. The reason why EURO code needs more computation budget is that its aging term has 3 parameters, while the ACI only has 2. Figure 5(b) shows that the sampling points tends to be much denser when approaching to the minima, and more diluted

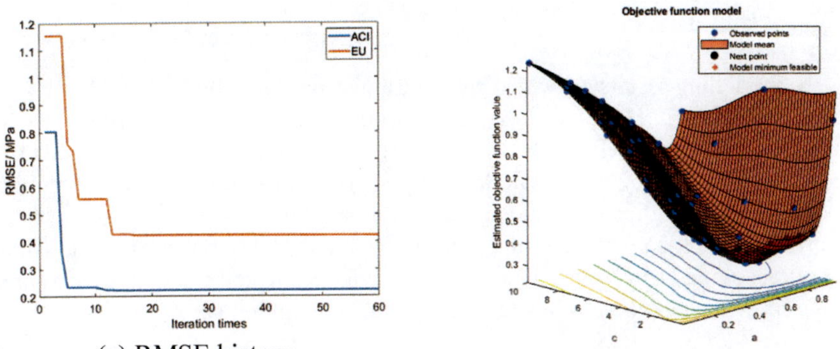

(a) RMSE history

(b) Response surface

Fig. 5. Optimization history

when approaching to the peaks, which guarantees the tradeoff between exploitation and exploration and therefore ensure the efficiency.

Accordingly, the modelling results with the adjusted aging terms are shown in Fig. 6. The results show obvious improvement comparing with results based on codes. Most of the improvement is on early-age stress. However, the modelling results are still not perfect, this can be explained by the following potential reasons: 1) the experimental curves are not smooth due to friction and temperature fluctuation. With the objective goal defined in Eq. (15), the model has to make compromise at every time points of the experimental curves; 2) the aging function proposed by the codes are different and may not suitable to fully reflect the aging pattern of concrete creep. As shown in the results based on original codes, the aging term creates gaps between first 150 h, while works well on a time-range after 150 h. However, in the inverse modelling process, the changes on parameters (a, b, c) change the whole aging function, which means that although the improvement at first 150 h can be made, the good performance of original code after 150 h is also compromised. The final results based on the specific function forms are the tradeoff described above and therefore are not perfect.

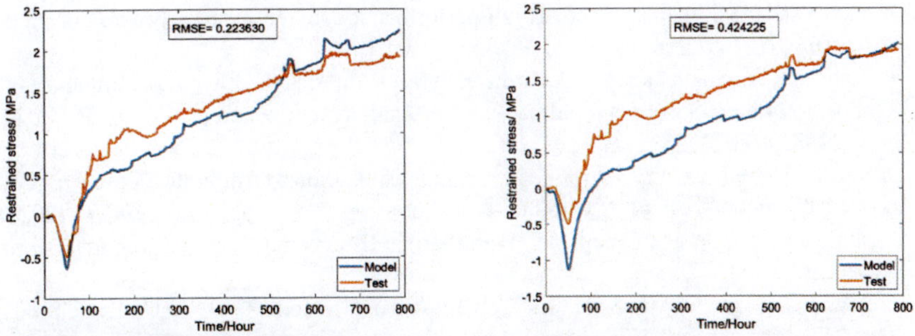

(a) model with ACI creep

(b) model with EURO creep

Fig. 6. Modelling results with adjusted aging term of creep

4 Conclusion

1) The modelling results show that creep formulas proposed by ACI and EURO codes performs well in later ages (after 200 h in this paper), which in turn validated the proposed FEM model in modelling the stress evolution of restrained concrete.
2) Due to the arbitrary settings in the aging of creep, modelling results based on both codes show significant gap at early ages before 200 h.
3) Inverse modelling based on Bayesian Optimization can guarantee efficiently find the optimal aging terms of creep compliance formulas, which can then used to predict the stress development of restrained concrete with higher accuracy.

References

1. Tianshi, L., Li, Z., Huang, H.: Effect of supplementary materials on the autogenous shrinkage of cement paste. Materials **13**, 3367 (2020)
2. Darquennes, A., Khokhar, M.I.A., Rozière, E., Loukili, A., Grondin, F., Staquet, S.: Early age deformations of concrete with high content of mineral additions. Constr. Build. Mater. **25**, 1836–1847 (2011)
3. Azenha, M., Kanavaris, F., Schlicke, D., et al.: Recommendations of RILEM TC 287-CCS: thermo-chemomechanical modelling of massive concrete structures towards cracking risk assessment. Mater. Struct. **54**, 135 (2021)
4. Irfan-ul-Hassan, M., Pichler, B., Reihsner, R., Hellmich, C.: Elastic and creep properties of young cement paste, as determined from hourly repeated minute-long quasi-static tests. Cem. Concr. Res. **82**, 36–49 (2016)
5. Spingenschmid, R.: Prevention of Thermal Cracking in Concrete at Early Ages. E&FN Spon, London (1998)
6. Semianiuk, V., Tur, V., Herrador, M.F.: Early age strains and self-stresses of expansive concrete members under uniaxial restraint conditions. Constr. Build. Mater. **131**, 39–49 (2017)
7. Briffaut, M., Benboudjema, F., D'Aloia, L.: Effect of fibres on early age cracking of concrete tunnel lining. Part I: Laboratory ring test. Tunnel. Underground Space Technol. **59**, 215–220 (2016)
8. Shen, D., Jiang, J., Shen, J., Yao, P., Jiang, G.: Influence of curing temperature on autogenous shrinkage and cracking resistance of high-performance concrete at an early age. Constr. Build. Mater. **103**, 67–76 (2016)
9. Markandeya, A., Shanahan, N., Gunatilake, D.M., Riding, K.A., Zayed, A.: Influence of slag composition on cracking potential of Slag-Portland cement concrete. Constr. Build. Mater. **164**, 820–829 (2018)
10. Di Luzio, G., Cedolin, L., Beltrami, C.: Tridimensional long-term finite element analysis of reinforced concrete structures with rate-type creep approach. Appl. Sci. **10**, 4772 (2020)
11. Bažant, Z.P., Jirásek, M.: Creep and Hygrothermal Effects in Concrete Structures. Springer, Dordrecht (2018)
12. International Federation for Structural Concrete (fib), Fib Model Code for Concrete Structures 2010. Ernst & Sohn, Wiley, Berlin (2013)
13. Schlangen, E., Liang, M., Šavija, B.: The influence of autogenous shrinkage and creep on the risk of early age cracking. In: RILEM International Conference Numerical Modelling Strategies for Sustainable Concrete Structures, Marseille (2022)

14. Liang, M., et al.: Stress evolution in restrained GGBFS concrete due to autogenous deformation: Bayesian optimization of aging creep. Constr. Build. Mater. **324**, 126690 (2022)
15. EN. EN 1992-1-1 Eurocode 2: Design of Concrete Structures—Part 1-1: General Rules and Rules for Buildings; CEN: Brussels, Belgium (2005)
16. American Concrete Institute Committee 209 (ACI). Guide for Modeling and Calculating Shrinkage and Creep in Hardened Concrete; ACI report 209.2R-08; ACI: Farmington Hills, MI, USA (2008)
17. Bažant, Z.P., Carol, I.: Viscoelasticity with aging caused by solidification of nonaging costituent. J. Eng. Mech. ASCE **119**, 2252–2269 (1993)
18. Bažant, Z.P., Xi, Y.: Continuous retardation spectrum for solidification theory of concrete creeps. J. Eng. Mech. ASCE **121**, 281–288 (1995)
19. Bažant, Z.P.: Creep and Shrinkage in Concrete Structures; Mathematical Modeling of Creep and Shrinkage of Concrete. Wiley, New York (1982)
20. Delsaute, B., Torrenti, J.-M., Staquet, S.: Modeling basic creep of concrete since setting time. Cement Concr. Compos. **83**, 239–250 (2017)
21. Liang, M., Chang, Z., Wan, Z., Gan, Y., Schlangen, E., Savija, B.: Interpretable ensemble-machine-learning models for predicting creep behavior of concrete. Cement Concr. Compos. **125**, 104295 (2022)
22. Mockus, J., Tiesis, V., Zilinskas, A.: The application of Bayesian methods for seeking the extremum. Towards Global Optim. **2**, 117–129 (1978)
23. Jones, D.R.: A taxonomy of global optimization methods based on response surfaces. J. Global Optim. **21**(4), 345–383 (2001)
24. Jones, D.R., Schonlau, M., Welch, W.J.: Efficient global optimization of expensive black-box functions. J. Global Optim. **13**, 455–492 (1998)

Modeling C-S-H Sorption at the Molecular Scale: Effective Interactions, Stability, and Cavitation

Fatima Masara(✉), Tulio Honorio, and Farid Benboudjema

Université Paris-Saclay, CentraleSupélec, ENS Paris-Saclay, CNRS, LMPS - Laboratoire de Mécanique Paris-Saclay, 91190 Gif-sur-Yvette, France
fatima.masara@ens-paris-saclay.fr

Abstract. The behavior of confined water molecules in C-S-H has a great influence on various physical and chemical properties of C-S-H gel, which further determine the macroscale behavior of cement-based materials such as creep, shrinkage, and cracking. Here, using molecular simulations, we investigate the effect of relative humidity (RH) on the behavior of C-S-H at the molecular scale taking as reaction path the interlayer distance (spanning interlayer pores up to small gel pores). The confining pressures, desorption isotherm, the potential of mean force (PMF), stable basal spacings, meta-stable domains, elastic modulus perpendicular to the pore surface, and cavitation of nano-confined water are analyzed. We evaluate these properties as a function of interlayer distance at various RH, ranging from (liquid) saturated (RH = 100%) to completely dried (RH = 0%) conditions at ambient temperature (300 K). From the PMF profiles and pressure isotherms, we can identify equilibrium basal spacings and meta-stable domains. We observe that the stable basal spacing decreases when the RH decreases, therefore interlayer pore shrinkage contributes to drying shrinkage of cement-based materials. We also show that cavitation of water in small C-S-H interlayer spaces is pore size-dependent. Each of these properties can be useful to explain the physical origins of the thermo-hygro-mechanical behavior of cement-based materials and provide a methodology to improve the performance of these materials.

Keywords: C-S-H · Desorption · Grand canonical Monte Carlo simulations · Molecular dynamics · Disjoining pressure · Cavitation · Hydration states

1 Introduction

Calcium silicate hydrate (C-S-H) is the main product of ordinary Portland cement hydration. C-S-H contributes significantly to the mechanical behavior and durability of cement-based materials. A fundamental understanding of C-S-H structure and its behavior is therefore crucial to understanding the macroscale behavior of the material. Despite the recent advances in understanding C-S-H atomic structure and behavior bottom-up, important questions remain answered regarding the behavior of this phase under RH changes and underlying mechanisms governing the drying shrinkage in cementitious materials [1].

P. Rossi and J.-L. Tailhan (Eds.): SSCS 2022, RILEM Bookseries 38, pp. 218–229, 2023.
https://doi.org/10.1007/978-3-031-07746-3_22

Experimental evaluations of the structure and the characteristics of C-S-H is challenging due to the fact that C-S-H structure is very complex due, for instance, to the variable chemical composition (with respect to the Ca/Si ratio) [2], possible co-existence of crystalline and glassy phases [3], and multiscale pore size [2].

Water sorption isotherms are one of the valued tools to assess the microstructure of the material. Since these techniques are sensitive to the microstructure and the surface chemistry of the studied material (e.g. hysteresis in adsorption-desorption isotherms is related to the microstructure of the material), significant information on the microstructure and the behavior of cement-based materials can be gained from the analysis of the water sorption isotherms. Several models have been proposed that have contributed to understanding sorption in cement-based materials (e.g. Kelvin-Laplace theory [4] and BJH model [5]). However, these models are well-suited to describe sorption in meso- and macro-pores only (e.g. they do not take into account the interaction of water molecules with the surfaces of both sides of the pore as in C-S-H interlayer slit-like pores) being restricted to pore larger than a few tens of nanometers [1].

In the last decades, molecular simulations have been an important tool for studying the various physical properties of the phases present in cement-based materials, also in investigating the structure, energetics, and behavior of cementitious materials at the atomic level [6]. Ji et al. [7] studied structural, mechanical, and dynamical properties of C-S-H using two different water models, the flexible SPC and the TIP5P water models. They showed that these models yield good results compared to the experimental values. Youssef et al. [8] showed that calcium counterions and defective silicate chains are the origins of the hydrophilic nature of C-S-H in very small interlayer pores. Bonnaud et al. [9] investigated the effect of RH on confined water in small C-S-H interlayer pores with pore sizes up to 10 Å. How the effective interactions of C-S-H and behavior of confined water in small C-S-H nanopores vary with the RH for equilibrium basal spacings are not yet studied.

In this work, we perform Grand Canonical Monte Carlo simulations combined with molecular dynamics to study the effect of RH on the behavior of C-S-H at the layer scale. For this purpose, we adopt a previously designed molecular model of calcium-silicate hydrate nano-slit pore developed by Kunhi Mohammed et al. [2], and for atom interactions, we use the Clay-FF interatomic potential [10]. Stable basal spacings and metastable domains, pressure isotherms, cohesive pressures, desorption isotherm, and elastic modulus perpendicular to the pore surface are analyzed. We also discuss the effect of C-S-H pore size on the cavitation of nano-confined water in C-S-H nanopores. Each of these properties can be helpful to model the thermo-hygro-mechanical behavior of cement-based materials using a multiscale approach.

2 Models and Methods

2.1 Atomic Structure

In this work, we adopt the C-S-H model of Kunhi Mohamed et al. [2]. The main features of C-S-H of this model are: an orthogonal structure made up of two micropores with an interlayer equilibrium distance of 13.76 Å. The exact chemical formula is $Ca_{1.67}SiO_{3.7}\cdot nH_2O$, for which the Ca/Si ratio is equal to 1.67. The same Ca/Si ratio is reported for a wide range of C-S-H encountered in hydrated Portland cement [3, 4]. Features such as the density [3, 5], the mean silicate chain length, and the concentration of protonated silanol groups [6] are consistent with experimental data.

2.2 Force Field

To describe atom interactions, we use the Clay-FF interatomic potential [10]. With this force field, the total potential energy of the system is represented using:

$$U_{total} = U_{coul} + U_{LJ} + U_{bonded} \tag{1}$$

For electrostatic interactions, using Coulomb potential, the Coulomb energy is given as:

$$U_{coul} = \frac{e^2}{4\pi\varepsilon_0}\sum_{i\neq j}\frac{q_i q_j}{r_{ij}} \tag{2}$$

where q_i and q_j are the partial charges of the interacting atoms, r_{ij} is the distance between particles i and j, e is the charge of the electron ($1.60217662 \times 10^{-19}$ C), and ε_0 is the dielectric permittivity of vacuum (8.85419×10^{-12} F/m).

The contribution of short-range interactions in the total potential energy using Lennard Jones potential is expressed as:

$$U_{LJ} = 4\,\epsilon_{ij}\left[\left(\sigma_{ij}/r_{ij}\right)^{12} - \left(\sigma_{ij}/r_{ij}\right)^{6}\right] \tag{3}$$

where ϵ_{ij} is the depth of the potential energy well, and σ_{ij} is the distance for which the potential energy between atoms is zero.

The bonded interactions ($U_{bonded} = U_{bond} + U_{angle}$) are modeled using the harmonic potential for only water molecules and hydroxides (the only type of bonded interactions explicitly defined in the force field). Bond and angle harmonic energies are respectively given by:

$$U_{bond}\left(r_{ij}\right) = \frac{1}{2}k_{ij}\left(r_{ij} - r_0\right)^2 \tag{4}$$

where r_0 is the equilibrium bond distance, k_{ij} is the force constant, and r_{ij} is the distance between the two particles i and j.

$$U_{angle}\left(\theta_{ijk}\right) = \frac{1}{2}k_{ijk}\left(\theta_{ijk} - \theta_0\right)^2 \tag{5}$$

In a similar way to the bond potential, θ_0 is the equilibrium angle, k_{ijk} is the force constant, and θ_{ijk} is the bond angle.

The bond stretch and the bond angle bend terms associated with water molecules and hydroxides, Lennard-Jones (LJ) parameters, and the partial charges for C-S-H and the fluid species present in the system are given in a previously published work [6].

2.3 Simulation Details

Hybrid GCMC-MD simulations are deployed in the Grand canonical ensemble (μVT) for water and NVT ensemble for other particles. The Grand canonical ensemble (μVT) is characterized by constant chemical potential, volume, and temperature while only the number of particles is allowed to fluctuate. In our C-S-H system, only water molecules can exchange between the simulation box and an external water reservoir at the same imposed chemical potential and temperature (GCMC phase). To increase the sampling efficiency, several MD steps are performed in-between the GCMC phases (NVT phase). A timestep of 1 fs is adopted. The Ewald summation method is used for computing the long-range component of electrostatic interactions. Tail corrections are deployed to deal with long-range LJ interactions. Periodic boundary conditions are imposed in all directions. The Nosé-Hoover thermostat is used to control the temperature during the simulation period. As initial configuration, we adopt a system at equilibrium at RH = 100% and T = 300 K for all the imposed interlayer distances (from approximately 10 to 26 Å). Then, at the same temperature using the same initial configuration, various simulations at lower RH from 90 to 0% are performed. To reduce the computation time, only water molecules, calcium counterions, and hydroxides are time-integrated. Five cycles are enough to equilibrate the system at each new RH (in terms of energy, number of particles, etc.), and five other cycles are used for production.

2.4 Stable and Meta-stable Domains

From the confining pressures and energy profiles, stable and meta-stable domains can be identified. The confining pressure can be computed from the thermodynamic potential using [6, 11]:

$$P_{pore} = \frac{1}{S_s}\left|\frac{\partial \Lambda}{\partial d}\right|_{\mu,T} \tag{6}$$

where S_s is the surface area (constant in our case), d is the basal distance, and Λ is the thermodynamic potential associated with the hybrid μVT-NVT ensemble. This thermodynamic potential is a function of the Helmholtz free energy F, the number of water molecules N_w, and the chemical potential $\mu(\Lambda = F - \mu N_w)$, and can be obtained from various μVT-NVT simulations with controlled interlayer distance d using the expression [6, 11]:

$$\Lambda(d) = \Lambda_0(d_0, \mu, T) - S_s \int_{d_0}^{d} P_{pore} d(d) \tag{7}$$

From this equation, the minimum in thermodynamic potential curve corresponds to an equilibrium basal spacing. The transition between equilibrium basal spacings indicates discrete water layering. In other words, each equilibrium basal spacing is associated with a given hydration state (0W, 1W, 2W, etc.). Unstable domains occur whenever the slope in the pressure isotherm is positive when plotted as a function of the interlayer distance [6]. The confining pressures and energy profiles are also useful to compute other properties like the cohesive pressures and Young's modulus that can be compared with data from the literature.

3 Results and Discussion

3.1 Pressure Isotherms, Cohesive Pressure, and Young's Modulus

Figures 1, 2 (A), and 2 (B) show the evolution of the disjoining pressure isotherms, cohesive pressures, and Young's modulus perpendicular to the pore surface as a function of RH, respectively. The disjoining pressure is defined as the difference between the pore pressure and the bulk pressure of the external water reservoir ($P_{dis} = P_{pore} - P_{bulk}$). The cohesive pressure is the pressure needed to separate two solid layers apart; it corresponds to the global minimum in the pressure isotherms. The out-of-plane Young's modulus is obtained from the ratio of the disjoining pressure to the basal distance ($E_{zz} = -d_i \frac{\partial P_{dis}}{\partial d}\Big|_{di}$) and is computed for equilibrium basal spacings at each RH where $d_i = d_{eq}$.

The disjoining pressure isotherms show a non-monotonic dependence of the disjoining pressure on the basal distance, an indication of water layering as the basal distance increases [6]. Each fluctuation is associated with a stable basal spacing. The increasing parts of the graphs correspond to unstable equilibrium; the system is thermodynamically unstable whenever the slope in the pressure isotherm is positive when plotted as a function of the interlayer distance. For 0% RH, and above 12.8 Å, the disjoining pressures are negative. Such negative pressures are due to the cohesive contribution of calcium counterions in the total disjoining pressure isotherm; calcium counterions possess negative pressures when plotted as a function of the basal distance [6, 9]. As the RH decreases from 100 to 0.1%, the cohesive pressure increases gradually. At 0% RH, a higher value of the cohesive pressure is observed (~2 GPa in absolute value), which suggests a larger cohesion of C-S-H in a dehydrated state. The cohesive pressure also shifts toward smaller basal distances as the RH decreases, indicating drying shrinkage. For Young's modulus, up to 40% RH, the value of E_{zz} is almost constant (~68 GPa), from 40 to 30%, a slight increase in E_{zz} is noticed. As for the cohesive pressure, at 0% RH, the value of E_{zz} is higher compared to other RH. The cohesive pressures and Young's modulus for simulated and experimental data from the literature are given in Table 1 respectively.

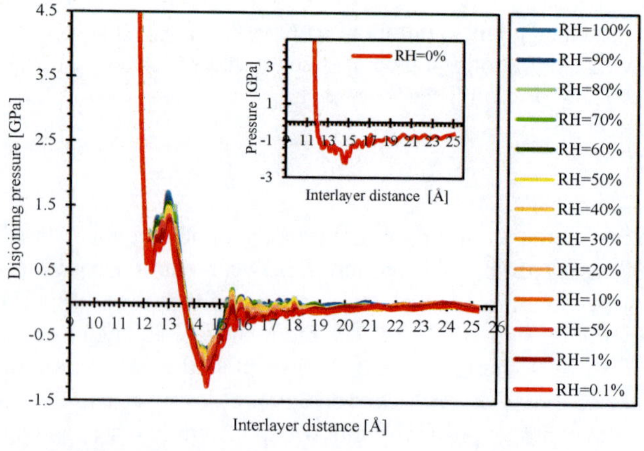

Fig. 1. Disjoining pressure isotherms as a function of the interlayer distance for different RH values.

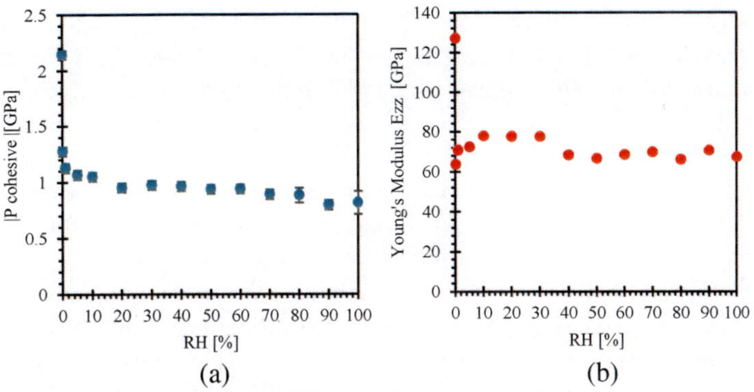

Fig. 2. (A) Cohesive pressures (in absolute value) as a function of RH. (B) Young's modulus in the c-direction for equilibrium basal spacings as a function of RH.

Table 1. Stable basal distance d_1 (d_2) (d_3), cohesive pressure P_{co}, and Young's modulus E_{zz} (perpendicular to the pore surface) for simulated and experimental data from the literature.

d_1 (d_2) (d_3) [Å]	P_{co} [Gpa]	E_{zz} [Gpa]	References
	0.93	–	[12][a]
12.5 (14.3) (16.1)	5.0	–	[13]
–	–	61.89	[14, 15][a]
11.4	0.931	42.28	[16]
10.4 (13.0) (15.7)	3.7	102.43	Ca/Si = 1.5 [17]
10.4 (12.9) (15.5)	3.0	81.74	Ca/Si = 1.7 [17]
12.6 (15.7)	4.17	61.5	[18][b]
13.5 (15.9)	6.5	77.6	[19][b]
12.7–13.7 (16.0)	0.82	67.4	[6]

[a] Experimental values
[b] Simulations on Tobermorite

3.2 Potential Energy Profiles, the Minimum of PMF, Drying Shrinkage, and Desorption Isotherm

The potential energy cprofiles and the global minimum of the potential energy well at saturated and undersaturated conditions are shown in Figs. 3 and 4, respectively. The global minimum of the potential energy well corresponds to the most stable basal spacing. Metastable domains exist whenever the energy of the C-S-H system (due to thermal fluctuations), on the order of kT (i.e., the Boltzmann constant times the temperature, which is characteristic energy associated with thermal fluctuations), is lower than the energy needed to overcome an energy barrier. Therefore, a C-S-H system with large energy barriers is prone to metastability.

As the RH decreases from 100 to 0.1%, the depth of the global potential energy well increases progressively. For 0% RH, the depth of global potential energy well is approximately six times larger than the values obtained for other RH. Therefore, as the

RH decreases, it becomes more difficult to overcome metastability. This also implies that the energy needed to disjoin two C-S-H layers increases whenever the RH decreases. The interaction of calcium counter ions with the charged C-S-H surface is the origin of such cohesive energies. Like the cohesive pressure, the global minimum of the potential energy well shifts toward lower hydration states with the decrease in RH. At low RH values, water in small C-S-H interlayer pores evaporates via drying, which causes shrinkage of C-S-H basal spacing. Figure 5 (A) shows the equilibrium basal distance versus RH obtained in this work compared to other experimental data from the literature with different Ca/Si ratios. Both our simulation results and the different experimental data show that the C-S-H basal spacing decreases due to water loss at low RH. According to previous experiments and simulations [20–27], the C-S-H basal distance decreases with increasing the Ca/Si ratio. The simulation results of Morshedifard [28] show that the removal of SiO_2 groups decreases the C-S-H basal distance with the Ca/Si ratio. On the contrary, the addition of $Ca(OH)_2$ increases the basal distance. The presence of $Ca(OH)_2$ groups in the interlayer C-S-H space of Kunhi Mohamed C-S-H model [2] might explain the larger basal spacings obtained in our simulations compared to the other experiments. Figure 5 (B) shows the desorption isotherm for equilibrium basal spacings taken at each RH.

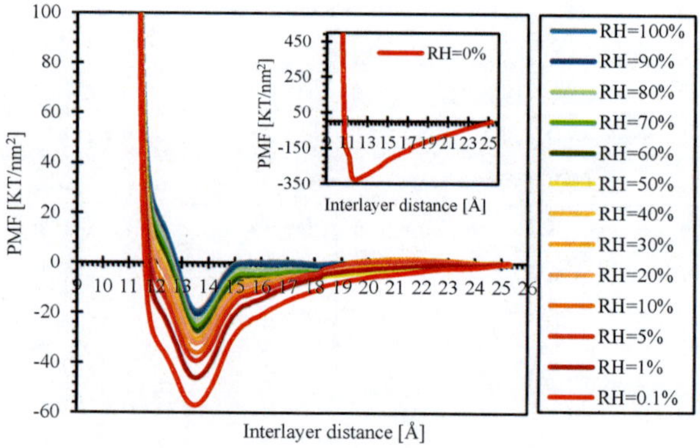

Fig. 3. Potential energy profiles as function of interlayer distance for different RH values.

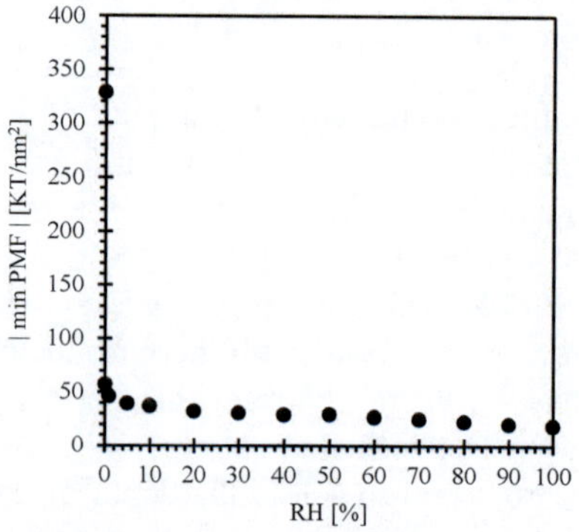

Fig. 4. Minimum of the global potential energy well (in absolute value) corresponding to equilibrium basal spacings as a function of RH.

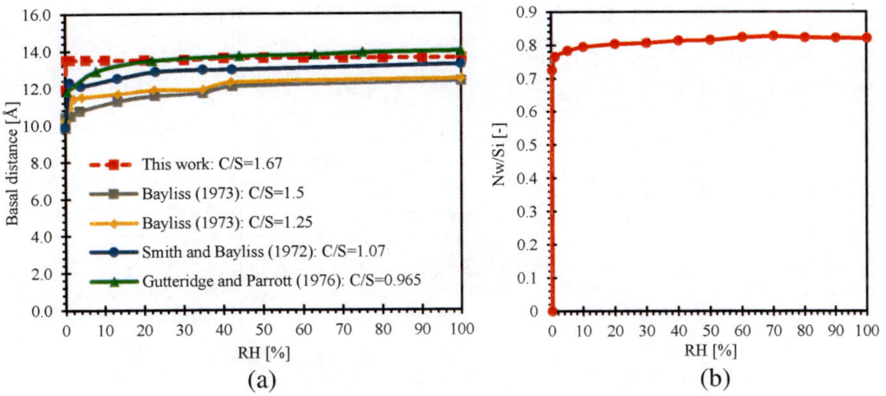

Fig. 5. (A) Simulated stable basal distance in this work against various experimental data from the literature for C-S-H with different Ca/Si ratios. (B) Desorption isotherm for equilibrium basal spacings obtained from the disjoining pressure and energy profiles.

3.3 Cavitation of Nano-confined Water in C-S-H

Desorption isotherms showing cavitation of water in C-S-H interlayer pores with a basal spacing between 10 and 26 Å are given in Figs. 6, 7, and 8, respectively. The rapid decrease in the amount of an adsorbed fluid is discussed in the literature as either being associated with cavitation or the ink-bottle effect (also known as the pore-blocking mechanism) [29, 30]. For small C-S-H interlayer pores, since these pores are surrounded with the gel porosity, and since the removal of water out of these pores occurs after the gel pores are completely empty, the pore blocking mechanism appears not to influence the removal of water in C-S-H interlayer pores, emptying the interlayer space, instead, is due to cavitation. Figure 6 shows the N_W/Si ratio as a function of the basal spacing according to different RH values. As shown in the figure, cavitation of water occurs in pores with a basal distance larger than 19 Å, for smaller basal distances, no cavitation is observed. The N_W/Si ratio for the range of basal spacings in which cavitation occurs is shown in Fig. 7. As the basal distance decreases, the point at which a rapid drop in the N_W/Si ratio shifts toward lower RH. The steepness of the slopes of the desorption curves for the region at which cavitation occurs decreases with the decrease in the basal distance. For nanometer-sized C-S-H interlayer pores, due to the hydrophilic C-S-H surface and the important disjoining pressures, water molecules begin to interact with the surface of the pore, affecting the cavitation of water. The Interlayer distances for which a sudden drop in the Nw/Si ratio occurs versus the RH are shown in Fig. 8 respectively. Figure 9 displays the atomic configurations for the upper (25.3 Å) and the lower bound (20.3 Å) of basal spacings in which cavitation occurs.

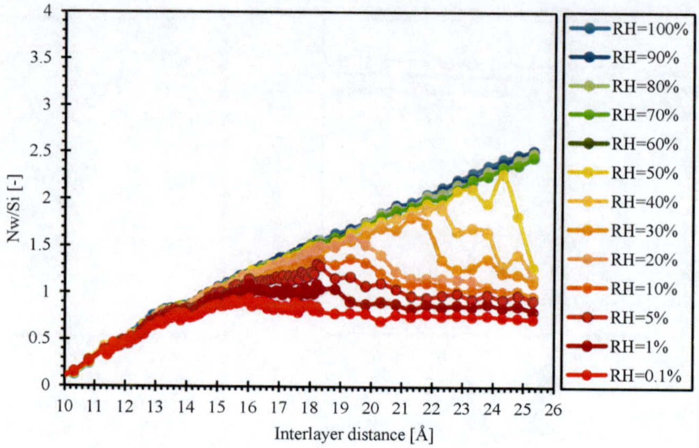

Fig. 6. Nw/Si ratio as a function of the interlayer distance for different RH.

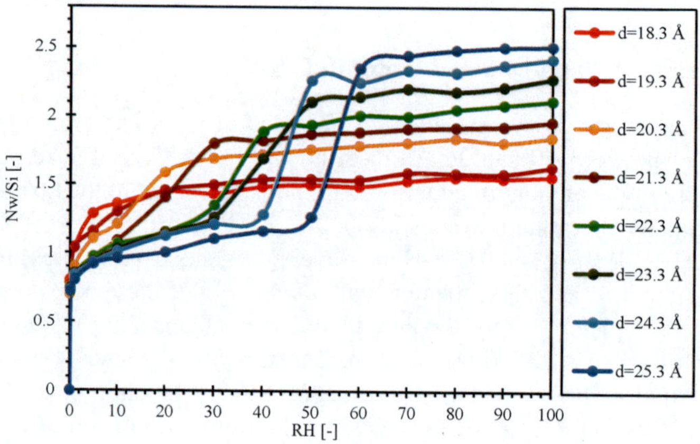

Fig. 7. Nw/Si ratio as a function of RH for interlayer distances between 18.3 Å and 25.3 Å for different RH.

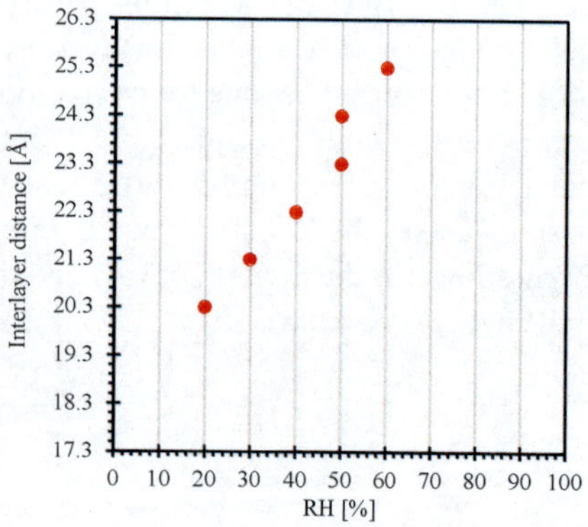

Fig. 8. Interlayer distances (between 20.3 Å and 25.3 Å) for which cavitation occurs versus RH.

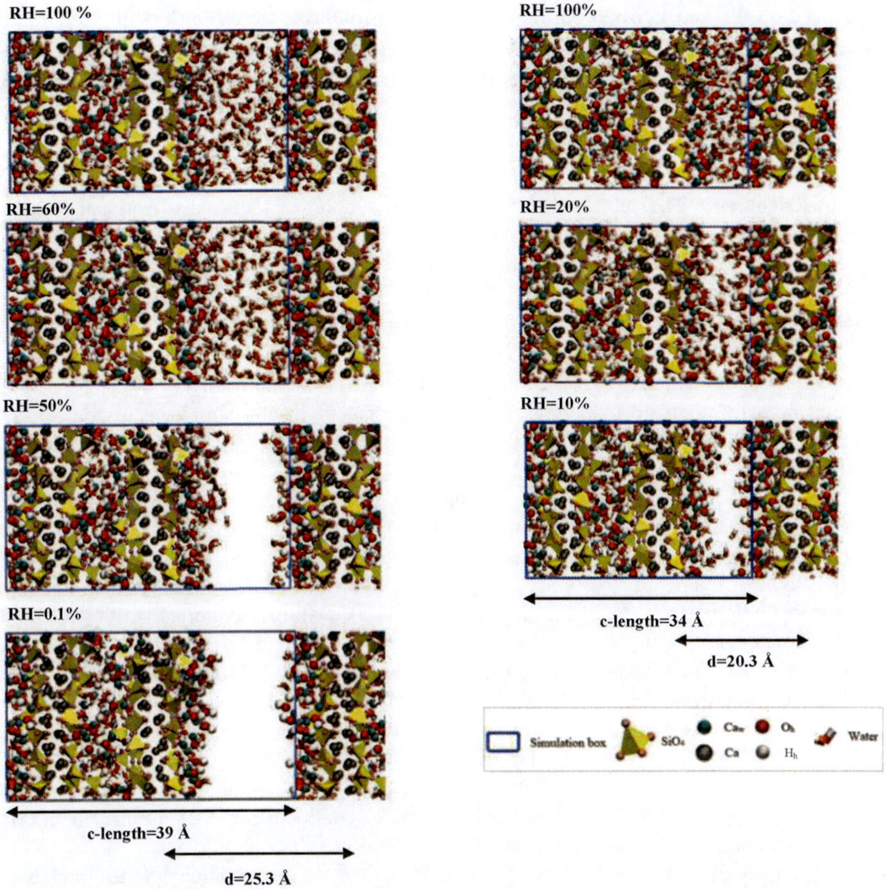

Fig. 9. Atomic configurations for the upper (25.3 Å) and lower limit (20.3 Å) of basal spacings for which cavitation occurs.

4 Conclusion

In this work, we use molecular simulations to study the effective interactions of C-S-H layers and desorption processes under RH change with atomic-level details. In function of RH, we study how water in small C-S-H interlayer pores behave. We determine the equilibrium basal spacing and metastable domains at a given RH from the energy profiles and pressure isotherms. As the RH decreases, we observe shrinkage of the C-S-H interlayer space. The cohesive pressures and the potential energy profiles are also RH dependent. Maximum cohesion is observed at low RH values due to calcium counterions in the interlayer space. We also study the cavitation of water confined in C-S-H nanopores and show that, unlike gel and capillary pores [29], cavitation of water in small C-S-H interlayer pores strongly depends on the size of the pore.

Based on this work, other valuables can be computed to better understand the behavior of water confined in C-S-H nano-pores. This includes the density profiles, adsorption isotherms, and thermal properties (heat capacity and thermal expansion coefficient). Using a multiscale approach, all these properties can be used as input data to model the behavior of water at a higher scale.

Acknowledgement. The authors thank the financial support of the French National Research Agency (ANR) through the project THEDESCO (ANR-19-CE22-0004-01).

References

1. Abdolhosseini Qomi, M.J., Brochard, L., Honorio, T., Maruyama, I., Vandamme, M.: Advances in atomistic modeling and understanding of drying shrinkage in cementitious materials. Cem. Concr. Res. **148**, 106536 (2021). https://doi.org/10.1016/j.cemconres.2021.106536
2. Kunhi Mohamed, A., Parker, S.C., Bowen, P., Galmarini, S.: An atomistic building block description of C-S-H - towards a realistic C-S-H model. Cem. Concr. Res. **107**, 221–235 (2018). https://doi.org/10.1016/j.cemconres.2018.01.007
3. Bauchy, M., Qomi, M.J.A., Ulm, F.-J., Pellenq, R.J.-M.: Order and disorder in calcium–silicate–hydrate. J. Chem. Phys. **140**(21), 214503 (2014). https://doi.org/10.1063/1.4878656
4. Ravikovitch, P.I., Neimark, A.V.: Density functional theory model of adsorption deformation. Langmuir **22**(26), 10864–10868 (2006). https://doi.org/10.1021/la061092u
5. Barrett, E.P., Joyner, L.G., Halenda, P.P.: The determination of pore volume and area distributions in porous substances. I. Computations from nitrogen isotherms. J. Am. Chem. Soc. **73**(1), 373–380 (1951). https://doi.org/10.1021/ja01145a126
6. Honorio, T., Masara, F., Benboudjema, F.: Heat capacity, isothermal compressibility, isosteric heat of adsorption and thermal expansion of water confined in C-S-H. Cement **6**, 100015 (2021). https://doi.org/10.1016/j.cement.2021.100015
7. Ji, Q., Pellenq, R.J.-M., Van Vliet, K.J.: Comparison of computational water models for simulation of calcium–silicate–hydrate. Comput. Mater. Sci. **53**(1), 234–240 (2012). https://doi.org/10.1016/j.commatsci.2011.08.024
8. Youssef, M., Pellenq, R.J.-M., Yildiz, B.: Glassy nature of water in an ultraconfining disordered material: the case of calcium−silicate−hydrate. J. Am. Chem. Soc. **133**(8), 2499–2510 (2011). https://doi.org/10.1021/ja107003a
9. Bonnaud, P.A., Ji, Q., Coasne, B., Pellenq, R.J.-M., Van Vliet, K.J.: Thermodynamics of water confined in porous calcium-silicate-hydrates. Langmuir **28**(31), 11422–11432 (2012). https://doi.org/10.1021/la301738p
10. Cygan, R.T., Liang, J.-J., Kalinichev, A.G.: Molecular models of hydroxide, oxyhydroxide, and clay phases and the development of a general force field. J. Phys. Chem. B **108**(4), 1255–1266 (2004). https://doi.org/10.1021/jp0363287
11. Honorio, T., Brochard, L., Vandamme, M.: Hydration phase diagram of clay particles from molecular simulations. Langmuir **33**(44), 12766–12776 (2017). https://doi.org/10.1021/acs.langmuir.7b03198
12. Plassard, C., Lesniewska, E., Pochard, I., Nonat, A.: Nanoscale experimental investigation of particle interactions at the origin of the cohesion of cement. Langmuir **21**(16), 7263–7270 (2005). https://doi.org/10.1021/la050440+
13. Pellenq, R.J.-M., Lequeux, N., van Damme, H.: Engineering the bonding scheme in C-S–H: the iono-covalent framework. Cem. Concr. Res. **38**(2), 159–174 (2008). https://doi.org/10.1016/j.cemconres.2007.09.026

14. Oh, J.E., Clark, S.M., Monteiro, P.J.M.: Does the Al substitution in C-S–H(I) change its mechanical property? Cem. Concr. Res. **41**(1), 102–106 (2011). https://doi.org/10.1016/j.cemconres.2010.09.010
15. Oh, J.E., Clark, S.M., Wenk, H.-R., Monteiro, P.J.M.: Experimental determination of bulk modulus of 14Å tobermorite using high pressure synchrotron X-ray diffraction. Cem. Concr. Res. **42**(2), 397–403 (2012). https://doi.org/10.1016/j.cemconres.2011.11.004
16. Bonnaud, P.A., et al.: Interaction grand potential between calcium–silicate–hydrate nanoparticles at the molecular level. Nanoscale **8**(7), 4160–4172 (2016). https://doi.org/10.1039/C5NR08142D
17. Masoumi, S., Zare, S., Valipour, H., Abdolhosseini Qomi, M.J.: Effective interactions between calcium-silicate-hydrate nanolayers. J. Phys. Chem. C **123**(8), 4755–4766 (2019). https://doi.org/10.1021/acs.jpcc.8b08146
18. Honorio, T.: Monte Carlo molecular modeling of temperature and pressure effects on the interactions between crystalline calcium silicate hydrate layers. Langmuir **35**(11), 3907–3916 (2019). https://doi.org/10.1021/acs.langmuir.8b04156
19. Masoumi, S., Valipour, H., Abdolhosseini Qomi, M.J.: Intermolecular forces between nanolayers of crystalline calcium-silicate-hydrates in aqueous medium. J. Phys. Chem. C **121**(10), 5565–5572 (2017). https://doi.org/10.1021/acs.jpcc.6b10735
20. Garbev, K., Beuchle, G., Bornefeld, M., Black, L., Stemmermann, P.: Cell dimensions and composition of nanocrystalline calcium silicate hydrate solid solutions. Part 1: Synchrotron-based X-ray diffraction. J. Am. Ceram. Soc. **91**(9), 3005–3014 (2008). https://doi.org/10.1111/j.1551-2916.2008.02484.x
21. Kalousek, G.L., Prebus, A.F.: Crystal chemistry of hydrous calcium silicates: III, Morphology and other properties of tobermorite and related phases. J Am. Ceram. Soc. **41**(4), 124–132 (1958). https://doi.org/10.1111/j.1151-2916.1958.tb13525.x
22. Matsuyama, H., Young, J.F.: Effects of pH on precipitation of quasi-crystalline calcium silicate hydrate in aqueous solution. Adv. Cem. Res. **12**(1), 29–33 (2000). https://doi.org/10.1680/adcr.2000.12.1.29
23. Kirkpatrick, R.J.: 29Si MAS NMR Study of the Structure of Calcium Silicate Hydrate, p. 13
24. **L'Hôpital, E., Lothenbach, B., Kulik, D.A., Scrivener, K.: Influence of calcium to silica ratio** on aluminium uptake in calcium silicate hydrate. Cem. Concr. Res. **85**, 111–121 (2016). https://doi.org/10.1016/j.cemconres.2016.01.014
25. Renaudin, G., Russias, J., Leroux, F., Frizon, F., Cau-dit-Coumes, C.: Structural characterization of C-S–H and C–A–S–H samples—Part I: Long-range order investigated by Rietveld analyses. J. Solid State Chem. **182**(12), 3312–3319 (2009). https://doi.org/10.1016/j.jssc.2009.09.026
26. Sugiyama, D.: Chemical alteration of calcium silicate hydrate (C–S–H) in sodium chloride solution. Cem. Concr. Res. **38**(11), 1270–1275 (2008). https://doi.org/10.1016/j.cemconres.2008.06.002
27. Taylor, H.F.W.: Relationships between calcium silicates and clay minerals. Clay Miner. **3**(16), 98–111 (1956). https://doi.org/10.1180/claymin.1956.003.16.06
28. Morshedifard, A., Masoumi, S., Abdolhosseini Qomi, M.J.: Nanoscale origins of creep in calcium silicate hydrates. Nat. Commun. **9**(1), 1785 (2018). https://doi.org/10.1038/s41467-018-04174-z
29. Maruyama, I., Rymeš, J., Vandamme, M., Coasne, B.: Cavitation of water in hardened cement paste under short-term desorption measurements. Mater. Struct. **51**(6), 1–13 (2018). https://doi.org/10.1617/s11527-018-1285-x
30. Coasne, B.: Multiscale adsorption and transport in hierarchical porous materials. New J. Chem. **40**(5), 4078–4094 (2016). https://doi.org/10.1039/C5NJ03194J

Contribution of Rib-scale Modelling to Study the Bond Mechanisms of Reinforcement in UHPFRC

Bruno Massicotte(✉), Fabien Lagier, Mohammadreza Zahedi, and Rémy Bastide

Group for Research in Structural Engineering, Polytechnique Montreal, Montreal, Canada
bruno.massicotte@polymtl.ca

Abstract. Using ultrahigh-performance fibre-reinforced concrete (UHPFRC) for reinforcement lap splice field connections simplifies construction details, reduces on site labour work and overall enhances connection performance and durability. Existing design equations for determining the required development length for normal concrete cannot be used for fibre-reinforced concrete. This paper presents the development of a refined nonlinear finite element model at rib-scale using the 3D concrete constitutive model EPM3D implemented in ABAQUS which enables explicitly expressing the bond performance of lap splices in UHPFRC according to the tensile properties of the concrete, the cover thickness and the bonded length. The numerical results show accurate simulation of the maximum strength, splitting failure mode, crack pattern, and steel stress distribution over the bonded length from different configurations of experimental bond tests. This methodology illustrates the value of nonlinear finite element analysis toward the harmonization of a bond test configuration and its contribution into the development of design guidelines for UHPFRC lap splice connections.

Keywords: Nonlinear analysis · Finite element · UHPFRC · Bond

1 Introduction

Ultrahigh-performance fibre-reinforced concrete (UHPFRC) has become an emerging material for the retrofitting of existing structures or for the construction of new ones. One of the most promising applications is for the strengthening deficient lap splice details [1, 2] or for connecting precast elements for accelerated bridge construction (ABC) projects [3, 4], as illustrated in Fig. 1. The exceptional bond characteristics are such that lap splice length in UHPFRC can typically be reduced by a factor of 5 when compared to normal concrete while secondary reinforcement detailing can be significantly simplified.

Several research projects carried out at Polytechnique Montreal on bond in UHPFRC are related to seismic applications. Tension hardening UHPFRC with 3% fibre content per volume was selected for strengthening existing bridge piers with deficient detailing where typically $24d_b$ lap splices without confinement were typically used at column-to-footing connections prior to the introduction of modern seismic design requirements. This technique allowed eliminating all bond failure mechanisms in the plastic hinge

P. Rossi and J.-L. Tailhan (Eds.): SSCS 2022, RILEM Bookseries 38, pp. 230–239, 2023.
https://doi.org/10.1007/978-3-031-07746-3_23

region. Although these choices conferred exceptional ductile behaviour to deficient columns [1], they are known as conservative and possibilities for optimizing these connections have been identified. In parallel to large specimen testing, in-depth research on the local behaviour of contact lap splices in UHPFRC was performed. Experimental investigations [5, 6] and numerical studies [7] have been carried out with the aim of getting a better understanding of the mechanisms involved in lap splice connections.

a) Precast column-footing joint for seismic applications [3]

b) Joints between precast bridge deck elements [4]

Fig. 1. Examples of lap splices in UHPFRC for ABC

This paper presents some of the numerical activities of an extensive research program aimed at developing guidelines for lap splice design of reinforcing bars in UHPFRC joints, for precast column-to-footing and precast abutment-to-footing connections meeting seismic requirements. The contribution of nonlinear finite element analysis (NLFEA) toward a better understanding of the bond mechanism is highlighted.

2 Adopted Approach

2.1 Seismic Performance Challenges

To meet the required column rotational ductility performances in plastic-hinge regions of bridges located in moderate to high seismic zones bring in important challenges for precast column-to-footing connections. The performance obtained in the strengthening of deficient bridge columns using UHPFRC in lap splice regions served as a basis for developing solutions for ABC. In such a situation, UHPFRC joints act as rigid elements connecting normal concrete members in which plastic deformations occur in the plastic hinge regions (L_p), as illustrated in Fig. 2a. For designing lap splices in UHPFRC, the length ℓ_s defined in Fig. 2a must be determined. This length depends on the UHPFRC tensile properties and the lap splice configuration. For seismic applications, the UHPFRC lap splice design must ensure that no bond failure occurs while the bars providing the rotational ductility (shown in blue) are subjected to repeated cyclic loading with tensile stress magnitude of the order of the bar ultimate strength (f_u). Determination of the embedment length ℓ_s according to the UHPFRC tensile properties and joint detailing to meet the desired joint performances constitute the main goal of the research program.

Bridge column geometry can be circular or rectangular. For large columns or abutments, long joints are expected. In all these situations, lap splice connections are characterized with simultaneously loaded bars subjected to high-amplitude cyclic tensile forces. Testing long joints with simultaneously loaded bars is a challenging endeavour. The relation between bond test on a single bar and the behaviour of simultaneously loaded bars in a structural joint must be determined. The adopted research strategy uses NLFEA to study the bond behaviour at two levels: in standard bond test specimens and in actual lap splice joints. Contact and non-contact lap splice configurations as illustrated in Fig. 2b are considered.

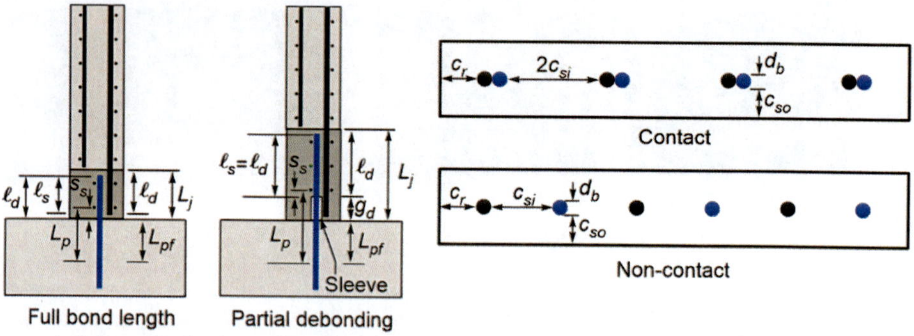

a) Column-to-footing connection b) Reinforcing bar configurations

Fig. 2. Geometrical parameter definitions for lap splice connections in UHPFRC

2.2 Rib-scale Modelling

Commercial finite element software Abaqus [8] was used to perform the rib-scale modelling, as presented in Fig. 3. Linear elastic properties for the reinforcing bar were adopted since the purpose of the study is to investigate the ultimate bond strength of different lap splice configurations. Tridimensional 8-node linear brick elements with reduced integration coupled with a relaxed hourglass stiffness method were used to mesh the bars and concrete. All the models were pre-processed, run and post-processed using a Python script. UHPFRC behaviour is modelled with concrete constitutive model EPM3D [9] considered as a user's subroutine in ABAQUS/Explicit. EPM3D is a general model for the nonlinear finite element analysis (NLFEA) of plain, reinforced and fibre-reinforced concrete structures. More details about the developed model strategy can be found in [7].

2.3 Research Strategy

Large-scale test results [3] showed that the full capacity of the lapped bars can be reached with a splice length ℓ_s equal to 10 d_b in joints with stirrups (see Fig. 1a) using a UHPFRC containing 2% by volume of steel fibres. The failure was governed by bar ruptures in tension at large strains. The multi-step approach adopted, uses a combination of nonlinear finite element modelling and the development of a standard bond test. Validated models offer a unique opportunity to observe the force transfer mechanism between lapped bars in UHPFRC. In the adopted strategy, analyses are coupled with experimental investigations at two levels: for standard bond tests and for actual joints between structural elements. In both cases analyses are first performed for designing the test specimens.

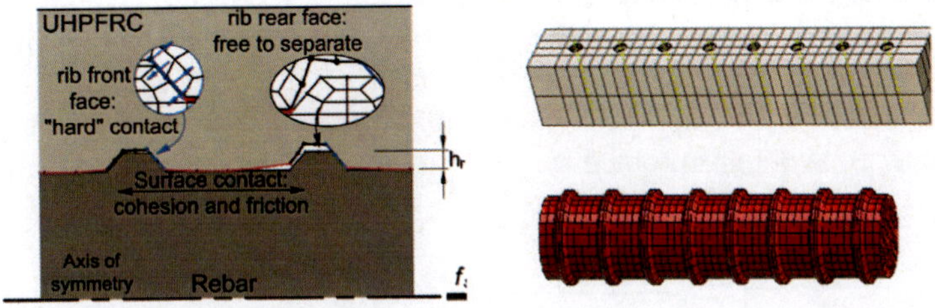

a) Interface properties [7] b) Model parts [7, 10]

Fig. 3. Rib-scale finite element modelling of lap splices

3 Standard Bond Test for UHPFRC

3.1 Defining Specimen Geometry

Challenges for defining a standardized bond test have been discussed for a long time for normal concrete and the same question remains for UHPFRC. This test should be relevant for correctly characterizing the main geometrical parameters. For UHPFRC, the casting procedure is important whereas there is not yet any standardized test for determining the tensile properties. In the process of defining the best geometry, several specimens were tested in a preliminary experimental campaign with contact and non-contact lap splice whereas extensive NLFEA was carried out [11, 12]. The two main conclusions indicate that DTB (Direct Tensile Bond) test (see Fig. 4c) is the most appropriate one, whereas preferential fibre orientation perpendicular to the bar orientation would be preferred.

DTB specimens were simulated using nonlinear finite element models to determine the impact of selecting the geometrical parameters (see Fig. 4c) including bar spacing (c_{si}), bar cover (c_{so} and c_r), embedment length of the support bar (ℓ_a vs. ℓ_d), bar gap (g) and overall specimen dimensions. For these analyses, isotropic reduced tensile properties presented in Fig. 4a were adopted. Actual measured tensile properties of four mixes from two suppliers used in recent experimental programs carried out at Polytechnique Montreal are shown in Fig. 4b for comparison purposes. The reduction of the maximum strength and strain values is based on previous calibration [7]. Reduced values in NLFEA account for the dispersion and variable orientation of fibres for random pouring and are assumed to be average conservative values.

Findings of a previous numerical study on continuous joints [11, 12] recommend using lateral bar length equal to the joint dimension, a condition adopted in the present study. Practically, longer lateral bars are used to ease the fabrication process. It is recommended to use for the lateral bars, the same bar diameter as the tested bar. Too large c_{si} values with a short anchorage length ℓ_d can create a beam-type behaviour of the top part of the DTB specimen. Indeed, tests on long strip joints containing several bars in which tension was applied on individual bars at the time [13] concluded the bond strength reduces while increasing bar spacing c_{si} (see Fig. 2b) which is counterintuitive. Numerical modelling using the approach describe above [11, 12] revealed that such testing method induces tensile bending stresses on the top of the specimen which add to the splitting stresses induced by the pulled bar anchorage, thereby reducing the bond strength for large bar spacing. Moreover, the analyses also indicated that for small bar spacing, the strip testing approach overestimates the bond strength leading to unsafe conclusions. To avoid this problem for DTB test specimens, it is recommended that $\ell_d \geq c_{si}$.

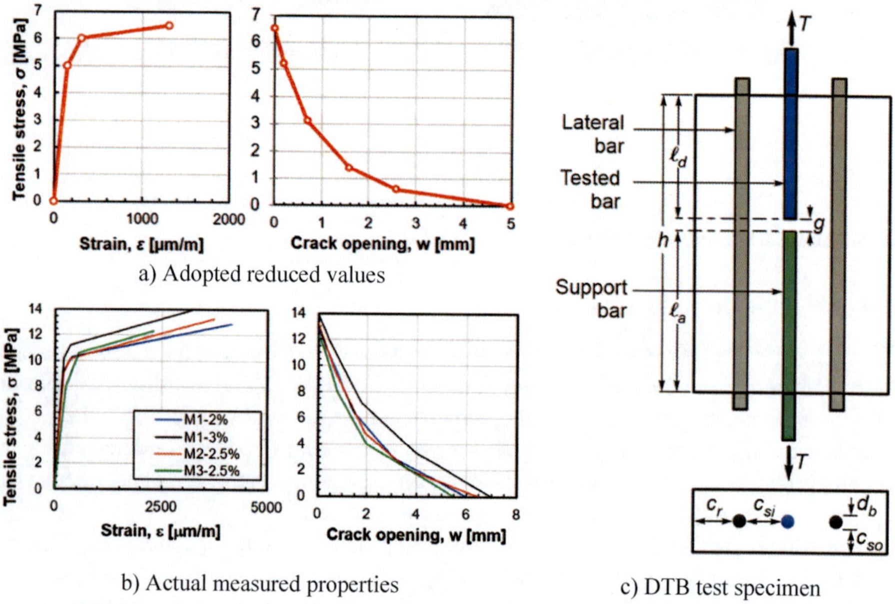

Fig. 4. UHPFRC tensile properties and proposed standard test specimen

NLFEA results suggested that $\ell_a \geq \ell_d + 2\, d_b$. It was also determined that the gap g has no influence on the specimen behaviour. Although DTB test specimens in Fig. 5 have very small end-cover c_r, further analyses indicated that larger covers increase the bond strength. It is therefore suggested to select $c_r = c_{si}$, justified by the failure mechanism, whereas $c_r \geq 2\, d_b$ and not less than 50 mm, both required to ensure proper fibre dispersion in the vicinity of the lateral bars which can be affected by the pouring process and the presence of the mould. Finally, both exploratory testing and numerical model results suggest for standard DTB test specimens using preferential fibre orientation, perpendicular to the bar orientation which was achieved by the filling of the moulds from one of the long sides.

Failure modes and maximum tensile stress for three specimens obtained using rib-scale modelling are illustrated in Fig. 5. Results indicate that for constant bar spacing and anchored length, varying the cover thickness from $2.16d_b$ to $6d_b$ changed the failure mode from a perpendicular one to a parallel one, with a mixed mode for an intermediate value. In all models, the initiation of splitting cracks always occurs before the maximum tensile force in the bars is reached. Hence, the contribution of the tensile softening part of UHPFRC plays an important role in the bond strength (see Fig. 4).

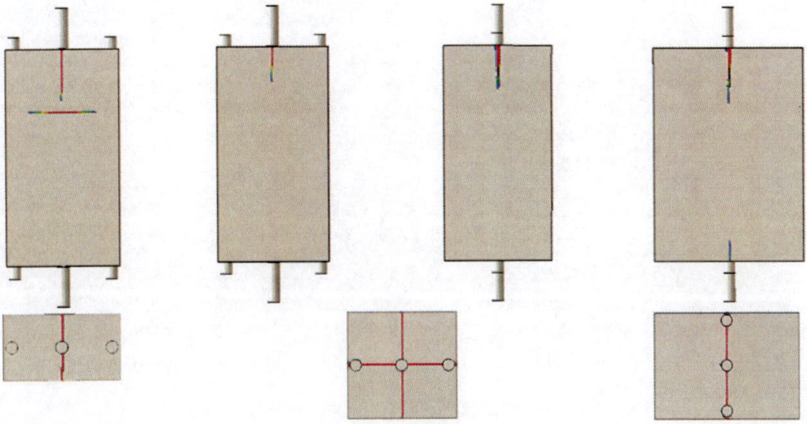

a) $c_{so}/d_b = 2.16$
$f_s = 354$ MPa

b) $c_{so}/d_b = 4$
$f_s = 607$ MPa

c) $c_{so}/d_b = 6$
$f_s = 695$ MPa

Fig. 5. Predicted failure modes for DTB test specimens with $c_{si} = 3d_b$ and $\ell_d = 6d_b$ [10]

3.2 Comparison with Experimental Results

Based on the specifications obtained from NLFEA, an extensive experimental campaign was successfully carried out. Observed experimental failure modes were similar to those obtained with the numerical model as presented in Fig. 6. Despite isotropic tensile properties were used, excellent correlation was obtained for the failure modes and trends with the varying geometrical parameters and material properties. Both NLFEA and tests indicated that stress localization, associated with UHPFRC softening behaviour and crack opening as shown in Fig. 4, occurs before the maximum tensile stress in reached in the pulled bar. Transverse splitting cracks form first at the specimen top side and progress toward the top of the specimen face then propagate downward along the bar. Horizontal cracks form at the end of the tested bar upon reaching the peak bar stress.

3.3 Bond Strength Mechanism

NLFEA was used to study the bond transfer mechanism. Variation of the concrete principal stresses and the bar stresses were examined at various locations along the embedded length as indicated in Fig. 7a. Factor RA indicates the bar stress ratio at various loading stages with respect to the maximum stress. One can see that for the modelled specimen with and embedment length of $10d_b$, nearly linear stress variations are observed in the bar up to 90% of the maximum stress whereas the shape of the stress distribution becomes nonlinear at failure, indicating a reduction of the bar confinement related to the UHPFRC that is in its crack opening softening response near the top of the specimen.

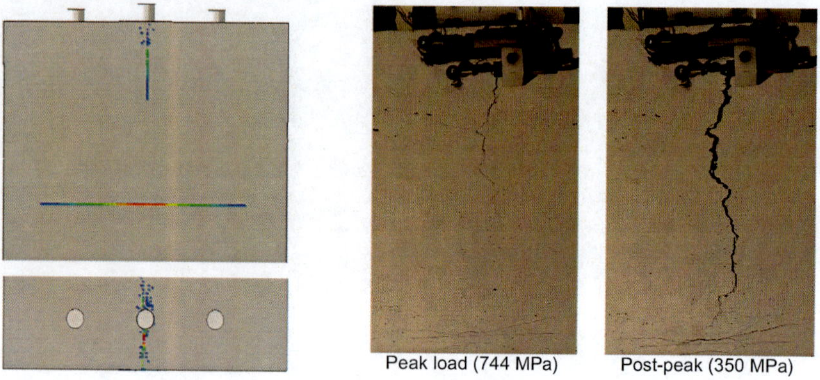

a) NLFEA crack pattern [14] b) Test specimen crack pattern [10]

Fig. 6. Typical DTB test specimen behaviour

The strain profile in the direction transverse to the bar orientation was also computed. The strain at which localization occurs is related to measured properties in the direct tensile test specimens shown in Fig. 4. The maximum strain is related to the test gauge length which typically varies from 70 mm to 140 mm. Different gauge lengths lead to different strains at peak stress. The average computed strain from simulations over the gauge length of the experimental strain gauge on the specimens, indicated in Fig. 7b and the displacement over the reference length of the potentiometers was used in the calibration process.

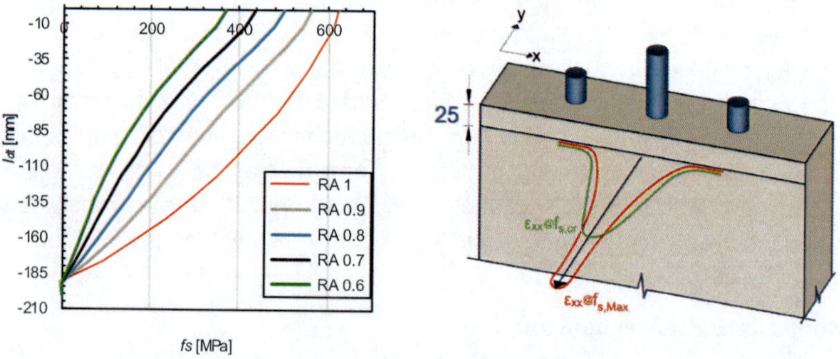

a) Reinforcement stress variation b) Specimen face strain distribution

Fig. 7. Modelled DTB test specimen response [14]

3.4 Model Calibration

DTB test specimens have preferred fibre orientation. Although the casting procedure is not representative of actual field-cast conditions, the calibration of the finite element model must be carried out with properties as close as possible to actual specimen properties. In that purpose, 90% of tensile proper of UHPFRC M-3 shown in Fig. 4 were adopted in the direction parallel to the fibre orientation whereas softening tensile properties without hardening characteristics were retained in the two other principal directions. The measured and modelled specimen behaviour are shown in Fig. 8.

Calibration was done for three identical specimens (#01, 02 and 03) with an embedment length equal to $10d_b$. On Fig. 8a, the measured and computed relative bar slip with respect of the top of the specimen are shown whereas the relative displacements measured by LVDT on each face of the specimen (see Fig. 6b and 7b) are compared in Fig. 8b with the model values. In both cases excellent agreements are obtained. The average maximum experimental bar stress is 600 MPa while 602 MPa is obtained with the model.

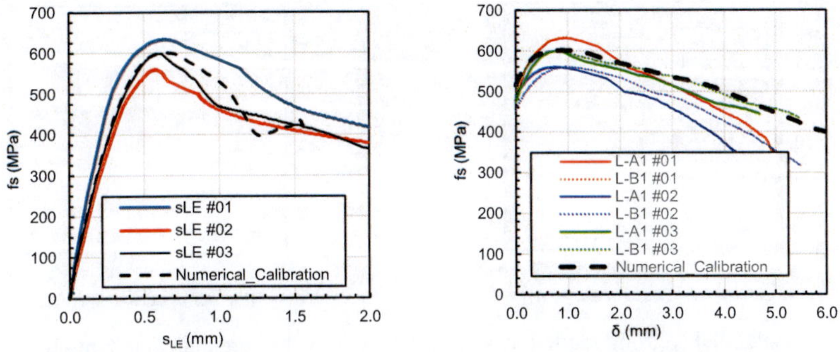

a) Bar slip b) Specimen face elongation

Fig. 8. Comparison of calibrated orthotropic properties with experimental results [14]

4 Continuous Joint Behaviour

For real field-cast joints, random properties are expected to be more representative of actual conditions. The behaviour of continuous joints was determined using the reduced properties given in Fig. 4a. Several joint configurations with different embedment length and bar numbers were considered. Typical failure modes are illustrated in Fig. 9 for two bar spacing leading to two different failure mechanisms.

Results of 10 models illustrated in Fig. 10 indicate that for small c_{si} values, failure occurs parallel to the joint. In this case, the number of bars in the joint does not affect noticeably the force supported by the bars. For larger bar spacing, the joint capacity increases with the number of bars pulled simultaneously, where both the average and maximum bar stresses are affected. These results indicate that test in continuous joint involving a few bars should lead to conservative results. The DTB test model with $c_r = c_{si}$ for the configuration of Fig. 9b leads to a strength of 431 MPa which corresponds to the average value of a joint with approximately 9 bars.

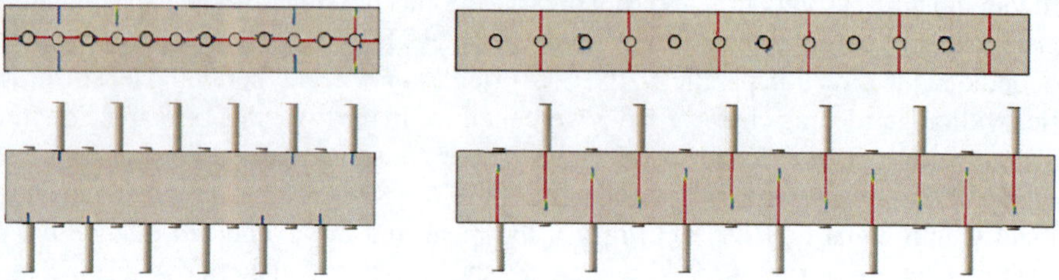

a) Parallel splitting mode for $n = 6$ $c_{so}/d_b = 2.16$ and $c_{si}/d_b = 1.5$

b) Perpendicular splitting mode for $n = 6$ $c_{so}/d_b = 2.16$ and $c_{si}/d_b = 3.0$

Fig. 9. Typical continuous joint failure modes with $\ell_s = 6d_b$ [10]

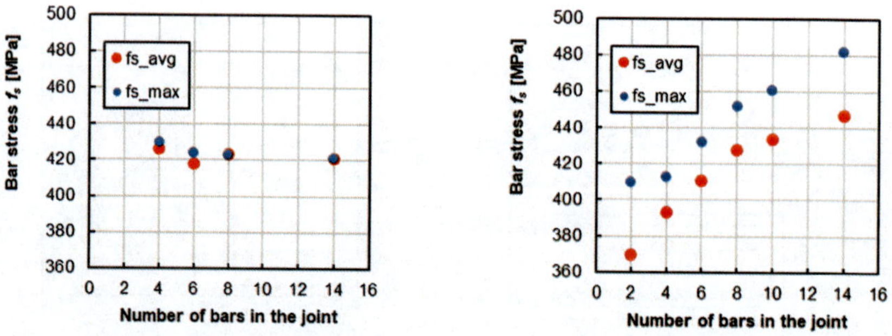

a) Parallel splitting mode for $c_{so}/d_b = 2.16$ and $c_{si}/d_b = 1.5$

b) Perpendicular splitting mode for $c_{so}/d_b = 2.16$ and $c_{si}/d_b = 3$

Fig. 10. Effect of the number of bars on the bar stress in continuous joints for $\ell_s = 6d_b$ [10]

5 Conclusion

Based on NLFEA, the Direct Tensile Bond (DTB) test emerged a potential candidate for standardization. The detailed nonlinear 3D FE model strategy with reinforcing bars modelled at rib-scale was used to highlight the importance of accurately selecting the UHPFRC tensile properties, accounting for fibre orientation and dispersion. More than a complement to experimental investigation, NLFEA has proved to be an essential component for planning experimental activities, for explaining experimental results, and ultimately contribute for developing design guidelines.

Acknowledgement. The authors would like to acknowledge for their financial support the Natural Sciences and Engineering Research Council of Canada (NSERC), the Center for Research on Concrete Infrastructures of Quebec (FRQNT - CRIB), and the Quebec Ministry of Transportation.

References

1. Dagenais, M.-A., Massicotte, B., Boucher-Proulx, G.: Seismic retrofitting of rectangular bridge piers with deficient lap splices using ultrahigh-performance fiber-reinforced concrete. J. Bridge Eng. **23**(2), 04017129 (2018)
2. Massicotte, B., Jolicoeur, O., Ben Ftima, M., Lagier, F.: Effectiveness of UHPFRC cover for the seismic strengthening of deficient bridge piers. In: Proceeding of the 39th IABSE Symposium, Vancouver, BC, Canada, 21–23 September 2017, pp. 2723–2730 (2017)
3. Darveau, P.: Development of precast columns with UHPFRC lap splice connections for seismic applications. M.A.Sc. Thesis, Polytechnique Montreal, Montreal, Canada (in French) (2017)
4. Marleau, B., Massicotte, B., Charron, J.-P.: Fatigue behaviour of UHPFRC joints between precast bridge deck panels. Report SR18-01, Group for Research in Structural Engineering, Polytechnique Montreal, Montreal, Canada (in French) (2018)
5. Lagier, F., Massicotte, B., Charron, J.-P.: Bond strength of tension lap splice specimens in UHPFRC. J. Construct. Build. Mater. **93**, 84–94 (2015)
6. Lagier, F., Massicotte, B., Charron, J.-P.: Experimental investigation of bond stress distribution and bond strength in unconfined UHPFRC lap splices under direct tension. Cem. Concr. Compos. J. **74**, 26–38 (2016)
7. Lagier, F., Massicotte, B., Charron, J.-P.: 3D nonlinear finite-element modeling of lap splices in UHPFRC. J. Struct. Eng. **142**(11), 04016087 (2016)
8. Hibbitt, H.D., Karlson, B.I., Sorensen, E.P.: ABAQUS version 6.14, finite element program. Hibbitt, Karlson and Sorensen, Providence, R.I. (2014)
9. Massicotte, B., Ben Ftima, M.: EPM3D-v3 – a user-supplied constitutive model for the non-linear finite element analysis of concrete structures. Report SR17-06, Group for Research in Structural Engineering, Polytechnique Montreal, Montreal, Canada (2017)
10. Bastide, R.: UHPFRC lap splice behaviour for precast abutment construction. M.A.Sc. Thesis, Polytechnique Montreal, Montreal, Canada (in French) (2020)
11. Rosini, A.: Design of field-cast UHPC connections for accelerated bridge construction in high seismic zones. M.A.Sc. Thesis, Polytechnique Montreal, Montreal, Canada (in French) (2018)
12. Lagier, F., Rosini, A., Ben Ftima, M., Massicotte, B.: Numerical simulation of bond strength in UHPFRC using high-resolution finite element modelling: effects of boundary conditions. In: Colombo, M., di Prisco, M., Ferrara, L. (eds.) SSCS2019, Lecco, Italy, 5–6 September 2019 (2019)
13. Yuan, J., Graybeal, B.: Bond of reinforcement in ultra-high-performance concrete. ACI Struct. J. **112**(6), 851–860 (2015)
14. Zahedi, M.: Numerical investigation of the bond behaviour of reinforcing bars in UHPFRC. M.A.Sc. thesis, Polytechnique Montreal, Montreal, Canada (2021)

Numerical Modeling of 3D Concrete Printing Wall Structure to Reliably Estimate the Failure Mechanisms

Meron Mengesha[1,2,4](✉), Albrecht Schmidt[1,3], Luise Göbel[1,3], and Tom Lahmer[1,2,3]

[1] Bauhaus-Universität Weimar, Weimar, Germany
[2] Institute of Structural Mechanics, Weimar, Germany
[3] Materials Research and Testing Institute (MFPA), Weimar, Germany
[4] Addis Ababa Institute of Technology (AAiT), Addis Ababa, Ethiopia
meron.wondafrash@aait.edu.et, Meron.Wondafrash@gmail.com

Abstract. 3D concrete printing technology (3DCP) have gained wide attention. They indicated their potential to become a serious supplement to conventional concrete casting in molds. The main reason for this is it directly addresses the challenges related to the sustainability and productivity of the construction industry. However, the current practice is based on the trial-and-error procedure, which makes the research of the 3DCP process expensive and time-consuming. One of the reasons is that there exist significant knowledge gaps regarding the relations between the design, material, and process parameters. Therefore, it is of vital importance to establish a relation between the process parameters and the printed product to avoid unreliability and failure. By implementing a numerical simulation of the 3DCP process, a more fundamental understanding of the relations between the printing process, the process parameters, and the properties of the printed product could be achieved.

In this study, layer-wise Finite Element Method (FEM) combined with a pseudodensity approach, known from in topology optimization is applied. Along with the progressing printing process, all material parameters vary spatially and temporarily due to the time dependency of the curing process. The numerical simulations allow to reliably estimate the failure mechanisms that might occur during the 3D concrete printing of a wall structure.

Keywords: 3D concrete printing · Numerical modeling · Pseudo-density approach · Time dependency · Printing velocity · Strength-based stability · Shape stability/geometry control · Overall buckling

1 Introduction

The extrusion of cement-based materials has recently received a lot of attention, especially since the advent of digital concrete construction techniques [1–4]. Extrusion-based digital construction, often known as 3D concrete printing, is one of the most extensively used digital construction processes [5].

P. Rossi and J.-L. Tailhan (Eds.): SSCS 2022, RILEM Bookseries 38, pp. 240–247, 2023.
https://doi.org/10.1007/978-3-031-07746-3_24

Reduced construction time, design flexibility compared to traditional construction methods, cost savings by avoiding formwork costs, reduced waste, reduced manpower, fewer injuries and fatalities on construction sites, and increased construction industry sustainability are just a few of the benefits of the technology [2, 6].

However, there are several challenges to overcome before the technology can be effectively implemented. Due to the complexity of material parameters and printing process parameters, a trial-and-error technique is typically used to find appropriate process parameters. As a result, 3D concrete printing (3DCP) is more expensive and time-consuming than it should be Which hamper the robustness of concrete printing [7].

Numerical simulations could help to achieve a better understanding of the relationships between the printing process, process parameters, and the properties of the printed product. They enable to study of a wide range of parameters such that dependencies of properties of the printed product on different influencing factors can be identified.

Wolfs et al. [5] rightly argue that the mechanical properties of the printed structure during the production process must be known and well defined in a recent paper. Disturbances in the printing process can cause inconsistencies and impair the product's quality [3]. Inadequate strength, stiffness, or stability may cause the structure to fail during the printing process.

The printing process factors, such as printing velocity, temperature, and nozzle diameter, as well as the concrete mixture, have a significant impact on these attributes. In this context, Van der Putten et al. [8] studied the effects of the linear printing speed and the time gap between two subsequent layers on the microstructure of printed concrete. Accordingly, the two parameters significantly influence the surface roughness, the compressive strength, and the interlayer bonding strength. To avoid unreliability and failure, it is critical to establish a relationship between the process parameters and the mechanical properties of the printed product [3, 4].

Therefore, information about a suitable printing velocity w.r.t. the failure mechanisms help steer the printing process optimally. The LayerWise Finite Element (FE) method is utilized in this study, together with a pseudo-density approach and analytical formulation, to find failure mechanisms in 3D printed concrete structures by changing one of the printing process parameters, namely printing velocity.

2 Methods

2.1 Spatial and Temporal Changes of the Material Parameters 3D Printable Concrete

To achieve a successful printing process and a reliable final product, numerous requirements must be met by the material when using the extrusion method. The rheological properties of the concrete vary during the extrusion process; it should be flowable during the pumping and extruding phases, and following deposition, it should gain rapid strength to maintain a stable shape and carry subsequent layers [2, 6].

The rheological properties for the extrusion-based printing process are shown in Fig. 1. At the pumping and extruding stage the fresh concrete is represented in light grey and after deposition, the concrete is represented in dark grey.

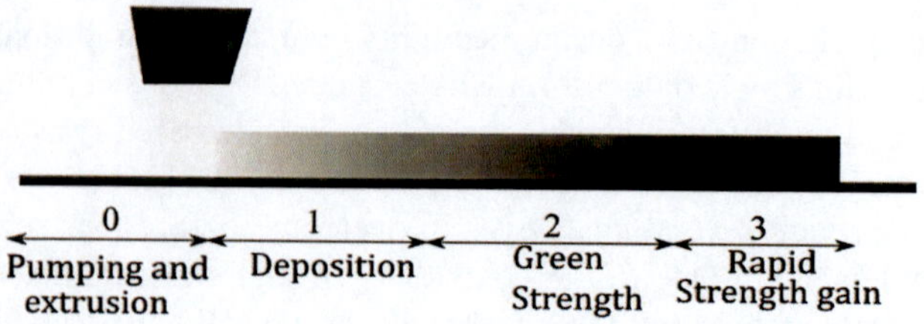

Fig. 1. Evolution of the static yield stress for the extrusion-based printing process according to [6].

The shape stability and the load-bearing capacity of the layers are linked to the rheology of the material, in particular to the yield stress. The yield stress and viscosity should be as low as possible during pumping and extrusion [2]. Following that, the structural build-up starts, and adequate yield stress should be achieved shortly after deposition [2, 6].

The goal is for each printed layer to maintain its shape after deposition and harden when poured while remaining liquid enough to bond with the one above it [9]. In addition, the time gap between consecutive layers should be limited to increase the bond strength between two consecutive layers [10].

Wolfs et al. [5] and Casdgrande et al. [11] observed this, and their findings reveal that all parameters change with time as a result of the hardening process and that the material parameters are both spatially and temporally variable. Based on these findings, the numerical model accounts for spatially and temporal changes in material parameters. For example, the time-dependent Young's modulus is given as:

$$E(x, t) = E_0(x) + E_1 t, \quad (1)$$

where $E_0(x)$ is initial the Young's modulus at the moment it leaves the printing nozzle, E_1 is the gradient of temporal increment, and t is the layer cycle time at a specific location.

2.2 Finite Element Model of a Printing Process

The layer-by-layer construction method of the concrete structure requires the Finite Element (FE) model to "grow." The aim is to work with a previously generated FE mesh in which the elements are activated sequentially to mirror the printing process. This has been addressed by introducing a novel modeling technique that includes FE with a pseudo density approach [14]. Similar to Bendse and Sigmund's [12] soft-killing approaches in topology optimization, the elements FE are endowed with a spatially varying pseudo-density $\rho(x) \in [0,1]$. This density is used to define the material properties of each FE. This method allows for a more efficient simulation while avoiding the computationally demanding meshing procedure.

Additionally, as all parameters do change with time according to the hardening process the material parameters are both spatially and temporarily varying, indicated by the space and time dependency in Eq. 1. During the simulation, i.e. the solution of a Finite Element system of the form [14]:

$$\boldsymbol{K}(\rho(x), t)\, \boldsymbol{u}(\rho(x), t) = \boldsymbol{F}(\rho(x), t), \quad (2)$$

where $\boldsymbol{K}$ is the global stiffness matrix, $\boldsymbol{u}$ is the vector of nodal displacements, $\boldsymbol{F}$ is the load due to self-weight and x and t denote the space and time variable, respectively.

The layer cycle time for each layer can be calculated using Eq. 3. when each layer is activated in the layer-wise Finite Element approach. The ith layer will have a layer cycle time of:

$$t = (i + (H(x,t)/h + 1))\left(L/v_p\right), \tag{3}$$

where v_p - printing velocity, H is the height, L is length and h is the thickness of one layer.

2.3 Failure Mechanisms in 3D Printed Concrete Structures

The simulation of the 3DCP process requires the selection of appropriate material models. For every "printed" element, and every "printed" layer, respectively, efficient analytical models estimating the time evolution of the mechanical properties induced by both the progressing hydration process and the thixotropic increase of the concrete need to be implemented. The mechanical properties of concrete, e.g. the yield stress and the elastic modulus, depend on the maturity of the material and are therefore all time-dependent.

As failure mechanisms, we consider the following types: strength-based stability, shape stability/geometry control, and overall buckling.

The strength-based failure of the structure is studied by comparing the temporal evolution of the yield stress with the increasing gravity-induced stresses caused by subsequently placed concrete layers. In this situation, the stress generated by the weight of the printed layers exceeds the printable concrete's yield strength, resulting in failure.

$$\sigma_y(x,t) \leq \sigma_p(x,t), \tag{4}$$

where $\sigma_y(x,t)$ is the compressive yield strength and $\sigma_p(x,t)$ gravity-induced stresses.

$$\sigma_y(x,t) = \sigma_{y,0} + \sigma_{y,1}t, \tag{5}$$

where $\sigma_{y,0}$ is the initial yield strength of the printable concrete at the moment it leaves the printing nozzle, $\sigma_{y,1}$ is the gradient of temporal increment, and t is the curing time at a specific location.

The gravity-induced stresses $\sigma_p(x,t)$ depend on the height of the wall to be printed. These gravity-induced stresses acting on any of the layers can be written as follows [15]:

$$\sigma_p(x,t) = \rho\, g\, H(x,t), \tag{6}$$

where ρ is the material density, g is the acceleration of gravity.

Another failure mode that might occur is the overall buckling of the printed structure, as also considered by Wolfs et al. [5]. For a slender vertical column structure subjected only to its self-weight, the critical height at which buckling is likely to be found, *H(x,t)* can be determined analytically as [2]:

$$H(x,t) = (8E(x,t)I/\rho g A)^{1/3}, \tag{7}$$

where I denote the second moment of area and A is the horizontal cross-sectional area.

By rearranging Eq. 7 and considering that the wall's width is the delta(δ), the time-dependent elastic Young modulus for the structure to be stable should be greater than a critical value equal to avoid overall buckling [2]:

$$E_0(x) + E_1 t > (3\rho g (H(x,t))^3/2\delta^2). \tag{8}$$

Gravity-induced stresses increase with the height of the printed product, whereby the maximum stress values occur in the bottom layer. Due to the increasing stresses, the layers exhibit progressively growing strains and may deform finally. Also for the cases, the yield stress and therewith strength-based failure is not reached, the amount of strain in each layer could influence the final geometry.

In the layer-wise Finite Element methods from Eq. 2, $\boldsymbol{u}$ the vector of nodal displacements can be determined from which further quantities like stresses and strain can be derived. The strain in each layer should be kept below the minimal critical strain to achieve a stable and load-bearing structure.

3 Numerical Example

3.1 Description

The wall is modeled as a two-dimensional structure and constructed in a layer-wise process to analyze the effect of changing the printing velocity on failure modes that may occur in the printed structures. The analysis is carried out using 4-node bi-linear finite elements. The numerical simulation's geometry includes a wall width of 43.5 mm, a layer thickness of 10 mm, a wall-length of 800 mm, and a Poisson ratio of 0.3 [5].The loading taken into account is the rising self-weight as the printing proceeds, with a mass density of 2020 kg/m^3 [5]. Because of friction on the printed bed, the modeled wall is fixed vertically and horizontally at the bottom edge [7]. The investigation is carried out by gradually increasing the printing velocity keeping the other parameters constant. Numerical simulation is performed using the commercially available programing language MATLAB.

As each layer is activated in numerical modeling, spatial and temporal variations in material properties are taken into account in the finite element formulation, as demonstrated by the time dependency in Eq. 2. The temporal evolution of Young's modulus can be seen in Fig. 2. An increase of $E_1 = 1.2$ kPa/min is chosen following [5] and the printing velocity of 2 cm/s.

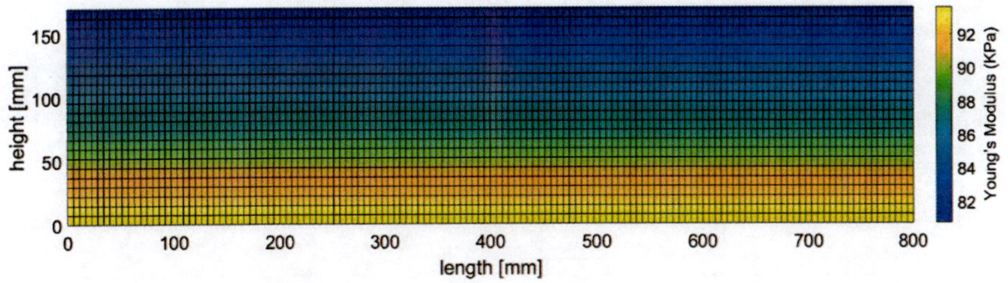

Fig. 2. Time-dependent Young's modulus at the end of the printing process.

3.2 Result

The principal strain of the bottom layer is shown in Fig. 3 from the layer-wise Finite Element analysis. As printing progresses, the stress in the underlying layers increases with each layer deposited, adding further strain.

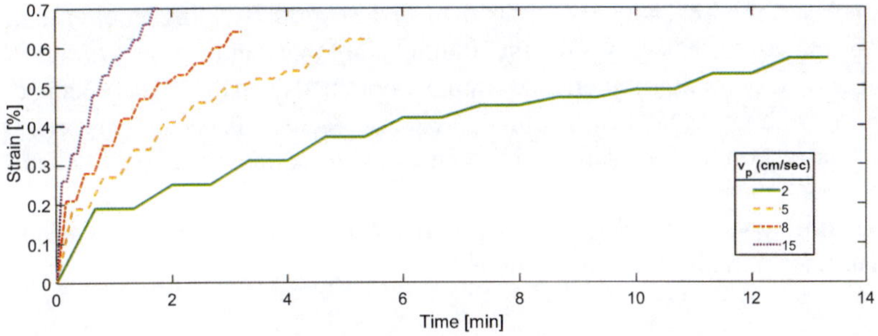

Fig. 3. Bottom layer strain as a function of time for various printing velocities.

In addition, the resting period between successive layers varies based on the printing speed. When printing at a faster speed, the resting time is shorter, resulting in more strain than when printing at a slower speed. As expected, lowering the printing velocity reduces strain; however, despite these benefits, we must keep the printing velocity within a certain range to avoid layer interface strength failure.

For this specific numerical example, raising the printing velocity from 2 cm/s to 15 cm/s resulted in a strain increase of 22.8%. The strain in each layer should be kept below the minimal critical strain to achieve a stable and load-bearing structure. The selected printing velocities satisfy the criteria, showing that the printed wall layers are intolerable deformation, based on [13] critical strain value of 2%.

When the printing velocity is increased, the other investigation is to determine the maximum height of the wall that can be built without strength-based failure or overall buckling. The numerical analysis output is displayed in Fig. 4 using Eq. 4 and Eq. 8. The findings are variable for the two failure scenarios, for the specified different printing velocities.

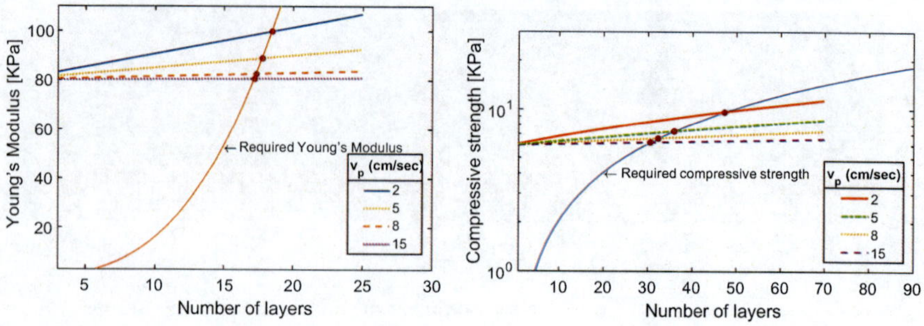

Fig. 4. Left - required elastic Young modulus and right- required compressive strength as a function of the number of printed layers, the failure was shown by the red dots.

As expected, depending on the printing velocity used, different heights can be built without failure. Results indicate that as most researchers agreed the 3D printed concrete wall structures are more susceptible to failure by buckling rather than plastic collapse.

For example, with a printing speed of 5 cm/s, the wall can be printed up to 0.36 m without plastic collapse, but only 0.18 m to meet the overall buckling requirement. Lower printing velocity also increases the maximum height that can be built without failure.

The geometry for this numerical example is obtained from the [7] in order to compare the findings. Choosing an 8 cm/s printing velocity results in 0.174 m that can be printed without elastic failure. This is in good agreement with the finding in [7], which is 0.179 m for the same geometry and printing velocity.

The printing velocity influences the failure mechanism of 3D printed wall structures, as can be shown in all of the cases studied.

4 Conclusion

The time-dependent printing process is efficiently simulated using a numerical modeling approach for a 3D concrete printed wall structure. In order to steer the printing process optimally, information on an appropriate printing velocity in relation to the failure mechanisms is useful.

The influence of printing velocity on the maximum height of the wall that may be built without failure is studied in a basic 2D numerical example.

The result indicates varying the printing velocity results in different heights of the wall to be built. Lowering the printing velocity enables the wall to be built to be greater in height while raising the printing velocity causes the wall to be lower in height for the selected printing velocities. The overall buckling is the governing failure mechanism for the 3D printed concrete wall, as numerous researchers and experimental programs have shown.

Further studies will include additional material and process parameters, and the effect of the printing velocity on layer interface strength and failure probability.

Acknowledgment. The work has been financially supported by different institutions which is highly acknowledged. Among them are: DAAD (Ethiopian - German Exchange of PhD candidates), DFG (German Research Foundation) priority program 1886 "Polymorphic uncertainty modelling for the numerical design of structures" and the Federal State of Thuringia, Germany.

References

1. Wangler, T., et al.: Digital concrete: opportunities and chlenges. RILEM Tech. Lett. **1**, 67 (2016)
2. Roussel, N.: Rheological requirements for printable concretes. Cem. Concr. Res. **112**, 76–85 (2018)
3. Buswell, R.A., Leal de Silva, W.R., Jones, S.Z., Dirrenberger, J.: 3d printing using concrete extrusion: a roadmap for research. Cem. Concr. Res. **112**, 37–49 (2018)
4. Wolfs, R.J.M., Bos, F.P., Salet, T.A.M.: Early age mechanical behaviour of 3d printed concrete: numerical modelling and experimental testing. Cem. Concr. Res. **106**, 103–116 (2018)
5. Perrot, A., Rangeard, D., Nerella, V.N., Mechtcherine, V.: Extrusion of cement-based materials - an overview. RILEM Tech. Lett. **3**, 91–97 (2019)
6. Marchon, D., Kawashima, S., Bessaies-Bey, H., Mantellato, S., Ng, S.: Hydration and rheology control of concrete for digital fabrication: potential admixtures and cement chemistry. Cem. Concr. Res. **112**, 96–110 (2018)
7. Suiker, A.S.J.: Mechanical performance of wall structures in 3d printing processes: Theory, design tools and experiments. Int. J. Mech. Sci. **137**, 145–170 (2018)
8. Van Der Putten, J., De Schutter, G., Van Tittelboom, K.: The effect of print parameters on the (micro)structure of 3d printed cementitious materials. In: Wangler, T., Flatt, R.J. (eds.) DC 2018. RB, vol. 19, pp. 234–244. Springer, Cham (2019). https://doi.org/10.1007/978-3-319-99519-9_22
9. Malaeb, Z., Hachem, H., Tourbah, A., Maalouf, T., el Zarwi, N., Hamzeh, F.: 3D concrete printing: machine and mix design. Int. J. Civil Eng. Technol. **6**, 14–22 (2015)
10. Le, T.T., et al.: Hardened properties of high-performance printing concrete. Cem. Concr. Res. **42**(3), 558–566 (2012)
11. Casagrande, L., Esposito, L., Menna, C., Asprone, D., Auricchio, F.: Effect of testing procedures on buildability properties of 3D-printable concrete. Constr. Build. Mater. **245**, 118286 (2020)
12. Bendsøe, M.P., Sigmund, O.: Material interpolation schemes in topology optimization. Arch. Appl. Mech. **69**(9), 635–654 (1999)
13. Roussel, N., Ovarlez, G., Garrault, S., Brumaud, C.: The origins of thixotropy of fresh cement pastes. Cem. Concr. Res. **42**(1), 148–157 (2012)
14. Mengesha, M., Schmidt, A., Göbel, L., Lahmer, T.: Numerical modeling of an extrusion-based 3D concrete printing process considering a spatially varying pseudo-density approach. In: Bos, F.P., Lucas, S.S., Wolfs, R.J.M., Salet, T.A.M. (eds.) DC 2020. RB, vol. 28, pp. 323–332. Springer, Cham (2020). https://doi.org/10.1007/978-3-030-49916-7_33
15. Perrot, A.: 3D Printing of Concrete, 1st edn. ISTE Ltd and John Wiley & Sons, Inc., Great Britain and the United States (2019)

Numerical Simulation of Concrete Fracture by Means of a 3D Probabilistic Explicit Cracking Model

Magno T. Mota[1](✉), Eduardo de M. R. Fairbairn[1], Fernando L. B. Ribeiro[1], Pierre Rossi[2], Jean-Louis Tailhan[2], Henrique C. C. Andrade[1], and Mariane R. Rita[1]

[1] Universidade Federal do Rio de Janeiro, Rio de Janeiro, Brazil
magnomota@coc.ufrj.br
[2] Université Gustave Eiffel, Paris, France

Abstract. The fracture process of concrete involves phenomena of considerable complexity, such as scale effect and softening behavior, which are directly linked to the heterogeneity of this material. These phenomena pose great challenges to the fracture modeling. Probabilistic models that deal directly with heterogeneity are a powerful tool to overcome these challenges, since they consider the natural variability of the mechanical responses. The present work refers to the simulation of concrete fracture by means of a 3D probabilistic finite element model in which interface elements are employed to represent the cracks explicitly. Friction between crack surfaces is also taken into account in the model. The three-dimensional modeling of cracks allows the fracture process to be analyzed in a more realistic way. Different sample sizes were considered to enable the assessment of the scale effect prediction, taking into account empirical reference data. The possibility of occurrence of different softening levels was also investigated.

Keywords: Concrete fracture · Probabilistic cracking model · Scale effect · Tensile failure · FEM

1 Introduction

The fracture process of concrete involves relevant phenomena, such as scale effect and softening behavior, which are directly linked to the heterogeneity and microstructural complexity of this material.

Studies have shown that probabilistic cracking models that directly deal with heterogeneity are a promising alternative for numerical simulation of concrete structures [1–10]. Through the statistical distribution of mechanical properties and the possible application of fracture mechanics concepts, the analysis of the failure process of structures can be performed in a more realistic and comprehensive way.

Rossi and Richer [1] proposed the probabilistic explicit cracking approach, which is based on the finite element method (FEM) and is primarily characterized by the use of interface elements to generate local discontinuities explicitly. This approach has advantages in terms of crack opening assessment and enables the development of strategies for

P. Rossi and J.-L. Tailhan (Eds.): SSCS 2022, RILEM Bookseries 38, pp. 248–257, 2023.
https://doi.org/10.1007/978-3-031-07746-3_25

directly handling phenomena that occur during the fracture process, such as the contact between crack surfaces. Limitations regarding the applicability of the approach refer to its high computational cost, mainly in 3D analyses.

The present work concerns the use of a 3D probabilistic cracking model based on the explicit cracking approach in order to simulate the failure of concrete specimens subjected to uniaxial tensile test. Some important characteristics of the applied model are as follows: employment of quadratic interpolation functions in the elements; use of the Weibull statistical distribution to take into account the material heterogeneity; and adoption of a friction model for handling the contact between crack surfaces. In order to deal with the high computational cost, a parallelized Monte Carlo method is applied.

2 Scale Effect Relations

Empirical relationships obtained by Rossi *et al.* [11] define correlations between concrete heterogeneity and scale effect, contributing to the validation of probabilistic models. The expressions depend on the V_T/V_A ratio, which is a parameter of heterogeneity, and the ultimate compressive strength f_c. The parameter V_T is the total volume of the specimen, and V_A is the volume of the coarsest aggregate, modeled by a sphere. In this way, the mean value of tensile strength or mean tensile strength (f_t^{mean}) and the coefficient of variation (CV), which is a measure of dispersion defined as the ratio between the standard deviation (SD) and the mean tensile strength, are determined by the following relations

$$f_t^{mean} = a(V_T/V_A)^{-\gamma} \tag{1}$$

$$CV = A(V_T/V_A)^{-B} \tag{2}$$

$$\gamma = 0.25 - 3.6 \times 10^{-3}\frac{f_c}{c_1} + 1.3 \times 10^{-5}\left(\frac{f_c}{c_1}\right)^2 \tag{3}$$

$$B = 4.5 \times 10^{-2} + 4.5 \times 10^{-3}\frac{f_c}{c_1} - 1.8 \times 10^{-5}\left(\frac{f_c}{c_1}\right)^2 \tag{4}$$

with $a = 6.5$ MPa, $A = 0.35$ and $c_1 = 1$ MPa.

Expressions (1)–(4) are applicable to any concrete, except fiber reinforced and lightweight concretes, with f_c ranging from 35 MPa to 130 MPa and V_T/V_A ratio varying from 10 to 6333.

3 3D Probabilistic Explicit Cracking Model

The probabilistic cracking model used in this work is outlined in Fig. 1. The three-dimensional treatment is given by the use of tetrahedral solid elements of 10 nodes and zero-thickness interface elements of 6 + 6 nodes geometrically idealized as wedges. Quadratic interpolation functions are adopted for both types of elements. The probabilistic character is established by the random assignment of tensile strength to the interface elements according to the Weibull probability distribution, whose parameters associated with mean and dispersion are known as scale and shape parameters. The interface elements follow a perfectly rigid and brittle behavior under tension, which is naturally associated with the Rankine failure criterion. This results in geometrical and material discontinuities when the stress reaches the tensile strength, leading the interface element to a cracked state. The solid elements have a linear elastic behavior. The acquisition of structural responses occurs by means of the Monte Carlo method, which corresponds to analyzing a significant number of computational samples having different random values of tensile strength generated from Weibull distribution defined by the same scale and shape parameters.

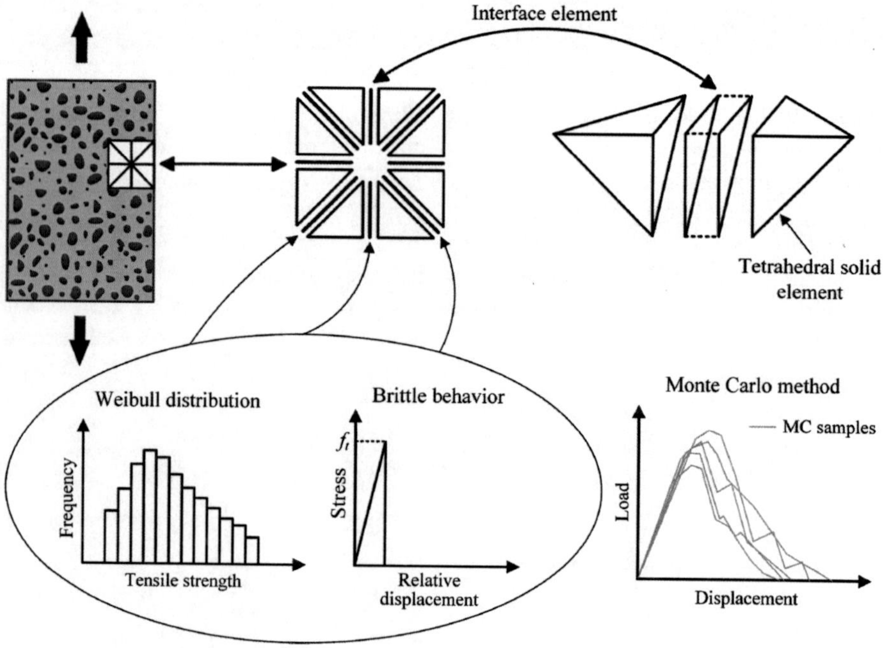

Fig. 1. General aspects of the 3D probabilistic model for explicit cracking of concrete.

Mota *et al.* [9] and Mota [10] provide more information about the FEM formulation, generation of random tensile strength, friction model, computational program, inverse analysis procedure and parallelization strategies.

3.1 Generation of Random Values of Tensile Strength

Regarding the Weibull distribution, the cumulative distribution function and the inverse cumulative distribution function related to a random variable x can be written respectively as

$$F(x, b, c) = 1 - e^{-\left(\frac{x}{c}\right)^b} = y \tag{5}$$

$$G(y, b, c) = c(-\ln(1-y))^{\frac{1}{b}} = x \tag{6}$$

in which $x \geq 0$, with this variable corresponding to the tensile strength in the present work; b is the shape parameter, associated with the coefficient of variation; and c is the scale parameter, directly related to the mean value of x.

The generation of random values of tensile strength is based on the inverse transformation method. The basic steps for generation and distribution of these values are summarized in Algorithm 1.

Algorithm 1 Basic steps for generation and distribution of random tensile strength

1. *n = 0; // interface element identifier*
2. ***do*** *(until the last interface element)*
3. $n = n + 1;$
4. *Generate a random number y between 0 and 1 from uniforme distribution;*
5. *Compute the random number x using Eq. (6);*
6. $f_t(n) = x;$ *// tensile strength of the n-th element*
7. ***end do***

3.2 Friction Model

The friction model elaborated for the contact between faces of cracked interface elements is illustrated in Fig. 2. It is based on the Mohr-Coulomb failure criterion without cohesion:

$$|\tau| - \tau_m \leq 0 \tag{7}$$

in which τ is the average shear stress, or the frictional stress; and $\tau_m = \sigma_c \tan\phi$, with ϕ and σ_c being the angle of friction and the average absolute value of the normal compressive stress in the cracked interface element, respectively.

The variables w_e, w_a, w_r, w_b, w_d and w_f are tangential relative displacements (TRD), and τ_e is the initial effective frictional stress. The parameters w_e and w_d are respectively the effective and elastic displacements, and w_r and w_f are the initial and final total residual displacements, respectively.

The friction model, in the condition represented in Fig. 2, can be mathematically summarized in the following expressions:

$$w_e = w_a - w_r \tag{8}$$

$$\tau_e = k_a w_e \tag{9}$$

$$w_d = \frac{\tau_m}{k_a} \tag{10}$$

$$w_b = w_e - w_d \tag{11}$$

$$w_f = w_r + w_b \tag{12}$$

with k_a being numerically equal to $\tan\theta_t$.

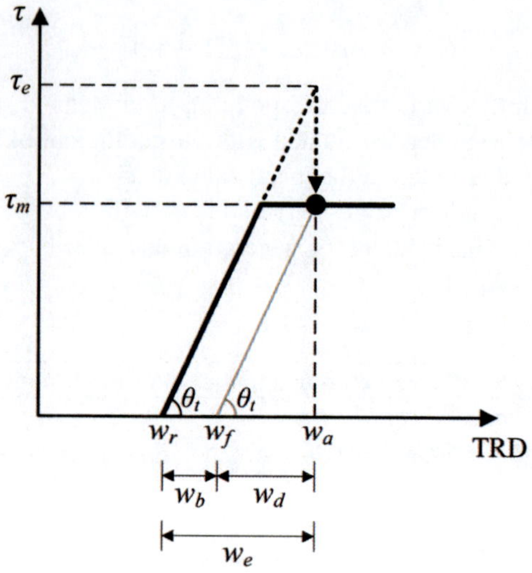

Fig. 2. Graphical representation of the friction model.

4 Numerical Modeling Procedures

Numerical simulations of tensile tests were performed based on the experimental investigation conducted by Rossi *et al.* [11]. The reasons that led to the choice of this experimental study were as follows:

- it refers to the concrete behavior in direct tensile test, in which the transition from diffuse to localized cracking is the most important phenomenon and the most difficult to capture;
- concretes with different mechanical properties were produced, allowing the evaluation of different scenarios;
- a considerable number of experimental samples were tested, implying statistical relevancy;
- experimental results are expressed by Eqs. (1)–(4) with great accuracy, so that it is entirely possible to use these equations to provide expected values according to the scale effect.

Prismatic samples (Fig. 3), namely P1 and P2, were simulated. In the mesh of both prisms, elements with approximately the same size were generated. Details relating to the mesh and geometry of P1 and P2 are displayed in Table 1, with $2Ve$ denoting the average volume of the two solid elements adjacent to the interface elements. Table 2 shows the compressive strength, the modulus of elasticity (E) and the volume of the coarsest aggregate of the considered concretes. The Weibull distribution parameters, determined by means of inverse analysis with the prism P1 [10], are presented in Table 3. Regarding the angle of friction, the adopted value was 36°.

Fig. 3. Geometry and mesh of the computational samples.

Table 1. Details relating to geometry and mesh of the computational samples.

Geometry	Height (dm)	Cross-section (dm^2)	Solid elements	Interface elements	Nodes	$2V_e(10^{-3}\ dm^3)$
P1	2.2	0.9503	837	1522	8370	5.0
P2	3.0	1.7671	2141	4010	21410	

Table 2. Properties of the simulated concretes.

Concretes	f_c (MPa)	E (GPa)	V_A (10^{-3} dm^3)
C1	35.0	39.8	4.2
C2	55.8	45.3	
C3	127.5	53.9	

Table 3. Weibull distribution parameters.

Samples	b (MPa)	c (MPa)
P1-C1 and P2-C1	1.17	11.35
P1-C2 and P2-C2	2.82	8.25
P1-C3 and P2-C3	23.98	7.64

5 Results

The results were obtained by simulating a total of 80 Monte Carlo samples. The values of mean tensile strength, standard deviation and coefficient of variation related to the numerical model, along with the corresponding empirical values, can be seen in Table 4. The parameter Δ denotes the discrepancy between numerical and empirical f_t^{mean}. By comparing the simulation results related to P1 and P2 for the same concretes, scale effect occurrences are observed in all cases. The absolute values of Δ for P1-C1, P1-C2 and P1-C3 are relatively very small due to the fact that the calibration was performed with these samples. In relation to P2-C1, P2-C2 and P2-C3, the values of Δ stay within an acceptable range. The numerical modeling results for *SD* and *CV* also show good similarity with the empirical ones.

Table 4. Simulation results.

Samples	Empirical			Numerical			Δ (%)
	f_t^{mean} (MPa)	*SD* (MPa)	*CV*	f_t^{mean} (MPa)	*SD* (MPa)	*CV*	
P1-C1	2.72	0.31	0.11	2.73	0.29	0.11	0.05
P1-C2	3.73	0.29	0.08	3.77	0.29	0.08	1.2
P1-C3	6.41	0.30	0.05	6.45	0.28	0.04	0.6
P2-C1	2.39	0.23	0.10	2.60	0.22	0.09	8.6
P2-C2	3.43	0.22	0.06	3.56	0.20	0.06	4.0
P2-C3	6.39	0.22	0.03	6.12	0.24	0.04	−4.3

Figure 4 presents the 80 complete load-displacements curves for the Monte Carlo samples of P2-C1, P2-C2 and P2-C3. In order to allow the global response of these samples to be examined in detail, Fig. 5 shows typical load-displacement curves for the analyzed concretes with displacement ranging from 0 to 1.5 × 10^{-3} dm. It is possible to notice that the lower the value of *b*, the higher the level of nonlinearity in the pre-peak response and the greater the prominence of the softening behavior. Considering that the values of *c* for P2-C2 and P2-C3 are relatively similar, it can be stated that the f_t^{med} value of samples P2-C3 is considerably greater than that of samples P2-C2 due essentially to the increase in the value of *b*.

The cracking process of the computational samples is illustrated in Fig. 6, which exhibits the typical cracking process of P2-C2 samples. The localized fracture process can be clearly noticed, indicating that the transition from diffuse to localized cracking is being captured.

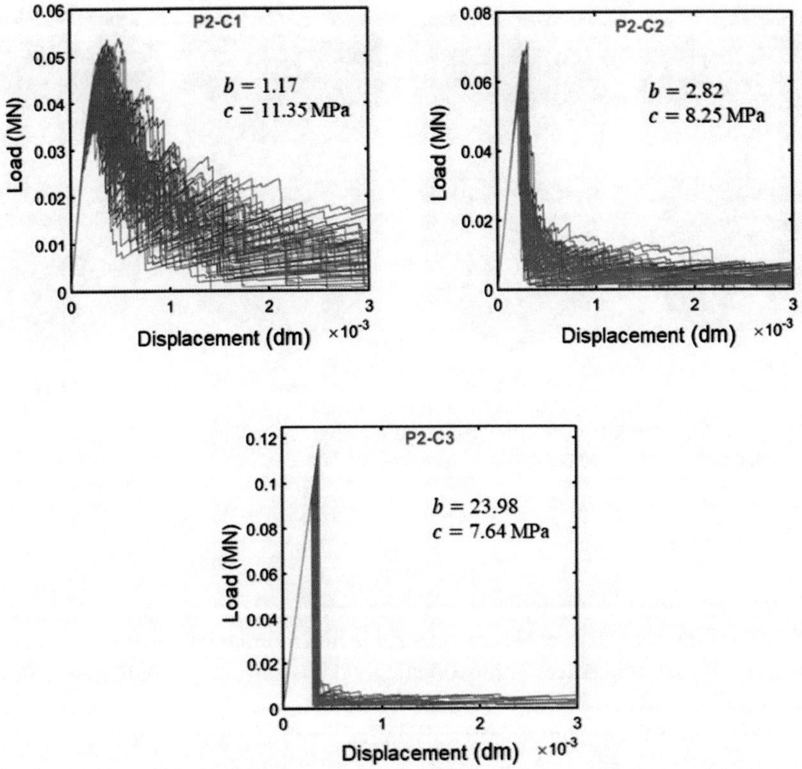

Fig. 4. Complete load-displacement curves of 80 Monte Carlo samples, for P2-C1, P2-C2 and P2-C3.

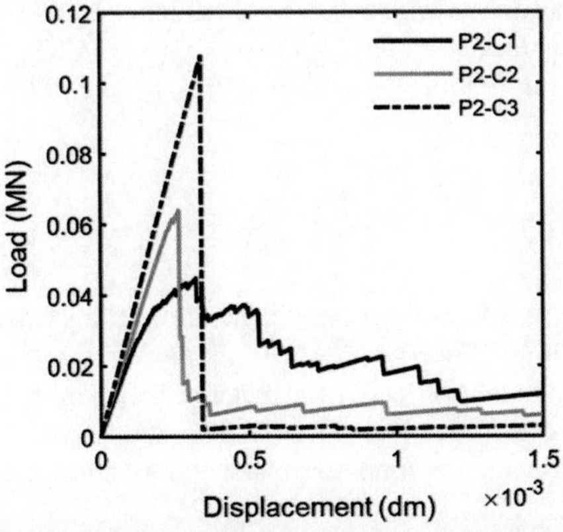

Fig. 5. Typical load-displacement curves for P2-C1, P2-C2 and P2-C3, with displacement ranging from 0 to 1.5×10^{-3} dm.

Fig. 6. Typical cracking process of P2-C2 samples. Crack pattern for different stages of the load-displacement curve: (a) peak; (b,c,d) post-peak.

6 Conclusions

A 3D probabilistic cracking model based on the explicit cracking approach was used to simulate the failure of concrete specimens subjected to uniaxial tensile test. The results demonstrate that the considered probabilistic model is capable of predicting scale effect occurrences in good agreement with empirical results.

The Weibull parameter b can strongly influence the fracture process. Variations in this parameter can significantly alter the mean peak load values and the nonlinearity in the pre- and post-peak behaviors of load-displacement diagrams of Monte Carlo samples.

With regard to the crack pattern exhibited by the computational samples, the important phenomenon of crack coalescence can be clearly observed in the post-peak branch of the load-displacement responses.

Acknowledgments. This study was financed by Brazilian scientific agencies, namely the *Conselho Nacional de Desenvolvimento Científico e Tecnológico* (CNPq), and the *Coordenação de Aperfeiçoamento de Pessoal de Nível Superior* (CAPES) - Finance Code 001.

References

1. Rossi, P., Richer, S.: Numerical modelling of concrete cracking based on a stochastic approach. Mater. Struct. **20**, 334–337 (1987)
2. Rossi, P., Wu, X.: Probabilistic model for material behavior analysis and appraisement of concrete structures. Mag. Concr. Res. **44**(161), 271–280 (1992)
3. Rossi, P., Ulm, F.-J., Hachi, F.: Compressive behavior of concrete: physical mechanisms and modeling. J. Eng. Mech. **122**(11), 1038–1043 (1996)
4. Fairbairn, E.M.R., Guedes, Q.M., Ulm, F.-J.: An inverse problem analysis for the determination of probabilistic parameters of concrete behaviour modeled by a statistical approach. Mater. Struct. **32**, 9–13 (1999)

5. Fairbairn, E.M.R., Ebecken, N.F.F., Paz, C.N.M., Ulm, F.-J.: Determination of probabilistic parameters of concrete: solving the inverse problem by using artificial neural network. Comput. Struct. **78**, 497–503 (2000)
6. Tailhan, J.-L., Dal Pont, S., Rossi, P.: From local to global probabilistic modeling of concrete cracking. Ann. Solid Struct. Mech. **1**, 103–115 (2010)
7. Tailhan, J.-L., Rossi, P., Daviau-Desnoyers, D.: Probabilistic numerical modelling of cracking in steel fibre reinforced concretes (SFRC) structures. Cem. Concr. Compos. **55**, 315–321 (2015)
8. Rossi, P., Daviau-Desnoyers, D., Tailhan, J.-L.: Probabilistic numerical model of cracking in ultra-high performance fibre reinforced concrete (UHPFRC) Beams Subjected to Shear Loading. Cem. Concr. Compos. **90**, 119–125 (2018)
9. Mota, M.T., et al.: A 3D probabilistic model for explicit cracking of concrete. Comput. Concr. **27**(6), 549–562 (2021)
10. Mota, M.T.: 3D adaptive probabilistic model for explicit cracking of concrete (in Portuguese). PhD thesis, Universidade Federal do Rio de Janeiro, Rio de Janeiro, Brazil (2021)
11. Rossi, P., Wu, X., Le Maou, F., Belloc, A.: Scale effect on concrete in tension. Mater. Struct. **27**, 437–444 (1994)

Incremental Formulation of Early Age Concrete in the Finite Strain Range for the Modelling of 3D Concrete Printing

Boumediene Nedjar(✉) and Zeinab Awada

MAST-EMGCU, Université Gustave Eiffel, 77454 Marne-la-Vallée, France
{boumediene.nedjar,zeinab.awada}@univ-eiffel.fr

Abstract. During a 3D concrete printing process, the mechanical charactesitics of the material at hand are continuously evolving due to hydration. From the theoretical point of view, the constitutive relations must be defined in rate form. We think that this restriction is mandatory and must be taken into account. On another hand, to predict eventual instabilities, the finite strain range is a priori assumed within the formulation where the kinematics must be adapted in adequacy. Here use is made of a multiplicative decomposition of the actual deformation gradient into its known intermediate part at an earlier time and the relative deformation gradient with respect to the configuration at that time. The incremental constitutive relations and evolution equations can then be ideally defined on the above intermediate configuration prior to be transported back to the reference configuration within a Lagrangian formulation. The early age creep is here introduced through an internal variable approach that is motivated by the generalized Maxwell model. A set of finite element simulations is performed to illustrate the efficiency of the proposed framework. In particular, a slump-test-like example is given where a method can be highlighted to help identifying, in future works, the material parameters relative to the early-age creep.

Keywords: Additive manufacturing · Incremental formulation · Early age · Large deformations · Instabilities

1 Introduction

Concrete cracking at early age is a major issue, especially for massive structures. By essence, a thermo-chemo-mechanical analysis is required where, for instance, autogenous strains and the exothermic reaction of the cement paste must be considered, e.g. [3]. Moreover, as the mechanical characteristics evolve with hydration, the reversible part of the stress-strain relationship is of the hypo-elastic type, i.e. incremental elasticity, see for example [1,4] among the abundant literature on early-age concrete. Within the assumption of small perturbations, the constitutive relation is therefore of the form,

P. Rossi and J.-L. Tailhan (Eds.): SSCS 2022, RILEM Bookseries 38, pp. 258–267, 2023.
https://doi.org/10.1007/978-3-031-07746-3_26

$$\Delta\boldsymbol{\sigma} = \boldsymbol{H}(\xi(t)) : \Delta\boldsymbol{\varepsilon} \equiv \boldsymbol{H}(t) : \Delta\boldsymbol{\varepsilon}, \tag{1}$$

where $\Delta\boldsymbol{\sigma}$ and $\Delta\boldsymbol{\varepsilon}$ are respectively the increments of stress and strain tensors, and the fourth-order tensor $\boldsymbol{H}$ is the Hooke-type elasticity that is time-dependent due to the evolving hydration $\xi(t)$. Let us stress out that Eq. (1) should not be confused with the direct form $\boldsymbol{\sigma} = \boldsymbol{H}(\xi(t)) : \boldsymbol{\varepsilon}$. They could become identical only for fixed elastic parameters, which is not the case for an ageing concrete.

Nowadays, the challenging 3D concrete printing technology is receiving considerable attention in civil engineering; the 'Construction 4.0' as termed in [2]. In general, a concrete mixing in pumped into a pipe connected to a nozzle, and with the help of a robotic mechanism, precise positioning control and deposition path can be achieved. After several layers have been deposited onto each other, the fresh (early-age) concrete must be able to retain its shape due to the lack of confining formworks, i.e. the fresh concrete must carry its self-weight. As the mechanical properties are initially low, possible collapse during the printing process can occur. A predictive theory must then be established within a geometrically nonlinear framework to be able to predict such instabilities. Hence, the above hypo-elastic form, Eq. (1), should be extended toward the finite strain range. On the one hand, the mechanical equilibrium equation is rewritten in an adequate form because, in this case, the stress is given in rate (evolutive) form and, on the other hand, the gravity cannot be incremented since it is imposed all at once. These peculiarities have been derived in [7] together with an adequate kinematical choice. For this latter, use is made of a multiplicative decomposition of the actual deformation gradient into its known part an earlier time and the relative deformation gradient with respect to the configuration at that time. This gives rise to an intermediate configuration on which incremental constitutive relations for stresses and evolution equations for early-age creep can be well defined. Herein, a Lagrangian formulation is adopted: the chosen incremental constitutive relations are then pull-back from the known intermediate configuration toward the reference (initial) configuration. We have shown that the above kinematics is well suited for this transport procedure.

Modeling examples are proposed that are inspired from nowadays well-known models widely used in the literature, such like a Saint-Venant- and a compressible neoHooke-like models [5]. For early-age creep, the concept of internal variables is adopted the evolution of which is motivated by the generalized Maxwell rheological model [8]. Here a stress-like internal variable is chosen that consists on an over-stress that has the same structure as the thermodynamically equilibrated stress tensor increment. So far, focus is made on purely mechanical aspects. Topics of thermo-hydric couplings will be addressed in future contributions.

2 Basic Equations

A material particle is identified by its position vector $\boldsymbol{X}$ in the reference configuration $\mathcal{B}_0$. At the current time t, the new position is traced as $\boldsymbol{x} = \boldsymbol{\varphi}_t(\boldsymbol{X})$ in the spatial configuration $\mathcal{B}_t$, where $\boldsymbol{\varphi}_t(\centerdot) \equiv \boldsymbol{\varphi}(\centerdot, t)$ denotes the deformation map at time t within a time interval $[0, T]$. The deformation gradient is defined as $\boldsymbol{F} \equiv \boldsymbol{F}_t = \nabla_{\boldsymbol{X}} \boldsymbol{\varphi}_t$, where $\nabla_{\boldsymbol{X}}(\boldsymbol{\cdot})$ is the material gradient operator with respect to $\boldsymbol{X}$. The determinant $J = \det[\boldsymbol{F}]$ is the Jacobian of the transformation with the standard convention $J > 0$.

2.1 Mechanical Balance

The mechanical balance is equivalently given by the following weak form in terms of the second Piola-Kirchhoff stress tensor $\boldsymbol{S} \equiv \boldsymbol{S}_t$:

$$\int_{\mathcal{B}_0} \boldsymbol{S} : \boldsymbol{F}^T \nabla_{\boldsymbol{X}} (\delta\boldsymbol{\varphi})\, \mathrm{d}V = \int_{\mathcal{B}_0} \rho_0 \bar{\boldsymbol{b}}.\delta\boldsymbol{\varphi}\, \mathrm{d}V, \tag{2}$$

that holds for any admissible variation $\delta\boldsymbol{\varphi}$ of the deformation. Here ρ_0 is the initial density, and $\rho_0\bar{\boldsymbol{b}}$ defines the body force vector due to the self-weight. Unlike classical constitutive relations where stresses are directly linked to strain measures, here the stress is restricted to be defined in incremental form, see Eq. (1). Within a typical time interval $[t_n, t_{n+1}]$, we then make the following choice:

$$\boldsymbol{S} = \boldsymbol{S}_n + \Delta\boldsymbol{S}, \tag{3}$$

where $\boldsymbol{S}_n$ is the *known* stress tensor at time t_n, and $\Delta\boldsymbol{S} \equiv \Delta\boldsymbol{S}_t$ is the second Piola-Kirchhoff stress tensor increment at the actual time $t \in [t_n, t_{n+1}]$. Thus, when replacing (3) into (2), we end up with the basic balance equation to be solved in our formulation [7]:

$$\int_{\mathcal{B}_0} (\boldsymbol{S}_n + \Delta\boldsymbol{S}) : \boldsymbol{F}^T \nabla_{\boldsymbol{X}} (\delta\boldsymbol{\varphi})\, \mathrm{d}V = \int_{\mathcal{B}_0} \rho_0 \bar{\boldsymbol{b}}.\delta\boldsymbol{\varphi}\, \mathrm{d}V. \tag{4}$$

This latter equation is valid for all classes of *incremental* constitutive relations through the definition of model-dependent stress increment $\Delta\boldsymbol{S}$ including, for instance, creep and/or plasticity.

2.2 Kinematics

In view of Eq. (4), the kinematical choice must be adapted as well. Let $\boldsymbol{F}_n$ be the deformation gradient at time t_n, its update $\boldsymbol{F}$ is given by the following multiplicative decomposition, see the sketch of Fig. 1,

$$\boldsymbol{F} = \tilde{\boldsymbol{f}} \boldsymbol{F}_n, \tag{5}$$

where $\tilde{\boldsymbol{f}} \equiv \tilde{\boldsymbol{f}}_t$ is the relative deformation gradient [11]. With this latter one can define the relative right Cauchy-Green tensor $\tilde{\boldsymbol{c}}$ with respect to the intermediate configuration $\boldsymbol{\varphi}_n$ as,

$$\tilde{\boldsymbol{c}} = \tilde{\boldsymbol{f}}^T \tilde{\boldsymbol{f}}, \tag{6}$$

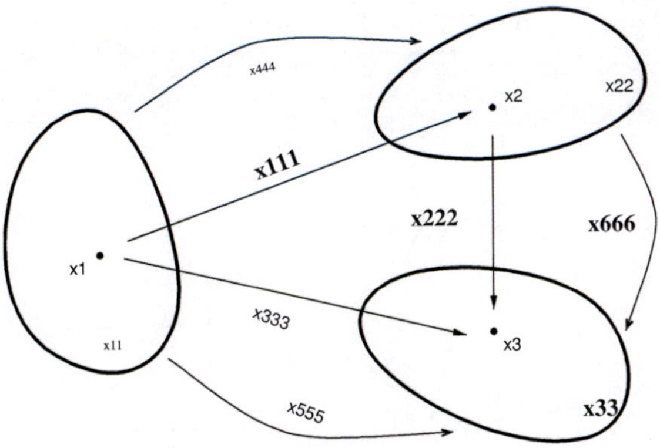

Fig. 1. Relative deformation gradient $\tilde{\boldsymbol{f}}_t$ connecting configurations $\mathcal{B}_n$ and $\mathcal{B}_t$. $\boldsymbol{F}_n$ is the total deformation gradient at time t_n, and $\boldsymbol{F}_t$ is its update at time $t \in [t_n, t_{n+1}]$.

which, in turn, is connected to the (total) right Cauchy-Green tensor $\boldsymbol{C} = \boldsymbol{F}^T\boldsymbol{F}$ through the relation,

$$\tilde{\boldsymbol{c}} = \boldsymbol{F}_n^{-T}\boldsymbol{C}\boldsymbol{F}_n^{-1}. \tag{7}$$

Formally, we first define the stress increment on the configuration $\boldsymbol{\varphi}_n$, i.e. the second Piola-Kirchhoff-type stress *increment*, that we denote by $\tilde{\boldsymbol{s}}$, and with the form,

$$\tilde{\boldsymbol{s}} \equiv \tilde{\boldsymbol{s}}(\tilde{\boldsymbol{c}}, \ldots), \tag{8}$$

in terms of the relative right Cauchy-Green tensor, and the dots refering to eventual internal variable. This stress increment is then transformed back to the reference configuration with the help of the pull-back procedure, e.g. [6],

$$\Delta\boldsymbol{S} = \boldsymbol{F}_n^{-1}\,\tilde{\boldsymbol{s}}\,\boldsymbol{F}_n^{-T}. \tag{9}$$

It is well adapted for incremental models inspired by many classical hyperelastic ones, for instance, Neo-Hooke, Mooney-Rivlin, Ogden ... e.g. [5,9,10]. Last but not least, Eq. (9) is to be replaced back into Eq. (4).

3 Incremental Viscoelastic Constitutive Relation

Motivated by the generalized Maxwel model, see Fig. 2, The second Piola-Kirchhoff stress $\boldsymbol{S}$ is additively split as,

$$\boldsymbol{S} = \boldsymbol{S}^{\infty} + \boldsymbol{Q}, \tag{10}$$

where $\boldsymbol{S}^{\infty}$ is the equilibrium part, and the introduced internal tensor variable $\boldsymbol{Q}$ may be interpreted as a non-equilibrium over-stress. Hence, within the time interval $[t_n, t_{n+1}]$, Eq. (3) becomes,

$$\boldsymbol{S} = \underbrace{\boldsymbol{S}_n^{\infty} + \boldsymbol{Q}_n}_{=\boldsymbol{S}_n} + \underbrace{\Delta\boldsymbol{S}^{\infty} + \Delta\boldsymbol{Q}}_{=\Delta\boldsymbol{S}}. \tag{11}$$

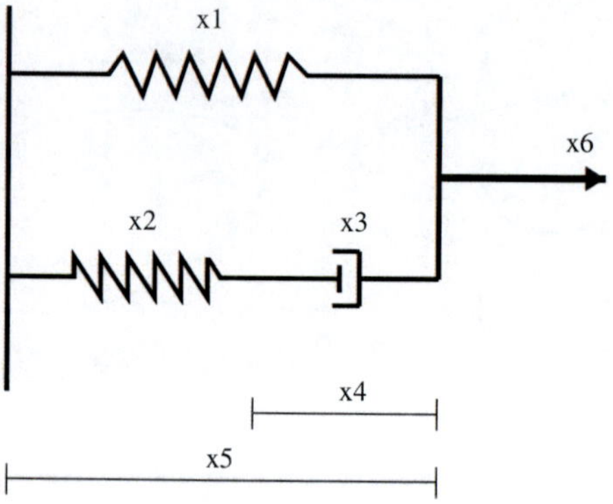

Fig. 2. Motivation: generalized Maxwell model.

We now have to define an incremental constitutive relation for the equilibrium part $\Delta\boldsymbol{S}^{\infty}$, and hence the update of $\boldsymbol{S}_{\infty}$. Then we will give a local evolution equation for the internal variable $\boldsymbol{Q}$, and hence the over-stress increment $\Delta\boldsymbol{Q}$. For the former, and inspired by a compressible version of the neoHooke model, we postulate the following relation for the equilibrated second Piola-Kirchhoff-type stress tensor increment[1], see [7] for more details:

$$\tilde{\boldsymbol{s}}^{\infty} = \lambda_{\infty}(t)\,\log[\tilde{j}]\,\tilde{\boldsymbol{c}}^{-1} + \mu_{\infty}(t)\,(\boldsymbol{1} - \tilde{\boldsymbol{c}}^{-1}), \tag{12}$$

where $\log[\boldsymbol{.}]$ is the natural logarithm function and $\tilde{j}$ is the jacobian of the relative deformation gradient, $\tilde{j} = \det[\tilde{\boldsymbol{f}}]$. Here and in all what follows, $\boldsymbol{1}$ is the second-order identity tensor with components δ_{ij} (δ_{ij} being the Kronecker delta). The time-dependent parameters $\lambda_{\infty}(t)$ and $\mu_{\infty}(t)$ are Lamé-like coefficients within the asymptotic infinitesimal limit. They are related to the time-dependent Young's modulus $E_{\infty}(t)$ and Poisson's ratio $\nu(t)$ of the equilibrium part as:

$$\lambda_{\infty}(t) = \frac{\nu(t)\,E_{\infty}(t)}{(1+\nu(t))(1-2\nu(t))} \qquad \mu_{\infty}(t) = \frac{E_{\infty}(t)}{2(1+\nu(t))} \tag{13}$$

[1] For the hyperelastic version of the model of Eq. (12), the strain energy function would be $W^{\infty} = \frac{1}{2}\lambda_{\infty}\log^2[J] - \mu_{\infty}\log[J] + \frac{1}{2}\mu_{\infty}(\boldsymbol{C} : \boldsymbol{1} - 3)$ in terms of the (total) right Cauchy-Green tensor $\boldsymbol{C}$. With the state law $\boldsymbol{S}_{\infty} = \frac{\partial W^{\infty}}{\partial \boldsymbol{C}}$, this gives $\boldsymbol{S}_{\infty} = \lambda_{\infty}\log[J]\boldsymbol{C}^{-1} + \mu_{\infty}(\boldsymbol{1} - \boldsymbol{C}^{-1})$.

Finally, the pull-back (9) applied to (12) gives the following equilibrium stress increment:

$$\Delta \boldsymbol{S}^{\infty} = \lambda_{\infty}(t) \log \left[\frac{J}{J_n} \right] \boldsymbol{C}^{-1} + \mu_{\infty}(t) \Big(\boldsymbol{C}_n^{-1} - \boldsymbol{C}^{-1} \Big), \tag{14}$$

with $J_n = \det[\boldsymbol{F}_n]$, $\boldsymbol{C}_n = \boldsymbol{F}_n^T \boldsymbol{F}_n$, and where use has been made of the kinematic relation $J = \tilde{j} J_n$.

Next, we need to specify an evolution equation for the internal variable $\boldsymbol{Q}$. This is motivated by the generalized Maxwell model of Fig. 2. Still within the time interval $[t_n, t_{n+1}]$, we have shown that the discrete update is given by, see [8] for full details:

$$\boldsymbol{Q}_{n+1} \;=\; e^{-\omega \Delta t} \boldsymbol{Q}_n \;+\; f \frac{1 - e^{-\omega \Delta t}}{\omega \Delta t} \Delta \boldsymbol{S}^{\infty}, \tag{15}$$

where $\Delta t = t_{n+1} - t_n$, and $\boldsymbol{Q}_n$ is the known (stored) value of $\boldsymbol{Q}$ at time t_n, while $\Delta \boldsymbol{S}^{\infty}$ has just been evaluated in Eq. (14). Here the adimensional parameter f is a scaling factor of the overstress contribution, chosen to be constant. The quantity ω is given by,

$$\omega \;=\; \frac{\dot{E}_{\infty}}{E_{\infty}} + \frac{1}{\tau} \tag{16}$$

in terms of the characteristic time τ and the rate of change of the early-age Young modulus $E_{\infty}(t)$.

Once $\boldsymbol{Q}_{n+1}$ is updated with the help of Eq. (15), the difference $\Delta \boldsymbol{Q} = \boldsymbol{Q}_{n+1} - \boldsymbol{Q}_n$ is to be evaluated and replaced into Eq. (11) for the (total) stress increment $\Delta \boldsymbol{S}$.

4 Numerical Example: Closed Wall

For illustrative purposes, we consider the closed wall described in Fig. 3 manufactured layer-by-layer, each layer with cross-section dimensions of 45 mm width and 10 mm height. For symmetry reasons, only one-fourth-wall is considered for the finite element discretization with adequate boundary conditions. We consider low mechanical properties so as to trigger structural buckling:

$$E_{\infty}(t) = 0.08 + 0.01t\,[\mathrm{MPa}], \quad \nu = 0.3, \quad f = 0.5, \quad \tau = 5\,\mathrm{min}, \tag{17}$$

where the time t expressed in minutes. The Young modulus E_{∞} is updated for each layer during the simulation. It enters into the definition of the Lamé-like coefficients of Eq. (13). For the gravity loading, we consider the density $\rho_0 = 2020\,\mathrm{kg/m^3}$.

At a printing speed of 0.3 min per layer, Fig. 4 shows selected deformed configurations after 20, 25 and 30 layers. Here the wall is straight in the though-the-layers' direction at least until 20 layers, then it starts to visibly deviate at around

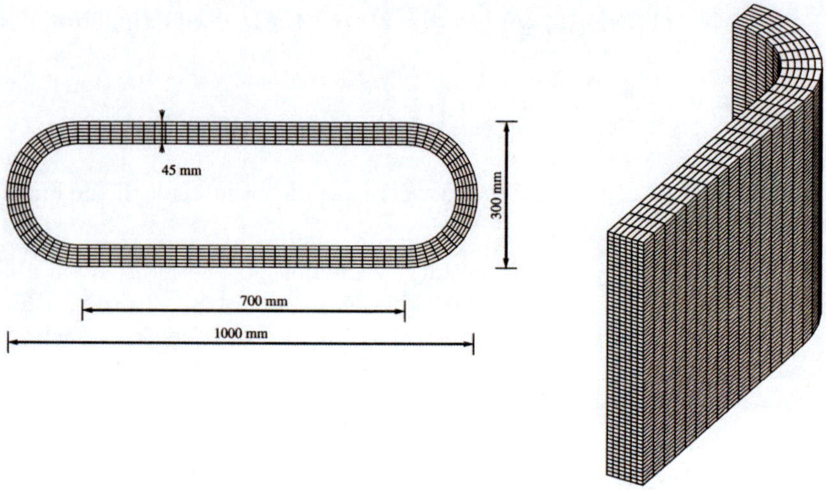

Fig. 3. Geometry of the closed wall (top view), and finite element discretization of 1/4-wall with up to 30 layers.

25 layers. The computed wall after 30 layers shows the very large deformation meaning collapse in bending mode.

For the sake of comparision, the same computation is now accomplished with a printing speed of 1.5 min per layer. As shown in Fig. 5, the wall remains almost straight until the end of the computation, here with 30 layers.

5 Numerical Example: A Slump-Test

As clearly shown, it is of major importance to correctly identify the mechanical parameters of the fresh concrete. This could certainly help optimizing the printing process for a given concrete formulation via numerical tools. A simple test would be for instance to use the very popular slump test. Indeed, at first view, it is obvious that different concrete mixes would slump differently. Hence, it would be advantageous to be able to numerically simulate such a test. This is the objective of this example where we show that the present theoretical and numerical framework can accomplish this task.

Let us consider a classical cone of 300 mm height, with a diameter of 200 mm at the bottom, and an upper diameter of 100 mm, see Fig. 6(a). The simulation is performed as an axisymmetric problem. A frictional contact is accounted for with the bottom horizontal surface. We consider the following very low early-age mechanical parameters (an almost liquid concrete):

$$E_\infty(t) = 0.0025 + 0.001t\,[\text{MPa}], \quad \nu = 0.3, \quad f = 2, \quad \tau = 0.075\,\text{min}, \tag{18}$$

with a density $\rho_0 = 2070\,\text{kg/m}^3$.

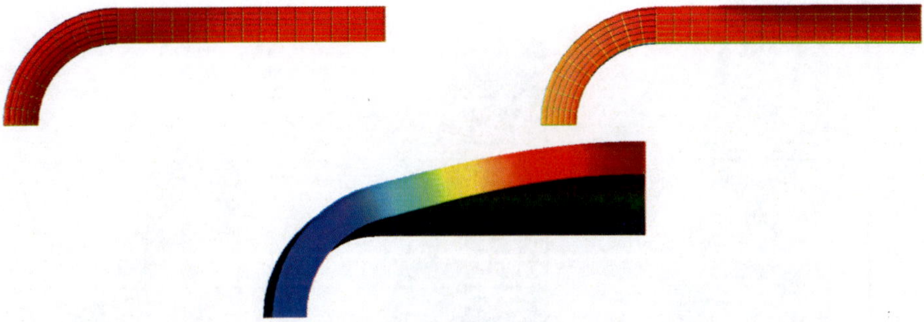

Fig. 4. Top views of deformed configurations at a printing speed of 0.3 mn/layer: after 20 layers (top left), 25 layers (top right), and 30 layers (bottom).

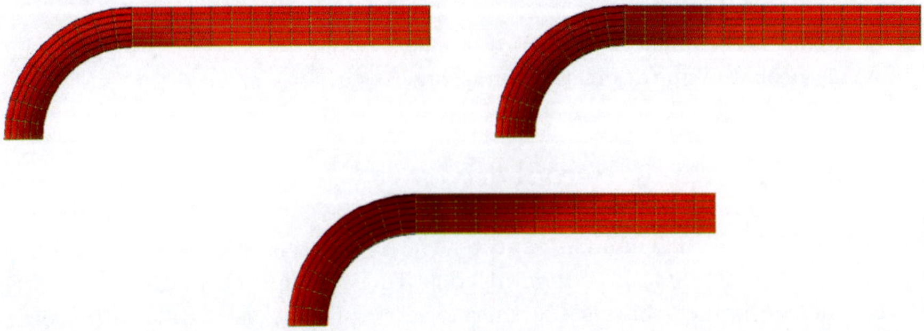

Fig. 5. Top views of deformed configurations at a printing speed of 1.5 mn/layer: after 20 layers (top left), 25 layers (top right), and 30 layers (bottom).

Figure 6(b) shows the deformed configuration as soon as the cone is removed. Here one can observe the instantaneous slump predicted by the model within the finite strain range. Next, the deformation still evolves due to the ealry-age viscoelasticity as illustrated in Fig. 6(c).

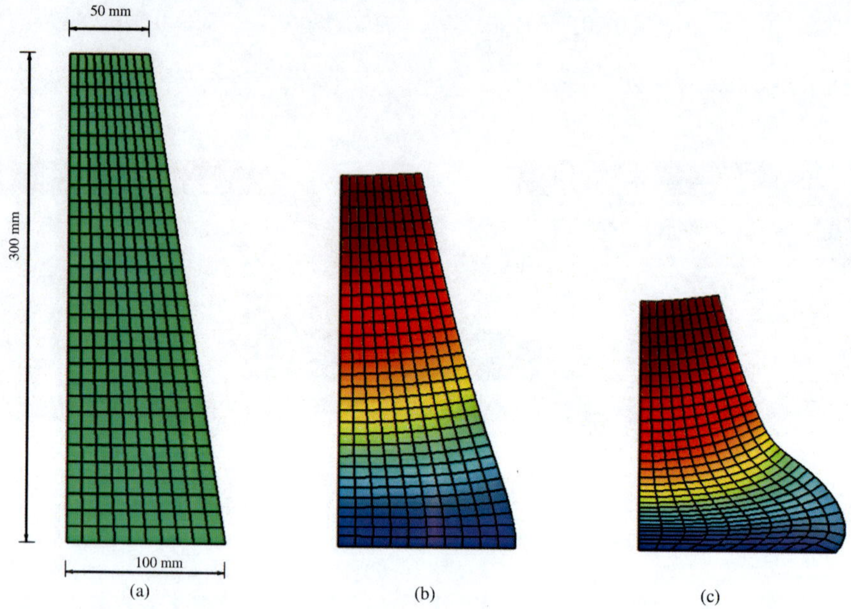

Fig. 6. Slump test simulation: (a) Axisymmetric geometry and finite element mesh, (b) instantaneous slump (c) delayed slump due to early-age creep.

6 Conclusion and Perspectives

Focusing on purely mechanical aspects, the fresh concrete has been described through time-dependent incremental constitutive relations in the finite strain range. The mechanical balance is therefore adapted so as to take into account this particularity. As a principal ingredient, the kinematical choice is based on the multiplicative decomposition of the deformation gradient into its known part at a precedent time and the relative deformation gradient. This latter constitutes the main strain quantity from which a constitutve relation must be built.

Furthermore, the early age creep has been motivated by the generalized Maxwell model written in rate form. Among others, only two parameters are introduced in addition to the incremental equilibrium elasticity: a characteristic time, and an adimensional factor representing the amplitude of the over-stress with respect to the equilibrium incremental elasticity.

Numerical examples have shown the efficiency of the whole procedure. In particular, eventual structural buckling can be captured, and hence circumvented in real tests. This feature could certainly help the optimization of the printing process through simulations that could limit the number of costly real experiments.

A future step will be the coupling with the hydration of concrete that, in turn, is strongly coupled to the exothermy of the hydration reaction.

References

1. Benboudjema, F., Torrenti, J.M.: Early-age behaviour of concrete nuclear containments. Nucl. Eng. Des. **238**(10), 2495–2506 (2008)
2. Craveiro, F., Duarte, J.P., Bartolo, H., Bartolo, P.J.: Additive manufacturing as an enabling technology for digital construction: A perspective on construction 4.0. Automation in Construction, vol. 103, pp. 251–267 (2019)
3. De Schutter, G., Taerwe, L.: Degree of hydration based description of mechanical properties of early-age concrete. Mater. Struct. **29**, 335–344 (1996)
4. Hauggaard, A.B., Damkilde, L., Hansen, P.F.: Transitional thermal creep of early age concrete. J. Eng. Mech. **125**, 458–465 (1999)
5. Holzapfel, G.A.: Nonlinear Solid Mechanics. A Continuum Approach for Engineering. John Wiley and Sons, Ltd., Chichester (2000)
6. Marsden, J.E., Hughes, T.J.R.: Mathematical foundations of elasticity. Prentice-Hall, Englewood-Cliffs (1983)
7. Nedjar, B.: On a geometrically nonlinear incremental formulation for the modeling of 3D concrete printing. Mech. Res. Commun. **116**, 103748 (2021). https://doi.org/10.1016/j.mechrescom.2021.103748
8. Nedjar, B.: Incremental viscoelasticity at finite strains for the modelling of 3D concrete printing. Comput. Mech. **69**, 233–243 (2022). https://doi.org/10.1007/s00466-021-02091-5
9. Nedjar, B., Baaser, H., Martin, R.J., Neff, P.: A finite element implementation of the isotropic exponentiated Hencky-logarithmic model and simulation of the eversion of elastic tubes. Comput. Mech. **62**(4), 635–654 (2017). https://doi.org/10.1007/s00466-017-1518-9
10. Ogden, R.W.: Non-linear Elastic Deformations. Dover, New York (1997)
11. Simo, J.C., Hughes, T.J.R.: Computational Inelasticity. Springer-Verlag, New York (1998)

Virtual Design Laboratory for Sustainable Fiber Reinforced Concrete Structures: From Discrete Fibers to Structural Optimization Under Uncertainty

Gerrit E. Neu, Vladislav Gudžulic, and Günther Meschke(✉)

Institute for Structural Mechanics, Ruhr University Bochum, Bochum, Germany
guenther.meschke@rub.de

Abstract. Concrete is one of the most used materials worldwide with a high environmental impact. Over past years, numerous attempts to minimize the associated effects on the environment by using more sustainable materials or by improving the performance of the material (e.g. high strength concrete, fiber reinforcement) have been introduced. The increase in material performance must be accompanied by better models and design approaches to take full advantage of the potential benefits. In this contribution, a discrete fiber and a multi-level model for the analysis of SFRC structures are used to assess the influence of a chosen fiber type, content, and orientation on the structural response. Zero-thickness cohesive interface elements capture the post-cracking behavior. The discrete fibers are modeled using truss elements. The bond between fibers and concrete is modeled using an elastoplastic bond-slip law, and the effects of fiber bending, friction, and matrix spalling are accounted for using a sub-model at the level of the interface element. The predictive capabilities of both models are validated and compared with fiber pull-out experiments. Finally, the prospects of applying complex FE models in conjunction with methods of optimization to design an SFRC tunnel lining segment are discussed. The objective is to minimize the total segment thickness and the fiber content while a constraint ensures that the required failure probability is retained.

Keywords: SFRC · Discrete crack models · Discrete fibers · Multi-level homogenization · Optimization · Uncertainty modeling · Reliability · Segmental tunnel lining

1 Introduction

Concrete is one of the most used materials globally, with an estimated yearly consumption of around 30 billion tons and continuously increasing demand [1]. With such massive consumption, naturally, the sustainability and environmental impact are questioned. The sustainability issue can be tackled in multiple ways, such as using recycled aggregates in concrete production (see e.g., [2] and [3] for an overview) or by improving the performance of the material, thus reducing consumption. In the past decades,

P. Rossi and J.-L. Tailhan (Eds.): SSCS 2022, RILEM Bookseries 38, pp. 268–277, 2023.
https://doi.org/10.1007/978-3-031-07746-3_27

significant effort has been invested in understanding and improving the lifetime performance of concrete. For example, setting forth novel recipes for high-performance and ultra-high-performance concretes [4, 5], introducing hybrid multiscale micro and macro fiber-reinforcements into the design [6, 7], have resulted in stronger, more ductile, and durable concretes allowing for optimized design both in regards to serviceability as well as ultimate load capacity.

Further, many developments regarding the use of advanced chemical admixtures have improved the rheology of fresh fiber-reinforced concrete, allowing for optimizing production quality [8–10], and opened the avenue for the application of 3D printing technology in fiber-reinforced concrete manufacture [11]. Naturally, the increase in material performance must be accompanied by better models and design approaches to take full advantage of the potential benefits. Many models over the past years have been developed to help researchers and engineers understand and design better structures with reduced material consumption. For instance, progress has been achieved in understanding the behavior of fiber-reinforced cementitious composites by devising analytical and semi-analytical models of crack bridging behavior of individual PVA [12] and steel fibers with and without hooks [13–16]. Further progress has been made by [17] and [18] in modeling fiber-reinforced specimens with the discrete representation of fibers and on the structural scale utilizing semi-analytical models and multi-level homogenization [16].

This contribution presents and compares two distinct approaches to the design of fiber-reinforced concrete: A concrete cracking model based on interface elements [19, 20] with a newly developed discrete fiber representation and a multi-level approach for the behavior of FRC structures [21]. In conjunction with optimization algorithms and uncertainty modeling, the multi-level approach is applied to design a tunnel lining, demonstrating how advanced numerical modeling approaches can be leveraged to reduce the environmental impact of concrete structures.

2 Multi-level Model for Steel-Fiber-Reinforced Concrete

This section aims to present two approaches for modeling steel-fiber-reinforced concrete, a model resolving fibers discretely and a multi-level modeling framework for simulation of fiber-reinforced concrete response [16, 19]. The emphasis is on assessing their suitability to model concrete structures with different levels of detail. Both approaches are shown to yield reasonable results on a single fiber pullout level, and the applicability of the second model to structures is discussed. The basis of both modeling approaches is discrete crack representation using zero-thickness cohesive interface elements. All material nonlinearities due to cracking are modeled by the cohesive-frictional traction-separation relationship, while the bulk of concrete is assumed to behave as a linear elastic material [21]. Zero-thickness interface elements are inserted between the solid finite elements in the part of the domain where cracking is expected, as illustrated in Fig. 3(right).

2.1 Discrete Fiber Model

This section outlines the single fiber model, which serves as a basis for the discrete fiber finite element and is the bottom-most level of the multi-level approach. The contributions to the pullout resistance from the bond stresses, fiber bending and plastification, concrete spalling and frictional sliding must be accounted for to capture the pullout response of inclined fibers accurately. The semi-analytical fiber pullout model formulation, used in the multi-level framework, is described in detail in papers by Zhan and Meschke [16, 22]. This model is extended to a 3D setting and implemented as a discrete fiber finite element in the FE program. The steel fibers are resolved explicitly, using the standard truss finite elements embedded into solid finite elements mesh with pre-inserted zero-thickness interface elements. Coupling between fibers and concrete is accomplished utilizing bond elements that transfer forces between concrete and fibers via penalizing the relative displacements (bond-slip) at the control points where the fiber element and the corresponding parent solid element are tied. The bond-slip is defined as $[\![\boldsymbol{u}]\!] = \boldsymbol{u}_S - \boldsymbol{u}_F$, where $\boldsymbol{u}_S$ and $\boldsymbol{u}_F$ are the displacements of the solid element and the fiber element at the control point. Based on the current value of the bond-slip, the resistance forces due to bond stress in the fiber direction are calculated by integration of the elastoplastic bond stress-slip law along the effective fiber-concrete contact length [19, 22] (see Fig. 1).

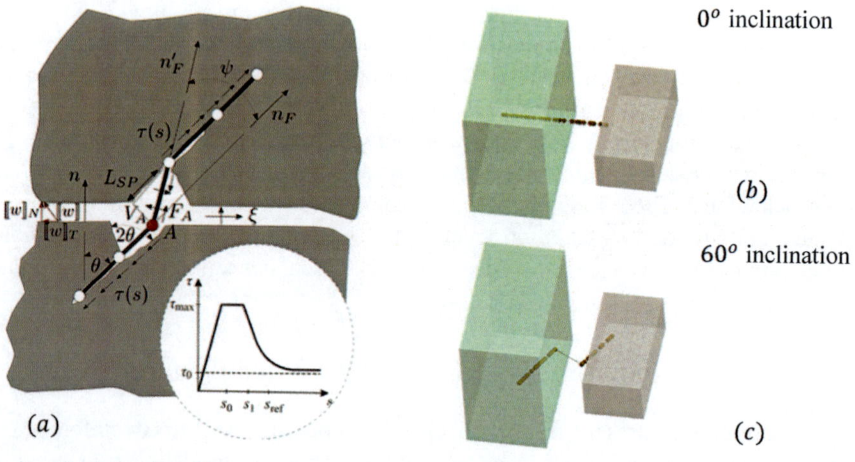

Fig. 1. (a) Illustration of discrete fiber model crossing the dominant crack. The related geometrical and model parameters are shown in the figure. (b, c) Finite element models of the inclined fiber pullout experiments by [23] with 0° and 60° inclinations to the crack.

As cracks open, further processes occur at the fiber exit point, such as fiber bending, matrix spalling, and frictional sliding [14, 15, 22]. These processes are specific only to the fiber segment that bridges the crack. When cracking occurs, among the interface elements crossed by the fiber, the one with the largest opening is identified and denoted as the dominant crack. The interface element then employs a sub-model to calculate additional forces acting on the crack faces due to fiber bending resistance and friction on the fiber exit point during the pullout.

The angle between the reference direction of fiber and the crack normal is calculated as $\theta = arccos(\boldsymbol{n}^T \boldsymbol{n}_F)$, where $\boldsymbol{n}_F$ is the reference direction of the fiber and $\mathbf{n}$ is the normal to the crack surface as illustrated in Fig. 1. The current inclination of the fiber $\boldsymbol{n}'_F$ depends on the crack opening $[\![\boldsymbol{w}]\!]$ and the spalling length L_{sp}, and is calculated as

$\boldsymbol{n}'_F = \frac{[\![w]\!]+s_f \cdot L_{SP} \cdot \boldsymbol{n}_F}{\|[\![w]\!]+s_f \cdot L_{SP} \cdot \boldsymbol{n}_F\|}$, where s_f is a factor that takes value 1 or 2 accounting for cases when spalling occurs only on one side of the crack or on both sides (Fig. 1). Finally, the deflection angle of the fiber is calculated as an angle between the reference and the current direction of the fiber $\psi = arccos(\mathbf{n}_F^T \boldsymbol{n}'_F)$. These angles are used for calculation of fiber bending forces according to the model in [22]. However, we assume a symmetric shape of the spalling cone with the apex angle 2θ (Fig. 1, left). The equilibrium between all internal forces due to bending and friction is ensured according to the procedure described in [22] and the normal (F_A) and transversal forces (V_A) at the point A in the Fig. 1 are calculated and added to internal force vector as $\boldsymbol{f}_{int}^{fib} = \boldsymbol{N}^T |_{\xi=\xi_f} \left(F_A \boldsymbol{n}'_F + V_A \left(\frac{[\![w]\!]}{\|[\![w]\!]\|} - \boldsymbol{n}'_F \right) \right)$, where $\boldsymbol{N}$ is the matrix containing shape functions of the interface element, evaluated at the isoparametric coordinates of the fiber exit point ξ_f.

2.2 Multi-level Model for Analysis of SFRC Structures

The second approach for modeling steel-fiber-reinforced structures is a multi-level approach described in detail in [16, 19]. This method is based on three primary levels of analysis. At the bottommost level, a semi-analytical model for the pullout of steel fibers [22] is used to obtain the pullout response of fibers under different inclinations relative to the dominant crack (Fig. 2 left). On the next level (Fig. 2 middle), the response of multiple fibers is homogenized, accounting for the distributions of fiber orientations and embedment lengths, yielding an effective traction-separation law. Finally, the traction separation law is applied as a constitutive relation for zero-thickness interface elements to simulate the structural response (Fig. 2 right). The procedure is illustrated in Fig. 2. Detailed steps are contained in [16, 19, 22].

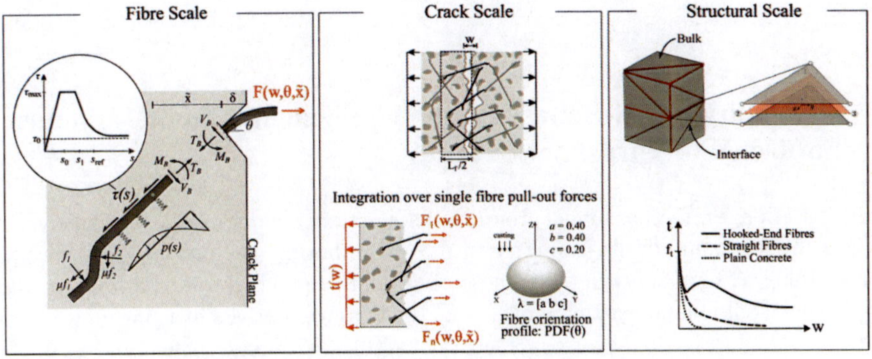

Fig. 2. Multi-level approach for modelling SFRC structures

2.3 Comparative Analysis of the Discrete and the Homogenized Approach for Modeling SFRC

In this subsection, we present the results of numerical analyses of two sets of fiber pullout experiments performed by Leung and Shapiro [23], differing only in the yield strength of the fiber material. The fibers of a total length of 22 mm between two Plexiglas blocks on one side and embedded in mortar block on the other side, with a 10 mm embedment length. The surface of the Plexiglas in contact with mortar was lubricated before casting to avoid cohesion. The parameters used in numerical simulations for both models can be found in [22]. Both approaches yield adequately accurate results, as shown in Fig. 3, and capture the main features of the pullout curves, which demonstrates the feasibility of applying the semi-analytical fiber pullout model as a bottom-most level of the multi-level approach. The finite element models for the discrete fiber simulations are shown in Fig. 1b, c. The multi-level approach has been validated in [16, 19].

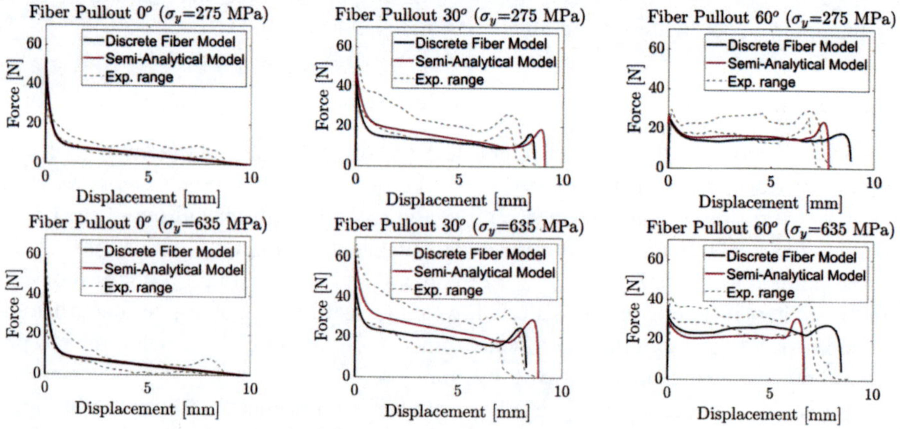

Fig. 3. Comparison of numerically obtained pullout response curves for two sets of fibers using two approaches with the experimentally obtained results [23].

3 Reliability Based Robust Design of Segmental Tunnel Linings Subjected to Thrust Jack Forces

In this section, a reference tunnel project is defined (inner diameter of 9.6 m, sandy soil), where the maximum thrust jack force of the tunnel boring machine F_{design} is specified as 5600 kN. A segment with a thickness of 500 mm and a conventional reinforcement Φ10-10 + additional rebars in high stressed regions) was derived by using conventional design procedures. This results in a steel content of ≈260 kg/segment and a concrete volume of 4.53 m^3 per segment. However, in this example, a hybrid fiber reinforced segment consisting of two SFRC layers (A and B) is designed, whose required thickness d and the fiber content c_F as well as the width ΔL of the strengthening SFRC layer B (see Fig. 4) is determined by solving an optimization task under consideration of uncertainties.

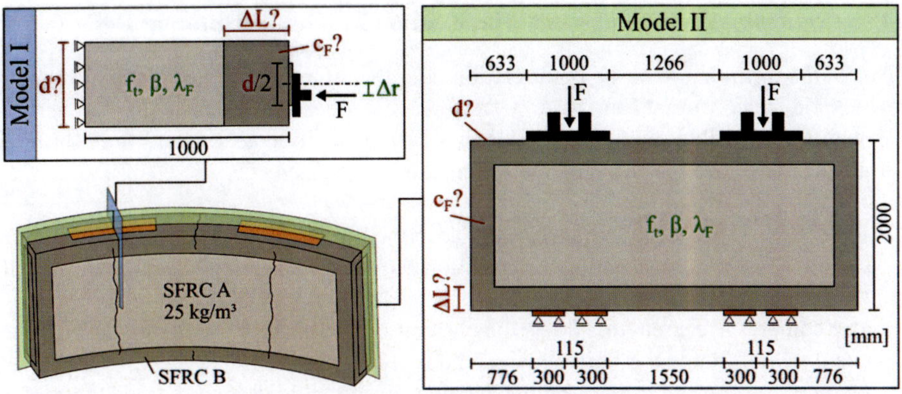

Fig. 4. Design variables (red), the material uncertainties and the jack eccentricity as uncertain a priori parameter (green) considered in two structural models used for the optimization task. The potential cracks due to the thrust jack loading are indicated in the bottom left.

In order to design the lining segment subjected to thrust jack forces, two 2D FE models, labeled as *Model I* (plane strain, element size between 2.5 and 10 mm) and *Model II* (plane stress, element size 25 mm) are used analogous to the design models in engineering practice (Fig. 4). A typical crack pattern is shown in Fig. 5(left).

3.1 Objective and Input of the Optimization Problem

The optimization task is formulated to minimize the design variables (segment thickness d, the fiber content c_F and the width ΔL of the strengthening SFRC layer) while material uncertainty (fiber orientation λ_F, tensile concrete strength f_t and shear strength β) and imperfections (thrust jack eccentricity Δr) are considered. It should be noted, that in this optimization problem the objective function to be minimized is specified as a weighted sum of the design variables d (80%), ΔL(10%) and c_F (10%), so that the minimum is at the constraint limit state defined by the four inequality constraints applied to both models I and II, respectively. All constraints are formulated in terms of accepted failure probabilities to account for the scatter through the considered a priori uncertain parameters and therefore providing a robust segment design (for the load bearing capacity $P_{f,ULS} = 10^{-5}$ and for the crack width control $P_{f,SLS} = 0.067$). For the failure criteria, the load bearing capacity F_{max} should be larger than the maximum thrust jack force $F_{design} =$ 5600 kN and in order to ensure a sufficient serviceability performance, the crack width between the thrust jack pads is restricted to 0.1 mm.

Due to the computational effort involved in solving an optimization task under uncertainty, Feed-forward Artificial Neural Networks (ANNs) with two hidden layers are used as surrogate models. The ANNs for both structural models are trained with back propagation using results of 1800 FE calculations each (design of experiment generated by Latin Hypercube).

3.2 Constraint Limit States and Optimization Results for Different Designs

Due to the formulation of the optimization task, the minimum of the objective function will be on the constraint limit state (CLS), which divides the design space into feasible as well as unfeasible regions and is defined by inequality constraints. Three different strategies for optimizing a segmental lining are investigated:

- **Reference:** The optimization task as described in Sect. 3.1 is solved to design a hybrid SFRC segment using a conventional C40/50 concrete. The concrete tensile strength is modeled as a lognormal distributed random variable ($\mu = 1.24$, $\sigma = 0.18$). The compressive strength f_c [24], the elasticity modulus Ec [24], the mode I fracture energy G_{fI} [25] and the ratio between mode II and mode I fracture energy κ [26] are assumed to be fully correlated with the lognormal distributed concrete tensile strength.
- **High-Performance:** This scenario should investigate how the use of a better performing concrete could further minimize the segment thickness. Therefore, the concrete tensile strength is modeled as a lognormal distributed random with the parameters $\mu = 1.51$ and $\sigma = 0.18$ (corresponds to C70/85).
- **Recycled:** In contrast, the use of a more environmentally friendly concrete (i.e. using recycled aggregates or modified cement paste) can lead to a lower concrete strength. Investigations in [27] showed that the use of 50% recycled concrete aggregates lowered the measured tensile strength by 18%. This strength reduction is applied by adopting the concrete tensile strength distribution parameters $\mu = 1.04$ and $\sigma = 0.18$.

Due to the monotonic behavior of all considered intervals, the CLS of the beforementioned strategies are evaluated for a single realization of the intervals ($\Delta r = 40$ mm, $\beta = 5.0$, $\lambda_F = 0.35$). More information regarding the uncertainty modeling, the parameter ranges and different realizations can be found in [28].

As can be seen in Fig. 5(right), the segment thickness d and the fiber content c_F are dominating the shape of the constraint limit states for each case, whereas the cap width ΔL has only a minor influence. When only using the minimum fiber content of 25 kg/m^3, a segment thickness of $d \approx 435$ mm ('high performance'), $d \approx 450$ mm ('reference') and $d \approx 465$ mm ('recycling') will lead to feasible designs. An increase of the fiber content up to 40 kg/m^3 enables a reduction of the segment thickness to $d \approx 400$ mm for the reference case. However, by further increasing the fiber content, only small reductions of the segment thickness can be achieved. The width of the SFRC cap ΔL is the least sensitive design variable. If a certain width is provided (tensile stresses occur within the outer SFRC layer), only small improvements with increasing cap width ΔL can be obtained (see reference case at higher fiber contents c_F in Fig. 5).

Finally, the evaluation of the objective function is performed for the above-mentioned cases and the optimum of each individual objective $R_{1,2,3}$ as well as the resulting design variables are summarized in Table 1.

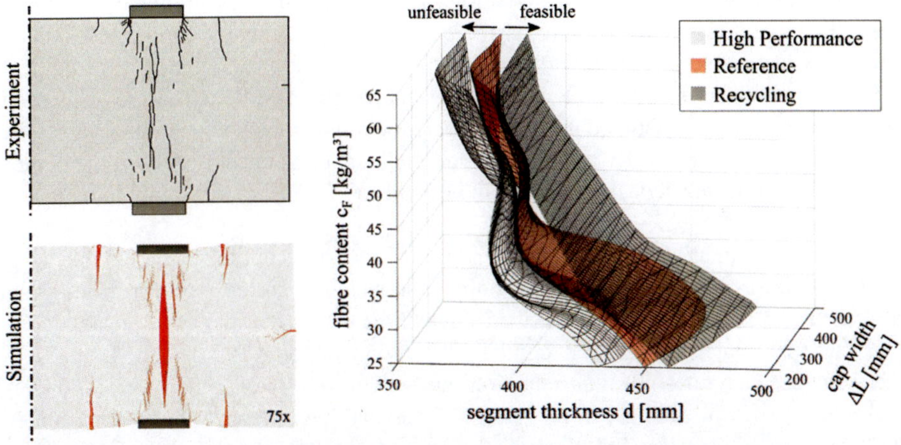

Fig. 5. Experimental and numerical crack pattern obtained through the re-analysis of [29] (left). Constraint limit states in the design space for the "reference", the "high performance" and the "recycling" case (right).

The optimization results in a 402 mm thick segment with a 200 mm wide SFRC cap containing 41 kg/m^3 of fibers for the reference case. When compared to the conventional RC segment design, the steel content is reduced by 59% (106 vs 260 kg/segment) and the concrete volume is reduced by 20% (3.61 vs 4.53 m^3/segment). The use of a better performing concrete can further reduce the segment thickness (3.2 vs. 3.61 m^3/segment). While a similar steel content in comparison to the reference case can be achieved (92 vs 106 kg/segment). As an alternative, the use of concrete with recycled aggregates still leads to a segment design with a lower concrete volume (3.83 vs. 4.53 m^3/segment) and a reduced steel content (108 vs. 260 kg/segment) in comparison to the conventional RC segment design.

Table 1. Design variables for each individual objective $R_{1,2,3}$ employed in the multi-objective optimization task for the investigated cases (results are rounded off).

	R1	R2	R3	d [mm]	ΔL [mm]	c_F [kg/m^3]
Reference	372	200	25	402	200	41
High Performance	358	200	25	391	200	38
Recycling	384	200	25	426	200	36

4 Conclusions

The results presented in this paper demonstrate the role of advanced numerical modeling approaches in the design of more sustainable fiber-reinforced concrete structures. The proposed discrete fiber model, which allows for an in-depth structural analysis of fiber distribution and orientation due to casting/printing, was compared with a less computationally expensive semi-analytical fiber pullout model and validated by pull-out tests of two types of inclined fibers. For the numerical design of SFRC structures, this semi-analytical model was included in a multi-level modeling strategy which allows assessing the influence of major SFRC parameters and the associated uncertainties in the context of finding an optimized design of FRC and hybrid RC-FRC structures with less material consumption.

As a representative application case, a reliability-based optimization for the design of fiber-reinforced segmental tunnel linings subjected to thrust jack forces has been performed. A segment design is derived, consuming up to 20% less concrete and reducing the steel content by 59% as compared to a conventional RC design. Using high-performance concrete, a further reduction can be achieved (steel content -13%, concrete volume -11%), while the use of recycled concrete has been shown to be an environmentally friendly alternative (58% less steel content and a 15% lower concrete volume in comparison to a conventional RC segment design).

Acknowledgement. Financial support was provided by the German Research Foundation (DFG) in the framework of project B2 of the Collaborative Research Center SFB 837 *Interaction modeling in mechanized tunnelling* (Project no.: 77309832) and the Priority Programme SPP 2020 *Cyclic deterioration of High-Performance Concrete in an experimental-virtual lab* (Project no.: 353819637). This support is gratefully acknowledged.

References

1. Monteiro, P.J., Miller, S.A., Horvath, A.: Towards sustainable concrete. Nat. Mater. **16**(7), 698–699 (2017)
2. Katz, A.: Properties of concrete made with recycled aggregate from partially hydrated old concrete. Cem. Concr. Res. **33**(5), 703–711 (2003)
3. Nedeljković, M., Visser, J., Šavija, B., Valcke, S., Schlangen, E.: Use of fine recycled concrete aggregates in concrete: a critical review. J. Build. Eng. **38**, 102196 (2021)
4. Walraven, J.C.: High performance fiber reinforced concrete: progress in knowledge and design codes. Mater. Struct. **42**(9), 1247–1260 (2009)
5. International federation for structural concrete (fib): fib Model Code for Concrete Structures 2010. Ernst & Sohn (2013)
6. Lawler, J.S., Zampini, D., Shah, S.P.: Microfiber and macrofiber hybrid fiber-reinforced concrete. J. Mater. Civ. Eng. **17**(5), 595–604 (2005)
7. Di Prisco, M., Plizzari, G., Vandewalle, L.: Fibre reinforced concrete: new design perspectives. Mater. Struct. **42**(9), 1261–1281 (2009)
8. Okamura, H., Ouchi, M.: Self-compacting concrete. Development, present use and future. In: Self-Compacting Concrete: Proceedings of the 1st International RILEM Symposium, pp. 3–14. Rilem Publications, Cachan Cedex, France (1999)

9. Grünewald, S., Walraven, J.C.: Parameter-study on the influence of steel fibers and coarse aggregate content on the fresh properties of self-compacting concrete. Cem. Concr. Res. **31**(12), 1793–1798 (2001)
10. Martinie, L., Rossi, P., Roussel, N.: Rheology of fiber reinforced cementitious materials: classification and prediction. Cem. Concr. Res. **40**(2), 226–234 (2010)
11. Li, V.C., et al.: On the emergence of 3D printable engineered, strain hardening cementitious composites (ECC/SHCC). Cem. Concr. Res. **132**, 106038 (2020)
12. Yang, E.H., Wang, S., Yang, Y., Li, V.C.: Fiber-bridging constitutive law of engineered cementitious composites. J. Adv. Concr. Technol. **6**(1), 181–193 (2008)
13. Fantilli, A.P., Vallini, P.: A cohesive interface model for the pullout of inclined steel fibers in cementitious matrixes. J. Adv. Concr. Technol. **5**(2), 247–258 (2007)
14. Laranjeira, F., Aguado, A., Molins, C.: Predicting the pullout response of inclined straight steel fibers. Mater. Struct. **43**(6), 875–895 (2010)
15. Laranjeira, F., Molins, C., Aguado, A.: Predicting the pullout response of inclined hooked steel fibers. Cem. Concr. Res. **40**(10), 1471–1487 (2010)
16. Zhan, Y., Meschke, G.: Multilevel computational model for failure analysis of steel-fiber–reinforced concrete structures. J. Eng. Mech. **142**(11), 04016090 (2016)
17. Schauffert, E.A., Cusatis, G.: Lattice discrete particle model for fiber-reinforced concrete. I: theory. J. Eng. Mech. **138**(7), 826–833 (2012)
18. Kang, J., Kim, K., Lim, Y.M., Bolander, J.E.: Modeling of fiber-reinforced cement composites: discrete representation of fiber pullout. Int. J. Solids Struct. **51**(10), 1970–1979 (2014)
19. Gudzulic, V., Neu, G., Gebuhr, G., Anders, S., Meschke, G.: Numerisches Mehrebenen-Modell für Stahlfaserbeton: Von der Faser- zur Strukturebene. Beton und Stahlbetonbau **115**(2), 146–157 (2020)
20. Snozzi, L., Molinari, J.F.: A cohesive element model for mixed mode loading with frictional contact capability. Int. J. Numer. Meth. Eng. **93**(5), 510–526 (2013)
21. Gudžulić, V., Meschke, G.: Multi-level approach for modelling the post-cracking response of steel fibre reinforced concrete under monotonic and cyclic loading. PAMM **21**(1), e202100194 (2021)
22. Zhan, Y., Meschke, G.: Analytical model for the pullout behavior of straight and hooked-end steel fibers. J. Eng. Mech. **140**(12), 04014091 (2014)
23. Leung, C.K., Shapiro, N.: Optimal steel fiber strength for reinforcement of cementitious materials. J. Mater. Civ. Eng. **11**(2), 116–123 (1999)
24. European Comittee for Standardisation, EN 1992 - Eurocode 2: Design of concrete structures (2005)
25. International federation for structural concrete (fib): fib Model Code for Concrete Structures 1990 (1993)
26. Naga Satish Kumar, C., Gunneswara Rao, T.D.: An empirical formula for mode-II fracture energy of concrete. KSCE J. Civ. Eng. **19**(3), 689–697 (2015)
27. Kang, T., Kim, W., Kwak, Y.-K., Hong, S.-G.: The choice of recycled concrete aggregates for flexural members. IABSE Congr. Rep. **18**(21), 726–731 (2012)
28. Neu, G.E., Edler, P., Freitag, S., Gudžulić, V., Meschke, G.: Reliability based optimization of steel-fibre segmental tunnel linings subjected to thrust jack loadings. Eng. Struct. **254**, 113752 (2022)
29. Hemmy, O.: Splitting of SFRC induced by local forces - investigations of tunnel segments without curvature. Sub-report (Annex 3) of the report of subtask 4.4. brite euram. Institut fuer Baustoffem Massivbau und Brandschutz, TU Braunschweig (2001)

Modeling Deterioration of RC Structures Due to Environmental Conditions

Joško Ožbolt(✉)

Institute of Construction Materials, University of Stuttgart, 70569 Stuttgart, Germany
ozbolt@iwb.uni-stuttgart.de

Abstract. Chloride-induced corrosion of steel reinforcement in concrete, creep and shrinkage and freezing-thawing are some of the major causes responsible for deterioration of reinforced concrete (RC) structures. The repair of damaged structures results in relatively high direct and indirect costs. Therefore, to predict their durability it is important to have a numerical tool, which is able to account for the above mentioned processes and their consequences for the structural safety. In the paper a recently developed coupled Chemo-Hygro-Thermo-Mechanical model for concrete is briefly discussed. The model is implemented into a 3D finite element code and it is aimed to model processes related to corrosion of steel reinforcement, freezing-thawing and creep and shrinkage of concrete. The macro-cell corrosion of steel reinforcement accounts for all relevant processes before and after depassivation of reinforcement and it is coupled with the mechanical model for concrete, which is based on the microplane model. Loading due to freezing-thawing of concrete is formulated in the framework of poromechanics. Drying creep of concrete is simulated based on the hygro-mechanical model at the meso scale. Investigated is the influence of the interaction between non-elastic deformations of cement paste (basic creep and shrinkage), load induced damage and heterogeneity of concrete. The application of the model is illustrated on several numerical examples.

Keywords: Reinforced concrete · Corrosion · Freezing-thawing · Drying creep · Fracture · Simulation · Finite elements

1 Introduction

RC structures exposed to aggressive environmental conditions, such as structures close to the sea or highway bridges and garages exposed to de-icing salts, very often exhibit damage due to corrosion. This damage is usually manifested in the form of cracking and spalling of concrete cover, caused by the expansion of corrosion products around steel reinforcement bar. The pitting corrosion is even more severe, it is not visible and it can cause strong degradation of reinforcement performance. Furthermore, freezing and thawing causes damage and accelerate corrosion of reinforcement. Moreover, close related to deterioration due to corrosion is also time dependent deformation of concrete structures due to creep and shrinkage (e.g. drying creep). To predict durability of RC structure it is important to have a reliable numerical tool, which is able to simulate these

P. Rossi and J.-L. Tailhan (Eds.): SSCS 2022, RILEM Bookseries 38, pp. 278–288, 2023.
https://doi.org/10.1007/978-3-031-07746-3_28

processes in order to better understand major phenomena and their consequences for the structural safety.

In the present contribution a brief overview of the recently developed coupled Chemo-Hygro-Thermo-Mechanical (CHTM) model for concrete is presented. In the first part of the paper a theoretical background is presented, with more details related to the modeling of corrosion of steel reinforcement, followed by the brief description of the approach employed for the modeling of freezing-thawing of concrete and modeling of drying creep. The application of the model is illustrated on several numerical examples.

2 Coupled Chemo-Hygro-Thermo-Mechanical Model

2.1 Modelling Corrosion of Steel Reinforcement in Concrete

A surface layer of ferric oxide covers and protects the steel in concrete. With this layer being damaged or depassivated the corrosion of reinforcement in concrete can be activated. Depassivation of the protective layer can occur by reaching a critical threshold concentration of free chloride ions near the reinforcement bar surface or as a consequence of carbonization of concrete [1, 2]. Here is discussed only macro-cell chloride type of corrosion that is the most severe one.

Processes Before Depassivation of Reinforcement. The corrosion of steel in concrete is an electrochemical process dependent on electrical conductivity of concrete, presence of electrolyte and the concentration of dissolved oxygen in the pore water near the reinforcement. The CHTM model couples the above mentioned physical and electrochemical processes with the mechanical behaviour of concrete (damage).

The transport of capillary water is described in terms of volume fraction of pore water in concrete by Richard's equation, based on the assumption that transport processes take place in aged concrete, i.e. the hydration of cement paste is completed:

$$\frac{\partial \theta_w}{\partial t} = \nabla \cdot [D_w(\theta_w)\nabla\theta_w] \tag{1}$$

where θ_w is volume fraction of pore water (m^3 of water/m^3 of concrete) and $D_w(\theta_w)$ is capillary water diffusion coefficient (m^2/s) described as a strongly non-linear function of moisture content. Transport of chloride ions through a non-saturated concrete occurs as a result of convection, diffusion and physically and chemically binding by cement hydration product:

$$\theta_w \frac{\partial C_c}{\partial t} = \nabla \cdot [\theta_w D_c(\theta_w, T)\nabla C_c] + D_w(\theta_w)\nabla\theta_w \nabla C_c - \frac{\partial C_{cb}}{\partial t} \tag{2a}$$

$$\frac{\partial C_{cb}}{\partial t} = k_r(\alpha C_c - C_{cb}) \tag{2b}$$

where C_c is concentration of free chloride dissolved in pore water (kgCl-/m^3 pore solution), $D_c(\theta_w, T)$ is the effective chloride diffusion coefficient (m^2/s) expressed as a function of water content and concrete temperature T, C_{cb} is concentration of bound chloride (kgCl-/m^3 of concrete), k_r is binding rate coefficient, $\alpha = 0.7$ is constant [3].

Assuming that oxygen does not participate in any chemical reaction before depassivation of steel, transport of oxygen through concrete is considered as a convective diffusion problem:

$$\theta_w \frac{\partial C_o}{\partial t} = \nabla \cdot [\theta_w D_o(\theta_w) \nabla C_o] + D_w(\theta_w) \nabla \theta_w \nabla C_o \tag{3}$$

where C_o is oxygen concentration in pore solution (kg of oxygen/m^3 of pore solution) and $D_o(\theta_w)$ is the effective oxygen diffusion coefficient [3], dependent on concrete porosity p_{con} and water saturation of concrete S_w.

Based on the constitutive law for heat flow and conservation of energy, the equation which describes temperature distribution of continuum reads:

$$\lambda \Delta T + W(T) - c\rho \frac{\partial T}{\partial t} = 0 \tag{4}$$

where λ is thermal conductivity (W/(m K)), c is heat capacity per unit mass of concrete (J/(K kg)), ρ is mass density of concrete (kg/m^3) and W is internal source of heating (W/m^3). More detail related to the strong and weak formulations of the processes up to the depassivation of reinforcement can be found in [3].

Processes after Depassivation of Reinforcement. The corrosion of steel is activated with depassivation of the steel reinforcement in concrete. The non-mechanical processes important for the propagation stage of steel corrosion in concrete are: (1) Mass sinks of oxygen at steel surface due to cathodic and anodic reaction, (2) The flow of electric current through pore solution and (3) The cathodic and anodic potential.

The oxygen consumption at the cathodic and anodic surfaces is the result of the following reactions:

$$2H_2O + O_2 + 4e^- \rightarrow 4OH^- \tag{5}$$

The transport of hydroxyl ions to the anode, where corrosion products forms:

$$Fe^{2+} + 2OH^- \rightarrow Fe(OH)_2 \tag{6}$$

$$4Fe(OH)_2 + O_2 + 2H_2O \rightarrow 4Fe(OH)_3 \tag{7}$$

It can be calculated as:

$$D_o(S_w, p_{con}) \frac{\partial C_o}{\partial n}\bigg|_{cathode} = -k_c i_c \quad k_c = 8.29 \times 10^{-8} \frac{\text{kg}}{\text{C}} \tag{8a}$$

$$D_o(S_w, p_{con}) \frac{\partial C_o}{\partial n}\bigg|_{anode} = -k_a i_a \quad k_a = 4.14 \times 10^{-8} \frac{\text{kg}}{\text{C}} \tag{8b}$$

where $\boldsymbol{n}$ is outward normal to the steel bar surface and i_c and i_a are cathodic and anodic current density (A/m^2), respectively. The constants k_c and k_a are calculated using the stoichiometry of chemical reactions (5–7) and Faraday's law.

According to Butler–Volmer kinetics, in the present model kinetics of reaction at the cathodic and anodic surface can be estimated from:

$$i_c = i_{0c}\frac{C_o}{C_{ob}}e^{2.3(\Phi_{0c}-\Phi)/\beta_c} \quad i_a = i_{0a}e^{2.3(\Phi-\Phi_{0a})/\beta_a} \tag{9}$$

where C_{ob} is oxygen concentration at surface of concrete element exposed to sea water (kg/m^3), Φ is electric potential in pore solution near reinforcement surface (V), i_{0c} and i_{0a} are the exchange current density of the cathodic and anodic reaction (A/m^2), Φ_{0c} and Φ_{0a} are the cathodic and anodic equilibrium potential (V), β_c and β_a are the Tafel slope for cathodic and anodic reaction (V/dec), respectively.

The electric current through the electrolyte is a result of motion of charged particles and, if the electrical neutrality of the system and the uniform ions concentration are assumed, can be written as:

$$\mathrm{i} = -\sigma(S_w, p_{con})\nabla\Phi \tag{10}$$

where σ is electrical conductivity of concrete.

The equation of electrical charge conservation, if the electrical neutrality is accounted for and the electrical conductivity of concrete is assumed as uniformly distributed, reads:

$$\nabla^2\Phi = 0 \tag{11}$$

Rate of rust production J_r (kg/m^{2}s) and mass of hydrated red rust per related surface (A_r) of rebar m_r (kg), respectively, are calculated as:

$$\begin{aligned} J_r &= 5.536 \times 10^{-7}\, i_a \\ m_r &= J_r \Delta t A_r \end{aligned} \tag{12}$$

where Δt is time interval in which the corrosion is taking place. The coefficient of proportionality between the anodic current density i_r and rate of rust production J_r is calculated using the stoichiometry of chemical reactions and Faraday's law [1, 2].

Recent experimental investigations [4] have shown that the penetration of corrosion products into the pores and their relatively large ingress through the radial cracks, generated around the bar, has a significant effect on the development of corrosion induced damage. The influence can be summarized as: (1) The distribution of rust and radial pressure over the anodic surface is not uniform and (2) Damage due to expansion of products is less pronounced. The distribution of corrosion product (red rust) R (kg/m^3 of pore solution) into the pores and through the cracks in concrete has been mathematically formulated as a convective diffusion problem:

$$\theta_w\frac{\partial R}{\partial t} = \nabla \cdot [\theta_w D_r \nabla R] + D_w(\theta_w)\nabla\theta_w\nabla R \tag{13}$$

in which D_r is the diffusion coefficient (m^2/s) of corrosion product. It is important to note that the Eq. (13) does not directly describe the transport of the red rust, but rather the distribution of the rust formed in the concrete pores and cracks as a consequence of soluble species, which can dissolve in concrete pore solution and subsequently migrate in pores and cracks, reacting with oxygen in the pore water [4].

Coupling with Mechanical Model. The microplane model for concrete with relaxed kinematic constraints [5] is applied in the mechanical part of the CHTM model. In the finite element analysis cracks are treated in a smeared way, i.e. smeared crack approach is employed. The governing equation for the mechanical behaviour of a continuous body in the case of static loading condition reads:

$$\nabla[D_m(u, \theta_w, T)\nabla u] + \rho b = 0 \tag{14}$$

in which D_m is material stiffness tensor, ρ_b is specific volume load and u is displacement field. In the mechanical part of the model the total strain tensor is decomposed into mechanical strain, thermal strain, hygro strain (swelling–shrinking-creep) and strain due to expansion of corrosion products.

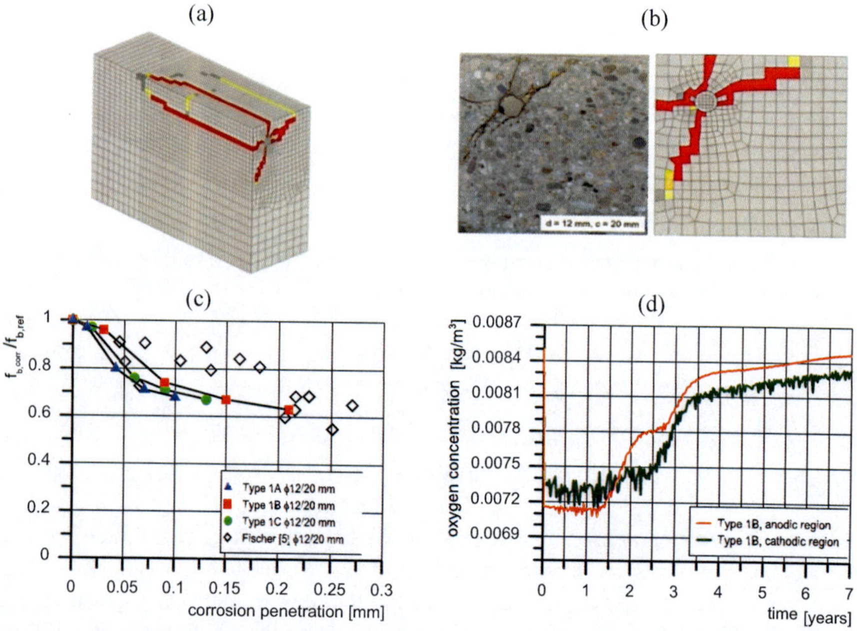

Fig. 1. Beam end specimen exposed to 7 years natural corrosion: (a) Crack pattern due to corrosion induced damage; (b) Crack pattern – experimental vs. numerical; (c) Reduction of the pull-out capacity due to corrosion induced damage and (d) Oxygen distribution at anodic and cathodic region after depasivation [6]

One-dimensional corrosion contact elements are used to account for the inelastic strains due to the expansion of corrosion products. They are placed radially around the bar surface and their main function is to simulate the contact between reinforcement and the surrounding concrete. These contact elements can take up only shear forces in direction parallel to reinforcement axis and compressive forces perpendicular to the adjacent surface of the reinforcement. The inelastic radial expansion due to corrosion

Δl_r is calculated as:

$$\Delta l_r = \frac{m_r}{A_r}\left(\frac{1}{\rho_r} - \frac{0.523}{\rho_s}\right) \tag{15}$$

where $\rho_r = 1.96 \times 103$ (kg/m^3) and $\rho_s = 7.89 \times 103$ (kg/m^3) are densities of rust and steel, respectively, 0.523 is the ratio between the mass of steel (m_s) and the corresponding mass of rust (m_r) over the related surface of reinforcement A_r that corresponds to the contact element. The stiffness of the rust layer is assumed to be $E_r = 100$ MPa. In the model it is represented by the axial stiffness of the corrosion contact elements. The shear resistance of the contact elements, defined by the bond-slip relationship, is used to model the bond between deformed steel reinforcement and concrete.

By using the finite elements to solve the partial differential equations implemented in the model, the strong form is rewritten into a weak form. The weak form of the system of partial differential equations is carried out by employing the Galerkin weighted residual method. The model is then implemented into a 3D FE code. The non-mechanical part of the problem is solved by using direct integration method of implicit type. To solve the mechanical part, Newton-Rapshon iterative scheme is used. To avoid mesh size dependency as a regularization method simple crack band approach is employed. Coupling between mechanical and non-mechanical part of the model is performed by continuous update of governing parameters during the incremental transient finite element analysis using staggered solution scheme. For more detail see [2, 3].

Numerical Example: Pull Out of Corroded Reinforcement. The application of the model is here demonstrated through numerical study of the pull-out of the reinforcement bar from the beam-end specimen [6]. The experiments were carried out under accelerated corrosion, which approximately corresponds to the severe splash natural conditions. The cross-section is 200 × 200 mm^2 and the diameter of the reinforcement bar is 12 mm with a concrete cover of 20 mm (ϕ12/20 mm). The specimen is first exposed to aggressive environmental conditions, which caused corrosion of embedded reinforcement bars. Subsequently the reinforcement is pulled out from the specimen and for the different levels of corrosion the numerical results are compared with test results [6]. As shown in Fig. 1, the model is able to realistically predict the corrosion induced damage, reduction of the pull-out resistance and distribution of oxygen in the active corrosion phase, i.e. for damaged concrete cover.

2.2 Modelling Freezing-Thawing of Concrete

When subjected to a uniform cooling below the freezing point a water-infiltrated porous material undergoes a cryo-deformation resulting from various combined actions: (i) the difference of density between the liquid water and the ice crystal; this density difference provokes the expansion of the solid matrix surrounding the crystal in formation, as well as the expulsion of some liquid water from the freezing sites towards the pores still filled by liquid water; (ii) the surface tension arising between the different constituents, which eventually governs the crystallization process in connection with the pore radius distribution; (iii) the drainage of the liquid water expelled from the freezing sites towards the air voids; (iv) the cryo-suction process, which drives liquid water towards the already frozen sites as the temperature further decreases; (v) the thermomechanical coupling between the pressurized pore space and the surrounding solid matrix, which governs the overall cryo-deformation [7].

The freezing behavior of cementitious materials is investigated through poromechanical approach. The material is taken as a porous medium saturated with water and subject to freezing. The involved thermodynamic laws are recalled to establish the constitutive equations for the phase change, mass transport and heat transfer processes. As a result, the pore pressure arising from freezing is converted to macroscopic effective stress through homogenization scheme. The established model is applied to predict the macroscopic freezing strain of a saturated cement paste.

The poromechanical approach is based on the poro-elastic assumption. To predict large deformation resulting from material nonlinearity and crack propagation during freezing, damage of concrete should be accounted for. Cracking can enlarge the permeability and diffusivity of concrete. On the other hand, it can lead to expansion and loss of mechanical properties, and ultimately, to the complete destruction of the concrete [8, 9].

Continuity Equation. Based on Darcy's law and the mass balance of the system, the continuity equations, with the consideration of the influence of coupling between deformation, ice formation, and heat and water migration, has been derived by Zuber and Marchand [10]:

$$\beta \frac{\partial p_w}{\partial t} = \nabla \cdot \left(\frac{D}{\eta} \nabla p_w \right) + S - b \frac{\partial \varepsilon_v}{\partial t} \tag{16}$$

where p_w = pressure in the water liquid; D = permeability of the porous media; η = fluid viscosity; ε_v = volumetric strain of the medium;

$$\begin{aligned} \beta &= \tfrac{nS_w}{K_w} + \tfrac{nS_i}{K_i} + \tfrac{b-n}{K_m} \\ S &= \left(\tfrac{1}{\rho_i} - \tfrac{1}{\rho_w} \right) \dot{w}_i + \overline{\alpha} \dot{T} - \tfrac{b-n}{K_m} \dot{X} - \tfrac{nS_i}{K_i} \dot{\kappa} \end{aligned} \tag{17}$$

where n = total porosity of a given mixture; S_i and S_w = proportions of the porosity filled with ice and liquid water, respectively; b = Biot's coefficient; K_w and K_i = compressibility modulus of water and ice, respectively; K_m = compressibility modulus of the solid matrix; ρ_i = mass density of ice; ρ_w = mass density of water; w_i = mass of frozen water; X = term that is related to the presence of ice in the frozen pores; κ accounts for the effects of the spherical interface between liquid water and ice; $\overline{\alpha}$ = volumetric thermal dilation of the system and T = temperature.

In Eq. (17) S denotes the source of pressure and it is composed of four terms. The first term corresponds to the pressures created by the formation of ice. The second term considers the pressure generated during temperature changes as a result of the difference between the thermal expansions of various phases. The last two terms correspond to the depression imposed on liquid water through the ice/water interface. Based on the homogenization approach [10], the effective pore pressure p^* of frozen continuum at the macro scale is:

$$p^* = p_w + X = p_w + X \tag{18}$$

where X is related to the pressure due to the presents of ice in pores. Note that this term (internal pore pressure) is responsible for the mechanical action and it comes into equilibrium Eq. (14).

Energy Conservation Equation. The differential equation for heat conduction takes the following form:

$$\rho C \frac{\partial T}{\partial t} = \nabla \cdot (\lambda \nabla T) + L \frac{\partial w_i}{\partial t} \tag{19}$$

where ρ = density of the system; t = time; λ, = thermal conductivity of the system, water, ice, and solid, matrix; C = heat capacity of the system, water, ice, and solid matrix, and L = latent heat of fusion of water. The term in Eq. (19) introduces a coupling between the heat transfer and the phase change and has an appreciable effect in freezing-thawing problems. Similar as in the case of corrosion, the differential Eqs. (16, 19) are rewritten into a weak form and solved using direct integration of implicate type. The governing variables are pressure in the water liquid (p_w), temperature (T) and strains. Note that the problem of freeze-thaw durability is in practice often coupled with reinforcement corrosion.

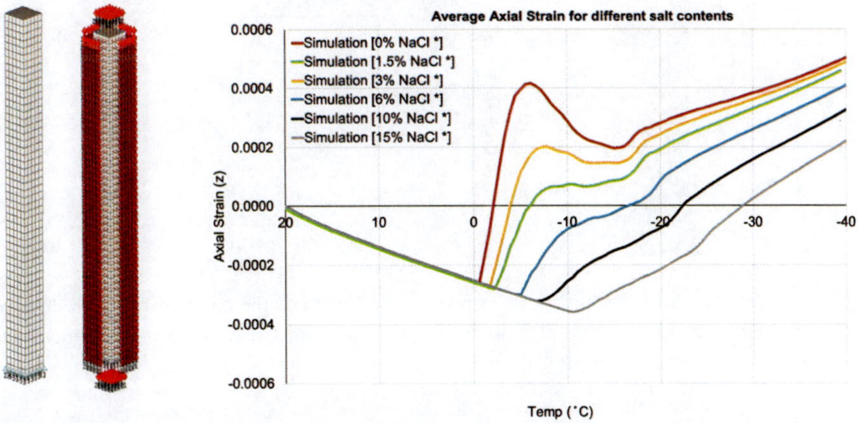

Fig. 2. Geometry and FE discretization with boundary conditions (left) and the influence of the chloride content on the axial surface strains (right)

Numerical Example: Influence of Chloride Content. To demonstrate the effect of chloride content on freezing, concrete specimen (see Fig. 2) was exposed to the linear cooling from 20 °C to −40 °C with the cooling rate of −20 °C/h. The content of chlorides in pore water was varied from 0% to 15%. Figure 2 shows (preliminary results) the average axial surface strain as a consequence of cooling. As can be seen, with increase of chloride content the surface temperature at freezing decreases the same as the maximum compressive strains at the surface specimen.

2.3 Creep and Shrinkage of Concrete

It is well known that creep and shrinkage of concrete have significant influence on the durability of concrete structures [11, 12]. To study draying creep of concrete, here is presented only a brief description of the approach. The interaction between the load-induced damage of concrete and its non-elastic time deformations, drying shrinkage and basic creep of mortar, is numerically investigated through a 3D meso-scale finite element (FE) simulation [13]. The transient numerical analysis is performed by employing two-phase meso-scale FE discretization of concrete (aggregate and mortar). The constitutive law for mortar is based on the hygro-mechanical model, which couples the Fickian moisture transport and the microplane-based mechanical model, similar as presented above. In the model the total strain is decomposed into mechanical, drying shrinkage and basic creep strains. Basic creep model is based on the rate type creep law and Maxwell chain model with eight aging units. The application of the model is illustrated on the numerical example, i.e. concrete cylinder loaded in compression. The main aim was to investigate the role of drying shrinkage of mortar and inhomogeneity of concrete on the drying creep of concrete.

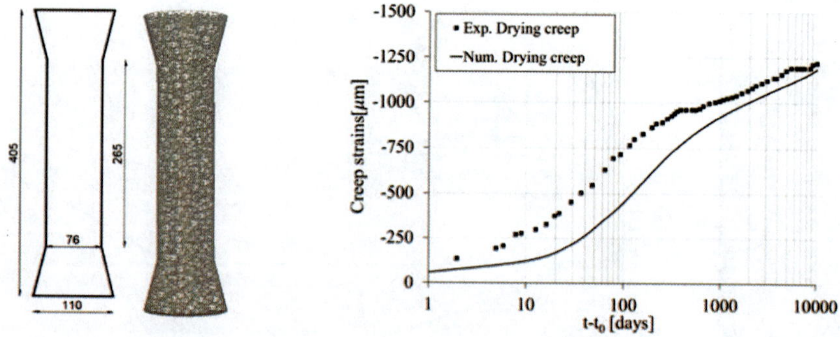

Fig. 3. Geometry and FE discretization at meso scale (left) and the comparison between experimental tests (compression, stress 30% of compressive strength) [14] and two-phase meso-scale FE prediction for drying creep (right)

Numerical Example: Simulation of Drying Creep. The aim of the simulation was to investigate whether the meso-scale model for concrete is able to replicate the long-term experimental tests for drying creep based only on the interaction between damage, drying shrinkage and basic creep, without the additional viscos kind of mortar strains due to drying. The numerical simulation [13] showed that there is a strong interaction between drying shrinkage, the load-induced damage of mortar and heterogeneity of concrete. This mainly contributes to drying creep of concrete and leads to the reduction of the uniaxial compressive strength to approximately 80% of short-term strength. For tensile load the interaction is even stronger and causes the reduction of the short term tensile strength to approximately 40%. The study also indicated that there is no strong interaction between the load-induced damage and basic creep of mortar. Figure 3 shows the specimen geometry, meso scale FE model and the test data for drying creep. It is interesting to observe that the meso-scale model is able to nicely replicate the long-term experimental tests [14], over almost 30 years, for drying creep without need for any additional viscos kind of mortar strains due to drying of cement paste.

3 Summary and Conclusions

In the present paper the recently proposed coupled 3D CHTM is briefly presented. The model can be employed for the modeling of macro-cell chloride induced corrosion of steel reinforcement in concrete, freezing-thawing of concrete and creep and shrinkage of concrete. It was implemented into 3D FE code and can be effectively used in studies of deterioration of concrete due to extreme environmental conditions. The application of the model is illustrated on several numerical examples, which shows its ability in realistic prediction of phenomena related to the influence of environmental conditions on durability of RC structures.

References

1. Ožbolt, J., Balabanić, G., Kušter, M.: 3D Numerical modelling of steel corrosion in concrete structures. Corros. Sci. **53**, 4166–4177 (2011)
2. Ožbolt, J., Oršanić, F., Balabanić, G., Kušter, M.: Modeling damage in concrete caused by corrosion of reinforcement: coupled 3D FE model. Int. J. Fract. **178**, 233–244 (2012)
3. Ožbolt, J., Balabanić, G., Periškić, G., Kušter, M.: Modeling the effect of damage on transport processes in concrete. Constr. Build. Mater. **24**, 1638–1648 (2010)
4. Wong, H.S., Zhao, Y.X., Karimi, A.R., Buenfeld, N.R., Jin, W.L.: On the penetration of corrosion products from reinforcing steel into concrete due to chloride-induced corrosion. Corr. Sci. **52**(7), 2469–2480 (2010)
5. Ožbolt, J., Li, Y.-J., Kožar, I.: Microplane model for concrete with relaxed kinematic constraint. Int. J. Solids Struct. **38**, 2683–2711 (2001)
6. Ožbolt, J., Oršanić, F., Balabanić, G.: Modeling pull-out resistance of corroded reinforcement in concrete, coupled three-dimensional finite element model. Cem. Concr. Compos. **46**, 41–55 (2014)
7. Coussy, O.: Poromechanics of freezing materials. J. Mech. Phys. Solids **53**(8), 1689–1718 (2005)
8. Yang, Z.F., Weiss, W.J., Olek, J.: Water transport in concrete damaged by tensile loading and freeze-thaw cycling. J. Mater. Civ. Eng. **18**(3), 424–434 (2006)
9. Pigeon, M., Pleau, R.: Durability of Concrete in Cold Climates, pp. 12–30. E&FN Spon, London (1995)

10. Zuber, B., Machand, J.: Predicting the volume instability of hydrated cement systems upon freezing using poro-mechanics and local phase equilibria. Mater. Struct. **37**(4), 257–270 (2004)
11. Bažant, Z.P., Qiang, Y., Li, G.-H.: Excessive Long-time deflections of prestressed box girders. I: record-span bridge in palau and other paradigms. J. Struct. Eng **138**(6), 676–686 (2012)
12. Bažant, Z., Wu, S.T.: Rate-type creep law of aging concrete based on Maxwell chain. Matériaux et Constructions **7**(37), 45–60 (1974)
13. Gambarelli, S., Ožbolt, J.: Interaction between damage and time-dependent deformation of mortar in concrete: 3D FE study at meso-scale. IOP Conf. Ser. Mater. Sci. Eng. **615**(1), 012013 (2019)
14. Brooks, J.J.: 30-year creep and shrinkage of concrete. Mag. Concr. Res. **57**(9), 545–556 (2005)

Development of a Thermo-Hydro-Mechanical Model of the Containment Vessel Vercors to Study Its Aging and Leak Tightness, Based on Specimen Tests and In Situ Measurements

Alessandro Perlongo[1(✉)], Nicolas Goujard[1], Mahsa Mozayan[1], and Farid Benboudjema[2]

[1] Ingérop, Rueil Malmaison, France
{alessandro.perlongo,nicolas.goujard,mahsa.mozayan}@ingerop.com
[2] LMT ENS Paris Saclay, Gif sur Yvette, France
farid.benboudjema@ens-paris-saclay.fr

Abstract. An important function of a reactor containment vessel is to provide sufficient leakage tightness during normal operation and severe accident condition. The purpose of this paper is to present a set of numerical models along with a study methodology to understand the mechanisms most responsible for aging and leakage tightness degradation of prestressed reinforced concrete containment vessels without liner. The solution proposed here is adapted to the study of full-size buildings while most research work dealing with this matter offers sophisticated modeling techniques mostly suitable for local studies.

The methodology is applied to study the thermo-hydro-mechanical behavior of Vercors mock-up [1]. Vercors is an experimental mock-up of a reactor containment building at 1/3 scale (with diameter of 7.7 m, height of 20 m and thickness of 0.4 m) built at EDF Lab "Les Renardières" near Paris in 2015 to analyze the possibility of the life extensions of French Nuclear Power Plants. The mock-up is finely instrumented, and its behavior is monitored from the beginning of the construction. In the numerical application of our modelling methodology to Vercors mock-up, the thermo-hydro-mechanical models are all calibrated by considering first the material test specimen and then the in-situ measurements, to have a representative digital clone of the mock-up. All necessary models and resolution strategy are implemented in Code Aster® (EDF) finite elements program.

Keywords: Ageing phenomena in concrete structures · Delayed strains · Damage and cracking of reinforced and prestressed concrete structures · Air transport in concrete

1 Notions of Ageing Phenomena in Concrete Structures

When talking about concrete materials, basically two ageing phases are to be distinguished:

P. Rossi and J.-L. Tailhan (Eds.): SSCS 2022, RILEM Bookseries 38, pp. 289–300, 2023.
https://doi.org/10.1007/978-3-031-07746-3_29

- Young age phase: by this term, we refer to the period during which the concrete hydrates and sets. It can vary from a few days after casting to a few months. In practice, it is usually limited to the first four weeks,
- Long-term phase: this corresponds to the behaviour of concrete in the long term once the main chemical phenomena have come to an end or their kinetics have slowed down drastically.

During the young age phase, several chemical phenomena take place, notably thermo-hydration, maturity and self-drying; while, during the long-term phase, phenomena of a physical nature such as desiccation take place.

All these ageing phenomena mean that the material is far from being considered as purely elastic. In addition to the elastic and thermal deformations of the material, delayed deformations, that are called shrinkage deformations in the absence of applied external mechanical loads or creep deformations under external stress, exist.

In turn, shrinkage deformations are decomposed into an endogenous part, i.e., the deformation observed at young age phase due to concrete hydration, and a desiccation one which corresponds to the part of long-term shrinkage linked to sorption/desorption phenomena. Creep deformations are also normally decomposed into a basic and a desiccation part, the latter being the surplus of deformations observed in the presence of water exchange with the external environment.

All delayed deformations are dependent on the water content within the concrete porosity as well as on the temperature of the concrete for some of them; hence, the requirement to perform a thermal plus a drying calculation prior to the mechanical one.

So, because of its porous structure and its potential for cracking, a concrete structure, even if pre-stressed, cannot ensure a perfect leak-tightness (delayed deformations in fact cause a loss of prestress, which leads to the appearance of tensile states). A global criterion on the maximum rate of outflow from the walls of the containment is normally set by the various nuclear safety regulations.

Three types of flow can be defined which contribute to the air mass passing through the containment in the event of a regulatory accident:

- Flow through the pore structure (Darcy flow). The amount of air that passes through the pores depends not only on the portion of voids in the volume but also on their connectivity (intrinsic permeability). This inherent permeability of the material is then influenced by the thermo-hydro-mechanical state of the structure. Essentially two state variables play a key role in modifying the intrinsic permeability:

  - The water content of the concrete. If we are only interested in the dry air flow, the apparent permeability of the material would be zero if all the porosity was filled with water and maximum for a dry porosity,
  - The level of damage to the structure since, even before the formation of a macro-crack, micro-cracks appear contributing to the increase in apparent permeability.

- The flow through the cracked areas (Poiseuille flow). The amount of air that passes through is undoubtedly a function of the opening of these cracks, but also of their traversing character, their connectivity and their internal roughness and tortuosity.

- Flow through singular zones. That is, areas such as concrete joints, concrete defects, areas in the vicinity of prestressing tendons,... where an increase in permeability is expected.

2 Numerical Modelling of VeRCoRS

2.1 VeRCoRs and the Third International Benchmark

To continue to use its power stations and extend its useful life, EDF ("Electricité de France") must be able to demonstrate the safety of its infrastructure. For this reason, a programme called VeRCoRs (French for "Realistic Verification of Reactor Confinement") was launched in 2013 (see [1]). This program consists of the construction and monitoring of a 1/3 scale mock-up of a power plant containment building and it has as its main aim is to demonstrate the degree of confinement of a nuclear power plant reactor building in the event of an accident.

The mock-up is strongly instrumented and subjected every year to a pressurization test that simulate the thermomechanical loading in case of an accident.

Vercors project has been the subject of several international benchmarks involving several engineering firms and university research centres over the years. The third and last benchmark (03/2021–03/2022), focus of this paperwork, has as its main objectives those of responding to two problems: the long-term prediction of the structural behaviour of the reactor building (mechanical state) and the prediction of air leakage in accidental situations (hydraulic calculation).

2.2 Computation Strategy

In the literature, there are mainly two families of approaches for carrying out thermo-hydro-mechanical modelling (see [5]): totally coupled approaches, which consider the complete coupling between the different physical phenomena, quite costly in terms of computational power required and only applicable for specimens; weakly coupled approaches, that are more suited for the thermo-hydro-mechanical computation of gas transport in unsaturated large-scale porous structures.

In this study, a weakly coupled approach was used. The following calculations were then carried out, in order:

- Thermal calculation to determine the temperature,
- Calculation of the drying of the concrete, to determine the internal relative humidity of the concrete (which influences the mechanical behaviour and the leak-tightness of the structure),
- Mechanical calculation,
- Calculation of damage (post-processing of the mechanical calculation),
- Flow calculation to determine the air leakage through the walls during the pressure tests of the vessel (Fig. 1).

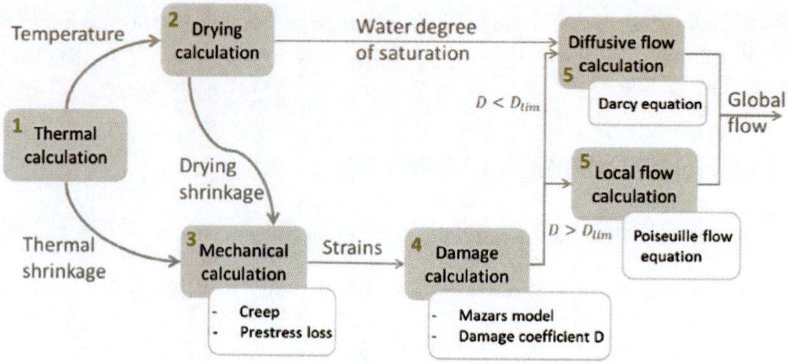

Fig. 1. Calculation sequence

2.3 Thermal Model

The heat equation was used to solve the thermal problem:

$$\rho c_p \frac{\partial T}{\partial t} = \nabla(\lambda \nabla T) \tag{1}$$

As material characteristics and boundary conditions, data provided in the benchmark input were used. Thermal results serve as input data for the drying calculation (Fig. 2).

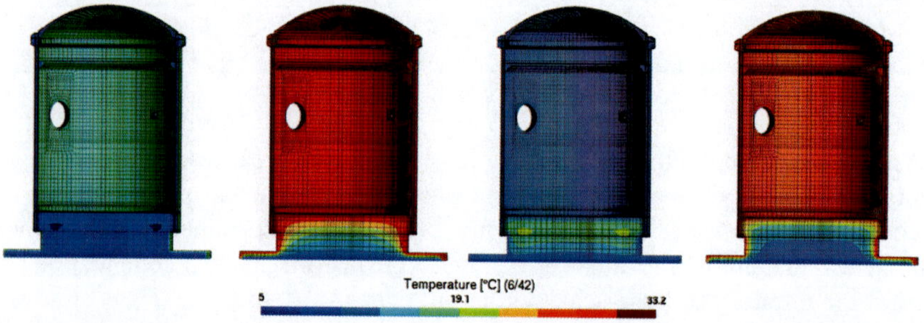

Fig. 2. Temperature profiles: a) at the start of the warm-up phase, b) at the end of the warm-up phase, c) during a typical ten-year test (heating shutdown), d) at the end of a typical ten-year test (restart heating)

2.4 Drying Model

The drying calculation was based on the theory developed by Granger in his doctoral thesis work [7]. The concentration of water in the material, defined as the ratio of the volume of water in the material to the total volume, verifies the following equation:

$$\frac{\partial C}{\partial t} = -div(-D(C, T)grad(C)) \tag{2}$$

where $D(C, T)$ is the non-linear diffusion coefficient can be expressed as:

$$D(C,T) = A\exp(B \times C)\frac{T}{T_0}\exp\left(\frac{Q_s}{R}\right)\left(\frac{1}{T} - \frac{1}{T_0}\right) \tag{3}$$

where A *and* B are material parameters while T_0 *and* Q_s/R represent respectively the reference temperature and the activation energy of drying.

All these coefficients were calibrated based on a concrete mass loss test. The latter, carried out by the TEGG laboratory in EDF, was performed by monitoring the mass loss of cylinder samples after an endogenous cure with a maturation of at least 90 days in a drying environment imposed in terms of temperature and relative humidity. The specimens, with a diameter of 16 cm and a height of 1 m, were then left to dry at a constant temperature T = 20 °C and a relative air humidity of 50% (Figs. 3 and 4).

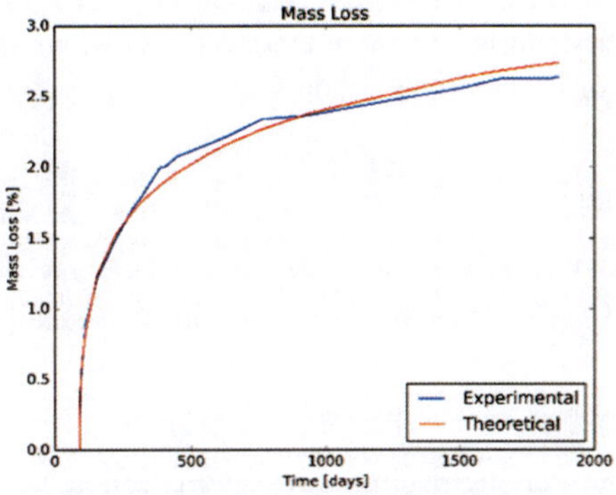

Fig. 3. Numerical simulation of mass loss

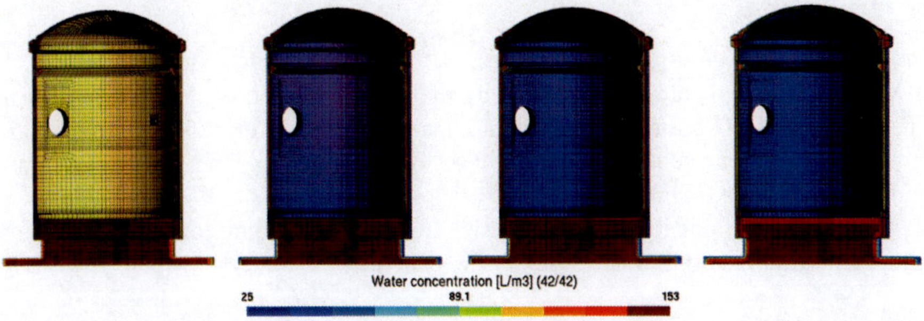

Fig. 4. Water concentration profile: a) before warming up b) at the start of the warm-up phase c) at the end of the warm-up phase d) at the final instant

3 Mechanical Model

3.1 General Assumptions

The mechanical model used to simulate the behaviour of concrete naturally considers elastic and thermal deformations but also delayed deformations characterising the ageing of the material (viscoelastic model).

The numerical simulation of the first days of life of the structure (young age) was not carried out, but still considered through the manual integration in damage results of the cracks found through visual inspection before the prestressing of the vessel ("young age cracks"). On the other hand, the three prestressing phases of the structure were simulated.

In the following sub-chapters, the theoretical model underlying each of the viscous phenomena involved in the modelling will be presented, followed by the methodology used to calibrate the parameters of these models. The latter was based on experimental tests on specimen at first, followed by a second phase of optimisation of parameters through local calculations on reduced models.

3.2 Drying Shrinkage

Since it is related to the loss of water mass, the desiccation shrinkage can be simply related to the variation of the internal relative humidity as (see [6]):

$$\underline{\varepsilon_{rd}} = k_d * \dot{h} * \underline{I_d} \tag{4}$$

where k_d *and* $\underline{I_d}$ represent the drying shrinkage coefficient and the identity matrix respectively.

3.3 Basic Creep

The model used to describe the basic creep is the so-called Burger model (see [6]). The set of Burger models are characterised by the decomposition of the deformation into a reversible part, whose associated rheological model is a Kelvin-Voigt chain, and an irreversible part represented by a non-linear viscous damper. The version used is the aging one, in which nonlinear viscosity varies linearly in time.

Moreover, the phenomenon of basic creep is by nature thermally activated: Arrhenius' law was used to describe this dependence.

3.4 Drying Creep

The commonly used model to describe drying creep is the one that relates the excess of strain during drying to the change in water concentration (see [6]). For this version, it is assumed that desiccation creep occurs with a Poisson's ratio of zero and its amplitude is proportional to that of the relative humidity taken as an absolute value. Furthermore, desiccation creep only occurs during the first time a lower relative humidity value is reached (the history of moisture loading is therefore considered). So, we have:

$$\underline{\dot{\varepsilon}_{dc}} = \frac{1}{\eta_{fd}} * \langle \dot{h} \rangle * \underline{\sigma} \tag{5}$$

where:

$$\langle \dot{h} \rangle = \begin{cases} \dot{h} & si\ \dot{h} < 0\ et\ h(t) < min_{\tau \in [0,t]} h(\tau) \\ 0 & otherwise \end{cases} \tag{6}$$

3.5 Calibration of Delayed Deformations

First Phase of Calibration of the Mechanical Model

Three experimental tests, carried out by the TEGG laboratory in EDF, were used to calibrate the ten viscoelastic parameters. These first three tests gave us a first play of parameters to use on reduced models.

The tests were all carried out on specimens 16 cm in diameter and 1 m in height, at a constant temperature of 20 °C. The specimens were also left to mature for 90 days under endogenous conditions before being tested.

- For the drying shrinkage coefficient, a drying shrinkage test was performed on a 2D axisymmetric specimen in drying condition and not mechanically loaded,
- To determine the 8 parameters of the basic creep model, a non-drying creep test was performed on a 2D axisymmetric specimen in a non-drying condition and at constant temperature, mechanically loaded to a level of 12 MPa on its upper face. As no radial deformations are available, an equivalence ratio between the spherical and deviatoric chains had to be imposed. A creep Poisson's ratio egal to the elastic one was used.
- Finally, to determine the desiccation creep viscosity, a drying creep test was carried out on an axisymmetric 2D specimen in the drying condition and mechanically loaded to 12 MPa on its upper surface (Fig. 5).

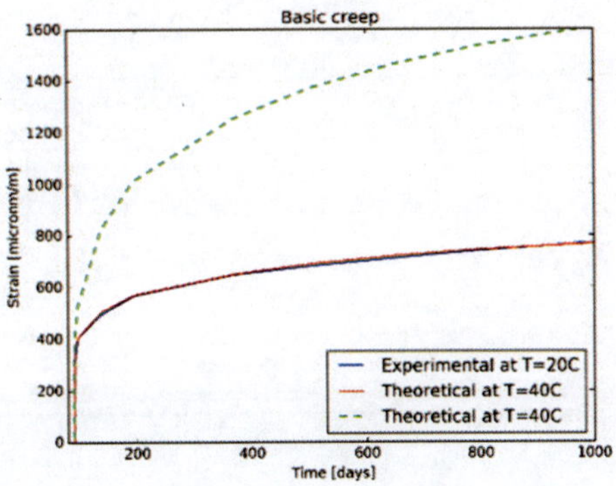

Fig. 5. Basic creep test under uniaxial stress

Second Phase of Calibration of the Mechanical Model
A second calibration phase on reduced models was carried out to verify the methodology used on the results of the laboratory tests and their real representativeness of the behaviour of the structure. The procedure used is as follows:

- A first global calculation of the vessel was carried out with the mechanical parameters obtained through the experimental tests,
- Two sets of sensors were selected for which vertical and radial strain measurements were available. These two sets (P1/P2 and H1/H2) belong to two different lifts, both located in the cylindrical part of the vessel,
- A reduced mesh around these points was extracted,
- From the global calculation, thermal and drying results as well as the nodal forces to be applied in the model were extracted,
- Finally, the optimisation algorithm was launched. The principle is then as follows: for a set of values of calibration parameters, a numerical calculation was carried out and, for a node at the location of each sensor, the value of total deformations at various representative moments of the life of the vessel was extracted. For these same instants and the sensor considered, the numerical values were compared with the experimental ones. The score (least squares method) was deduced and then the calculation was repeated with a new set of parameters (Fig. 6).

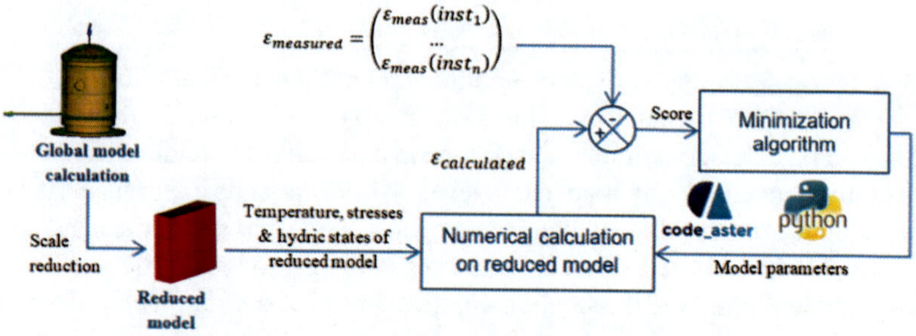

Fig. 6. Structure of the numerical setting on a reduced model

The figure shows the results of the optimisation (Fig. 7):

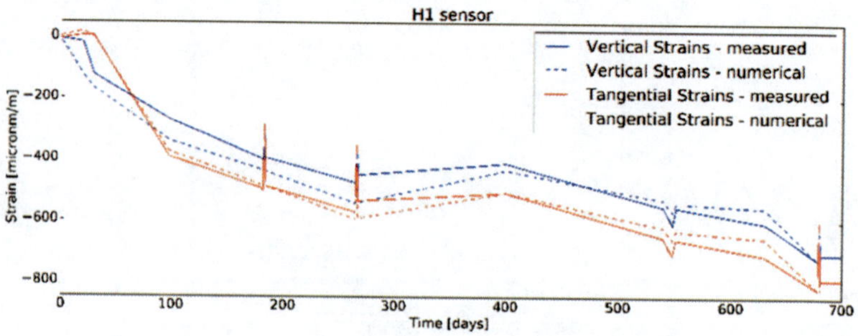

Fig. 7. Strains at sensor H1 after second optimisation

4 Damage Model, Cracking Model, and Integration of Early Age Cracks

The damage level of the structure was then estimated a posteriori of the mechanical calculation through the Mazars model (see [3]). Mazars model was not integrated directly in the global calculation because of the size of the model and the important computational time.

Three levels of damage were defined in this study: sound concrete; moderately damaged concrete, containing micro-cracks; and heavily damaged concrete, containing macro-cracks.

Depending on the level of damage, the flow law used through the concrete for leakage was different. If the concrete was sound, micro-cracked or macro-cracked but with a closed crack, Darcy's law was used for the flow calculation; however, if the concrete was macro-cracked, Poiseuille's law was preferred.

The calculation of macro-cracking was based on the value of the damage coefficient by comparing it to a threshold value D_{lim}. If $D > D_{lim} = 0.15$, there were enough micro-cracks to create a macro-crack, and the characteristics of this crack were calculated. The calculation of crack opening was based on a post-treatment of principal deformations, developed in [5].

However, as we chose not to model the young age behaviour and we knew from [2] that many cracks appear from the first days after pouring the concrete, it was necessary to complete the post processing routine to be able to manually define an initial state with already cracked elements (Fig. 8).

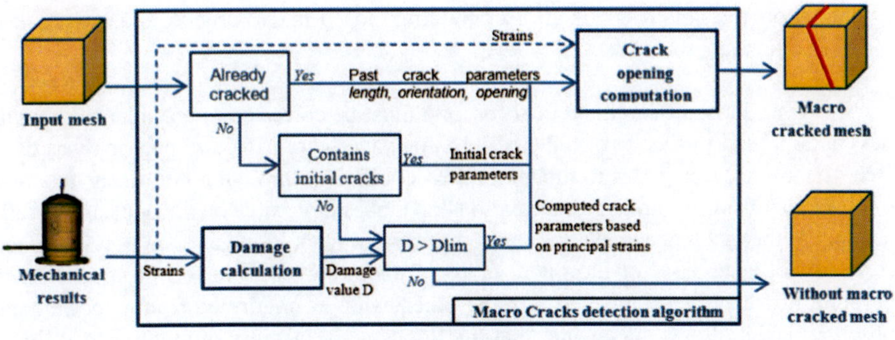

Fig. 8. Macro Cracks detection algorithm

Crack openings were used as basis for the hydraulic calculation (Fig. 9).

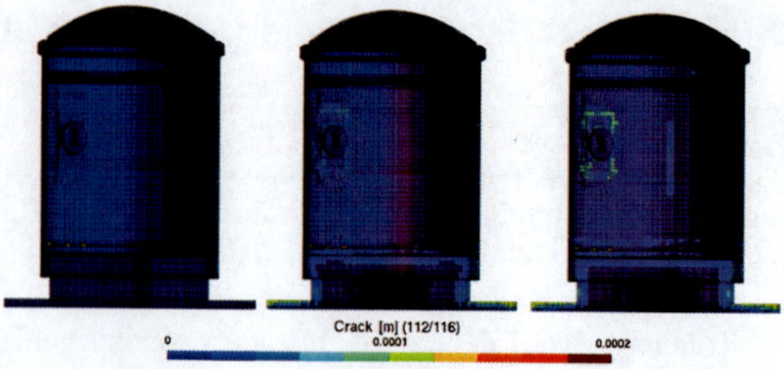

Fig. 9. Cracks profile: a) at the beginning: only "young age" cracks b) during a "normal" pressure test (4.2 bar): cracks in the gusset + superficial cracks in equipment hatch and dome c) during the "ultimate" pressure test (6 bar)

5 Hydraulic Model

The hydraulic calculation consisted in evaluating the air flow through the concrete during the pressurisation of the vessel at each ten-year test. The calculation of the air flow through the concrete depends on all the previous results:

- The temperature because, considering air as a perfect gas, the volume occupied by air is proportional to the absolute temperature. In addition, the temperature influences the drying, the mechanical calculation and therefore the damage and the hydraulics,
- The drying of the vessel because the wetter the concrete, the tighter it is,
- The mechanical condition of the concrete to obtain the damage,
- The damage, according to which different flow laws will be used: Darcy's law through sound concrete, micro-cracked or containing closed macro-cracks, and Poiseuille's law through macro-cracks.

However, Poiseuille's theoretical formula must be corrected to consider the roughness and internal tortuosity of the crack. In this paperwork Mivelaz approach was used (see [4]). Mivelaz proposes to model the loss coefficient through a non-linear function of the crack opening. A non-linear relationship makes it possible on the one hand to take account of the fact that a crack does not present significant losses as soon as it is formed (it is in fact necessary for the micro cracks to come together to form a macro crack before a flow can be established); on the other hand, as confirmed from experimental observations, it allows to consider that the flow cannot increase infinitely and it limits the loss coefficient to a maximum value of 0.3.

5.1 Numerical Results

One of the main objectives of the VeRCoRs model was to be able to predict the leakage rate during pressurisation tests, to numerically ensure that the vessel meets the safety criterion $Q < Q_{\mathrm{lim}}$, and even more so to be able to locate the origin of leaks and prevent excessive leakage.

The global (Darcy + Poiseuille corrected) numerical flux out of the outer structure surface was compared with the measurements available up to the third 10-year test, VD3.

As can be seen from Fig. 10, except for the first two operational tests, the numerical model was able to predict the values measured experimentally. The same prediction was found for the local flow (Fig. 11).

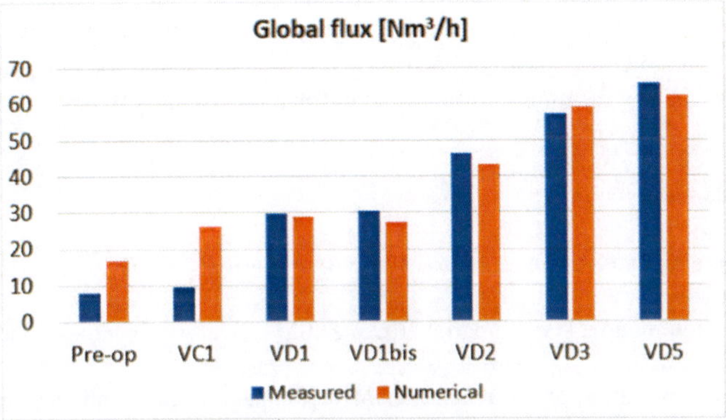

Fig. 10. Global flux results up to VD3

In our model, as well as experimentally confirmed from [2], it is the gusset that contributes almost all the local flow. Mivelaz's formula reflected well the experimentally measured value from the first ten-year test VD1.

For the operational tests (VO1 and VC1), the values were overestimated: this could be since we included the surface opening of the crack and not the in-depth one in our model; but "young age cracks" may during these first two tests not yet be fully traversed. The prediction of first tests in this case would certainly be wrong. If we then consider the local flow to be almost zero for the first two service tests, our flow prediction would be almost correct everywhere (with margins of error due to the necessary numerical refinement of the model and to experimental measurement errors).

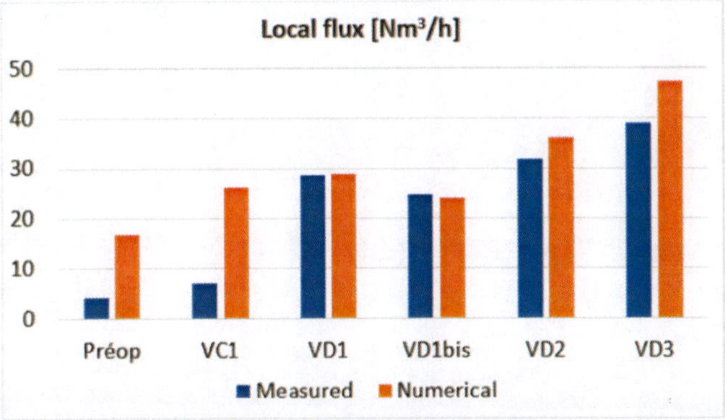

Fig. 11. Local flux results up to VD3

6 Conclusion

A partially coupled thermo-hydro-mechanical model adapted to study the long-term behavior of large-scale structures was developed. Different parameters of the model were calibrated with the Vercors laboratory tests, and then optimized with in-situ measures. The reasonable computation time, the stability of the numerical model and the quality of results, which follow the in-situ measures of strains, global and local leakage, constitute the assets of this tool and methodology to perform industrial studies.

References

1. EDF website of VeRCoRs benchmark. https://fr.xing-events.com/EDF-vercors-project.html?page=1426249
2. Corbin, M.: International Benchmark VeRCoRs 2015 – Overview, synthesis and lessons learnt. EDF Speten (2016)
3. Mazars, J.: Application of the mechanics of the damage to the nonlinear behavior and the rupture of the structural concrete, Ph.D. thesis, University Paris 6, Paris (1984)
4. Mivelaz, P.: Etanchéité des structures en béton armé, Ph.D. thesis, EPFL, Switzerland (1996)
5. Mozayan, M.: Une méthodologie de modélisation pour l'évaluation de l'étanchéité des enceintes de confinement des centrales nucléaires, Ph.D. thesis, ENS Cachan, France (2013)
6. Relation of behavior BETON_BURGER for the creep of the concrete, R7.01.35, Code_Aster
7. Granger, L.: Comportement differé du béton dans les enceintes nucléaries: analyse et modélisation. Ph.D. thesis, ENPC, France (1995)

Modeling of the Behavior of Concrete Specimens Under Uniaxial Tensile Stresses Through the Use of a 3D Probabilistic Semi-explicit Model

Mariane R. Rita[1(✉)], Eduardo de M. R. Fairbairn[1], Fernando L. B. Ribeiro[1], Jean-Louis Tailhan[2], Pierre Rossi[2], Henrique C. C. de Andrade[1], and Magno T. Mota[1]

[1] Universidade Federal do Rio de Janeiro, Rio de Janeiro, Brazil
mariane_rita@coc.ufrj.br

[2] Université Gustave Eiffel, Paris, France

Abstract. This paper presents the numerical modeling of plain concrete specimens subjected to uniaxial tensile stresses. The simulations are performed using a three-dimensional macroscopic probabilistic model for semi-explicit concrete cracking. As it is well known, concrete structures are largely sensible to the scale effects that can be attributed, among other reasons, to the heterogeneous nature of the material. The model used herein, which is developed in the framework of the finite element method, considers the material heterogeneity through the assumption that each finite element represents a volume of heterogeneous material, with mechanical properties of tensile strength and fracture energy being randomly distributed over the mesh according to the Weibull and lognormal distributions, respectively. The cracks are created with different energy dissipation according to an isotropic damage law. The results are obtained through Monte Carlo simulations using a parallelization strategy with OpenMp to allow feasible 3D simulations of real structures in a viable computational time. With the purpose of modeling the uniaxial tensile test and verifying the prediction of the scale effects, simulations of three prismatic plain concrete specimens with different sizes are performed.

Keywords: Numerical modeling · Probabilistic cracking model · Scale effects · Uniaxial tensile test · Finite element method

1 Introduction

Concrete is a heterogeneous composite material whose mechanical behaviour is subjected to the so-called scale effects. This phenomenon is related to the dependence of the global response of a given concrete structure to its size or volume. Some reasons can be pointed out to explain this phenomenon: the heterogeneous

Supported by the Brazilian Scientific Agencies CNPq, CAPES and FAPERJ.

P. Rossi and J.-L. Tailhan (Eds.): SSCS 2022, RILEM Bookseries 38, pp. 301–310, 2023.
https://doi.org/10.1007/978-3-031-07746-3_30

nature of concrete; the physical and chemical changing during its production and hardening; the micro-cracking due to drying; and the porosity resulting from the presence of water [8,10].

As stated by [7], the size effect is related to two levels: (1) the material level that determines intrinsic constitutive relations for concrete cracking, and (2) the structural level to account for size effect in design methods and finite element analysis. The material heterogeneity and size effect are strictly correlated aspects and should be taken into account when dealing with concrete structures modeling.

In this work, a 3D macroscopic probabilistic model for semi-explicit concrete cracking is applied to numerical simulations of the behavior of plain concrete specimens under tensile stresses. The purpose of these simulations is to investigate the model's capability to reproduce the phenomenon of scale effect on concrete tensile strength. The model considers the heterogeneity of the material through a probabilistic approach, and a Monte Carlo (MC) procedure is used to ensure the accuracy of the results. The MC procedure is implemented using a parallelization strategy, reducing the computational time.

2 Semi-explicit Probabilistic Model for Concrete Cracking

The semi-explicit probabilistic model is developed in the context of the finite element method; its main principle is to incorporate the concrete heterogeneity in its formulation. For this, it is assumed that each finite element represents a volume of heterogeneous material, with its heterogeneity degree (r_e) evaluated by the following ratio: finite element volume (V_e) divided by coarsest aggregate volume (V_a). Therefore, to describe the material heterogeneity, the tensile strength (f_t) and fracture energy (G_c) are randomly distributed for each finite element according to its respective r_e.

In this modeling, it is considered that the creation and propagation of one crack within the element itself induces some local dissipation of energy. The element is considered damaged when the total amount of energy that it can consume is reached. The evolution of this dissipative process is mathematically represented through a probabilistic isotropic damage law [2]. At a macroscopic level, the creation and propagation of a crack is the consequence of the elementary failure of successive elements that randomly appear and can coalesce to form the macroscopic cracks. In that context, the model does not deal with crack propagation laws in the sense of fracture mechanics [3,10].

The stress-strain relation of the material in a stage of damage can be expressed in terms of the undamaged stress-strain relation, as described in Eq. (1), where $\tilde{\mathbf{E}}$ and $\mathbf{E}_0$ are, respectively, the elastic modulus of the damaged and undamaged material and D is the damage variable. The damage evolution can be given by Eq. (2), where $\tilde{\varepsilon}_0$ represents the damage initialization strain; $\tilde{\varepsilon}_{fi}$ represents the maximum critical strain and $\tilde{\varepsilon}^k$ stands for the equivalent strain.

$$\sigma = \tilde{\mathbf{E}}\varepsilon, \quad \tilde{\mathbf{E}} = \mathbf{E}_0(1 - D), \quad (0 \leq D \leq 1) \tag{1}$$

$$D = 1 - \frac{\tilde{\varepsilon}_0}{\tilde{\varepsilon}^k}\left[1 - \frac{(\tilde{\varepsilon}^k - \tilde{\varepsilon}_0)}{(\tilde{\varepsilon}_{fi} - \tilde{\varepsilon}_0)}\right] \tag{2}$$

This constitutive law is completely defined by the tensile strength and volumetric density of dissipated energy (g_c). This latter can be evaluated considering the use of an energetic regularization technique [1], taking into account the material fracture energy, as follows: $g_c = G_c/l_e$; where l_e represents the elementary characteristic length and is evaluated as: $l_e = (V_e)^{1/3}$.

2.1 Random Distribution of the Material Properties

The tensile strength of the material is distributed according to the Weibull distribution [11,12]. Its probability density function for a random variable x is described in Eq. (3), where $b > 0$ is the shape parameter and $c > 0$ is the scale parameter of the distribution.

$$f(x, b, c) = \frac{b}{c}\left(\frac{x}{c}\right)^{b-1} e^{\left(-\frac{x}{c}\right)^b} \tag{3}$$

The mean μ and the variance σ^2 of the distribution can be seen in Eq. (4) and Eq. (5), where Γ is the Gamma function given by $\Gamma(\eta) = \int_0^\infty x^{\eta-1}e^{-x}dx$. If η is a positive integer then $\Gamma(n+1) = n!$ what means that $\Gamma(n) = (n-1)!$.

$$\mu = c\Gamma\left(1 + \frac{1}{b}\right) \tag{4}$$

$$\sigma^2 = c^2\Gamma\left(1 + \frac{2}{b}\right) - \mu^2 \tag{5}$$

The fracture energy of the material is distributed according the lognormal distribution. Its probability density function is defined by $f(x, b, c) : x \in (0, \infty] \rightarrow \mathcal{R}$, as can be seen in Eq. (6), where, μ is the mean and σ the standard deviation of the variable's natural logarithm.

$$f(x, \mu, \sigma) = \frac{1}{\mu\sigma\sqrt{2\pi}} e^{-\frac{(\ln(x)-\mu)^2}{2\sigma^2}} \tag{6}$$

The expected value $E(X)$ and variance $Var(X)$ are given by (Eq. (7)) and (Eq. (8)). The standard deviation is considered as the dispersion measure of the distribution and is defined as $d_{log} = \sqrt{Var(X)}$.

$$E(X) = e^{\mu + \frac{\sigma^2}{2}} \tag{7}$$

$$Var(x) = \left(e^{\sigma^2} - 1\right) e^{2\mu+\sigma^2} \tag{8}$$

2.2 Parameters Estimation

The Weibull distribution parameters are estimated through an iterative numerical procedure developed to solve a non-linear system of equations. The system combines the equations of the mean and standard deviation of the distribution (Eq. (4) and Eq. (5)) with the analytical scale law proposed by [9]. This formulation comes from an experimental investigation correlating concrete heterogeneity and scale effect. This law, applied here to the elementary level, estimates the expected values of the mean and standard deviation of a given volume of concrete. With the procedure, a pair of (b, c) is obtained for each element as a function of its volume, maximum aggregate size, and compressive strength (f_c). More details of the analytical expressions as well as the description of the iterative procedure implementation can be found in [5].

The lognormal distribution has two parameters; however, as it is assumed that the fracture energy is an intrinsic material property with a constant mean value, the only parameter that must be determined is its standard deviation d_{log}. The fracture energy mean value can be taken equal to the experimental value obtained by [6] ($G_c = 1.3141 \times 10^{-4}$ MN/m). Thus, an inverse analysis procedure was carried out for estimating d_{log}, based on several simulations of a macrocrack propagation test on a very large double cantilever beam specimen (DCB), modeling the experimental test performed by [6]. From this procedure, a function to define the value of the parameter for each mesh element related to its heterogeneity degree (Eq. 9) was proposed. The full description of the inverse analysis procedure is beyond the scope of this work, but more details can be found in [4].

$$d_{log}(r_e) = (A \ln(r_e) + B) \times G_c, \qquad r_e \in [1, 3000] \tag{9}$$

where, $A = -8.538$ and $B = 70.88$.

3 Modeling of the Uniaxial Tensile Test

Simulations using different volumes of plain concrete are performed to verify the presence of scale effects on the direct tensile strength of concrete in cubic and prismatic specimens. Since the model is macroscopic and its main objective is to treat the macrocrack propagation and not macrocrack initialization, a difference was expected between the numerical and experimental results concerning the tensile strength values. Therefore, the following simulations are performed:

1. Simulations of four concrete cubes with different volumes to verify the accuracy of the model and the set of estimated parameters.
2. Simulations of the four concrete cubes after calibrating the evaluation of mean and standard deviation of tensile strength, proposing an adjustment function to be used in the cases of simulation of concrete specimens under uniaxial tensile stresses.
3. Simulations of three prismatic plain concrete specimens with the purpose of validation of the proposed adjustment function.

The concrete properties are the following: Young's Modulus $E = 36\,\text{GPa}$; Poisson's ratio $\nu = 0.2$ and volume of the largest aggregate $V_a \approx 9 \times 10^{-4}\text{dm}^3$. The meshes are composed of linear tetrahedrons. The number of elements was fixed to ensure that the elements' failure has the same impact on the global response of the problem; this means that the ratio V_t/V_e is kept the same in all cases. The boundary and loading conditions of the cubes and prisms are consistent with the simulated direct tensile test, with incremental displacements applied in the longitudinal direction and increments corresponding to $\delta = 0.1 \times 10^{-4}$ dm. The measures of the specimens are in decimeters. For each Monte Carlo simulation, 400 samples were run to ensure a statistically consistent result.

3.1 Simulation of Cubic Specimens - First Evaluation

The simulations will be carried out on four cubes with different volumes whose geometry and mesh characteristics can be seen in Fig. 1. The four meshes are composed of 96 elements. More details about the cubic specimens are reported in Table 1, where the following information is presented: the simulation reference (REF), the specimen height and cross-section, the total volume of the cubes (V_T), the ratio V_t/V_a, the heterogeneity degree (V_e/V_a) and the empirical (theoretical) values of mean and standard deviation of tensile strength related to each analyzed specimen.

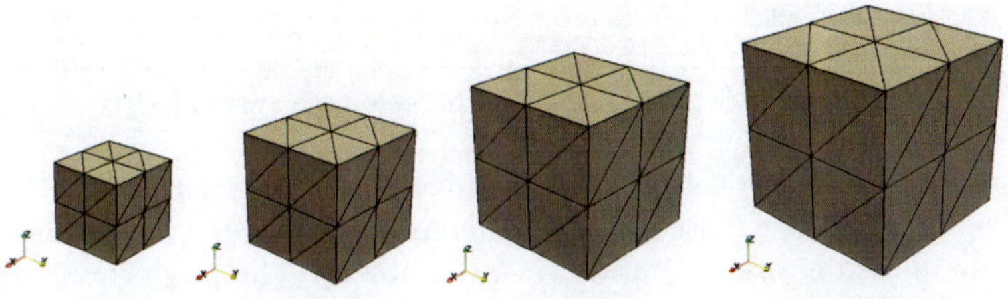

(a) C1_V1. (b) C2_V3. (c) C3_V6. (d) C4_V10.

Fig. 1. Geometry and meshes characteristics of cubic specimens.

The results of these simulations are displayed in Table 2, where Δ_{f_t} and Δ_{SD} are the comparison between numerical results and experiments, obtained by the respective division of numerical by empirical values of f_t^{mean} and SD (standard deviation). Notice that as the volume of the cubes increases, the difference between the expected and numerical tensile strength values increase. A reason for this behavior is the model's characteristics and purpose, i.e., it is a macroscopic model responsible for providing fine information about localized macrocrack propagation at a structural level. Thus, the model is not formulated to reproduce the crack creation (initialization) as its principal feature. Therefore, the obtained difference between the results is understandable.

Table 1. Detailed information about the cubic specimens.

REF	Height	Cross-section	V_T	V_t/V_a	V_e/V_a	Empirical	
						f_t^{mean}(MPa)	SD(MPa)
C1_V1	1.0000	1.0000	1	1111	11.52	3.17	0.23
C2_V3	1.4423	2.0801	3	3333	34.57	2.83	0.16
C3_V6	1.8171	3.3019	6	6666	69.15	2.64	0.13
C4_V10	2.1544	4.6416	10	11111	115.16	2.50	0.11

Table 2. Results of the first simulation on the cubes

Reference	Numerical		Comparison	
	f_t^{mean}(MPa)	SD(MPa)	Δ_{f_t}	Δ_{SD}
C1_V1	3.454	0.1849	1.09	0.80
C2_V3	3.273	0.1505	1.16	0.94
C3_V6	3.157	0.1335	1.20	1.03
C4_V10	3.155	0.1468	1.26	1.33

3.2 Simulation of Cubic Specimens - Second Evaluation

The simulations presented in this section aim to minimize the difference between numerical and expected results. For this purpose, a parameters calibration is proposed employing a proportional decrease in the elementary values of the mean and standard deviation of the tensile strength, provided by the analytical expressions describing the scale effect. Thus, the coefficient of variation of tensile strength theoretically obtained remains unchanged through this strategy. The adjustment is based on the difference between numerical and empirical results measured by the ratio Δ_{f_t} (see Table 2). The proportional decrease percentage, defined as $df_t(\%)$, is given in Table 3. This table also presents the numerical results of the second group of simulations and their comparison with the empirical data.

Table 3. Results of the second simulation on the cubes

Reference	$df_t(\%)$	Numerical		Comparison	
		f_t^{mean}(MPa)	SD(MPa)	Δ_{f_t}	Δ_{SD}
C1_V1	−10	3.097	0.1665	0.98	0.70
C2_V3	−13	2.854	0.1329	1.01	0.83
C3_V6	−17	2.618	0.1097	0.99	0.84
C4_V10	−20	2.526	0.1119	1.01	1.02

As can be seen, the mean value of the tensile strength is precisely achieved, whereas its standard deviation became more accurate only for the case of C4_V10. A level of discrepancy related to standard deviation was already expected and can be justified considering the following aspects: a) the decrease is proposed concerning Δ_{f_t}; b) it is a complex task to reproduce coefficients of dispersion of probabilistic parameters [9]; c) there is a substantial difference between the number of MC samples (400 analyzes) and the number of tests (around 13 experiments per concrete type and specimen size).

From these results, a function is proposed to estimate the percentage of decrease of the mean and standard deviation of tensile strength related to the mesh heterogeneity degree. The objective is to apply this function in the specific cases of direct tensile tests simulations and use the calibrated values for estimating the (b, c) parameters of the Weibull distribution. This function is defined as $df_t\left(\frac{V_e}{V_a}\right)$, and is described in Eq. 10.

$$df_t\left(\frac{V_e}{V_a}\right) = -4.8\left(\frac{V_e}{V_a}\right)^{0.3} \tag{10}$$

3.3 Simulations of Prismatic Specimens - Validation

For validation purposes, numerical analyzes on three prismatic specimens with different sizes, defined as P1, P2, and P3, are performed. Its geometry and mesh characteristics can be seen in Fig. 2. The number elements is equal to 837. More details about the prismatic specimens are reported in Table 4. The results are presented in Table 5 showing that there is a difference between numerical and analytical values of f_t^{mean}. However, the maximum discrepancy is around 14% for the case of P3, and the scale effect phenomenon is verified. Therefore, the results can be considered satisfactory. Besides, it is essential to highlight the mesh size distinction regarding to the cubic specimens; in this case, the number of elements is almost ten times increased.

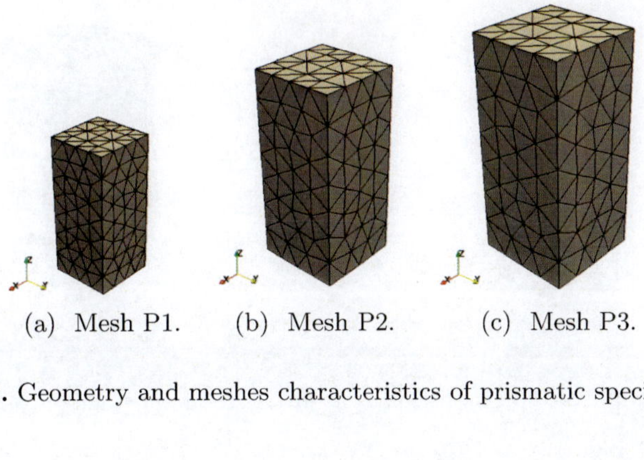

(a) Mesh P1. (b) Mesh P2. (c) Mesh P3.

Fig. 2. Geometry and meshes characteristics of prismatic specimens.

Table 4. Detailed information about the prismatic specimens.

REF	Height	Cross-section	V_T	V_t/V_a	V_e/V_a	Empirical	
						f_t^{mean}(MPa)	SD(MPa)
P1	2.20	0.9506	2.10	2333	2.78	2.94	0.1795
P2	3.00	1.7689	5.32	5911	7.06	2.67	0.1324
P3	3.52	2.4336	8.58	9533	11.38	2.54	0.1132

Table 5. Results of the prismatic specimens simulations.

REF	Numerical		Comparison	
	f_t^{mean}(MPa)	SD(MPa)	Δ_{f_t}	Δ_{SD}
P1	3.196	0.1144	1.09	0.64
P2	3.004	0.1176	1.12	0.89
P3	2.907	0.1006	1.14	0.89

A general overview of the global mechanical behavior of the performed simulations is presented through the load-displacements curves of each MC simulation, displayed in Fig. 4(a–c). Only twenty samples are reported in the graphics to allow clear visualization of the typical profile of $P - \delta$ curves. A comparison in terms of $(\sigma \times \varepsilon)$ curves is presented in Fig. 4(d). A more significant distinction between P1 and P2 mean curves are observed due to their volumes' larger difference. Moreover, the maximum discrepancy between numerical and empirical values of f_t^{mean} is related to P3. The typical cracking pattern of the analyzed prismatic specimens is illustrated in Fig. 3, where is presented the results of the damage variable at the final stage of the simulations for one of the MC samples of each specimen. In these figures, the light grayish blue elements represent the damaged elements. The dark gray represents the undamaged ones. The others with intermediate colors represent elements in the damaging process.

(a) Crack P1. (b) Crack P2. (c) Crack P3.

Fig. 3. Example of typical cracking pattern at the final stage of the simulations.

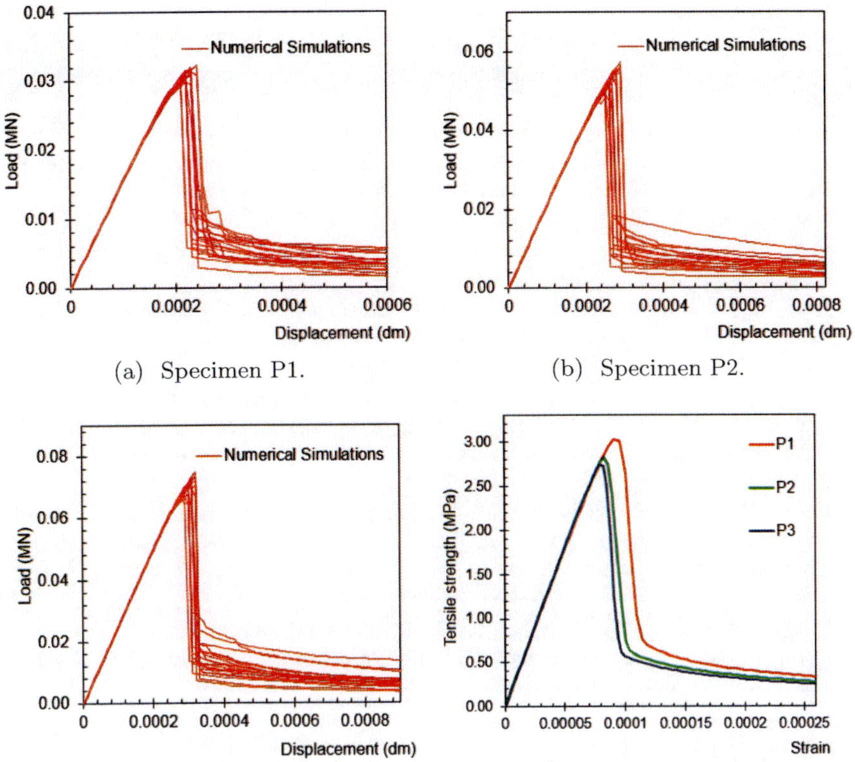

(a) Specimen P1.

(b) Specimen P2.

(c) Specimen P3.

(d) Comparison - ($\sigma \times \varepsilon$) mean curves.

Fig. 4. Global response of Monte Carlo simulations of the prismatic specimens.

4 Conclusions

In this paper, a 3D probabilistic semi-explicit cracking model is applied to simulate the behavior of concrete specimens under tensile stress in order to reproduce scale effects. After calibrating the parameters, the mean value of tensile strength was precisely achieved in the cubic specimens simulations, whereas its standard deviation became less accurate. However, the results are promising since the main objective was to achieve its mean value. Through these results, an adjustment function to be used in this cases of uniaxial tensile stress simulations was proposed to calibrate the parameters, and validation simulations were performed. The scale effect was verified in the prismatic specimens, although a slight difference between numerical and analytical values was observed. However, as the maximum discrepancy is around 14% and, considering that the results are concerning a very complex phenomenon, they can be considered satisfactory.

References

1. Bazant, Z., Oh, B.: Crack band theory for fracture of concrete. Mater. Struct. **3**, 155–177 (1983)
2. Rastiello, G.: Influence de la fissuration sur le transfert de fluides dans les structures en béton. Stratégies de modélisation probabiliste et étude expérimentale. Ph.D. thesis, Université Paris-Est, IFSTTAR, Paris, France (2013)
3. Rastiello, G., Tailhan, J.L., Rossi, P., Dal Pont, S.: Macroscopic probabilistic cracking approach for the numerical modelling of fluid leakage in concrete. Ann. Solid Struct. Mech. **7**, 1–16 (2015)
4. Rita, M.R.: Implementation of a 3D macroscopic probabilistic model for semi-explicit concrete cracking. Ph.D. thesis, Universidade Federal do Rio de Janeiro, UFRJ, Rio de Janeiro, Brazil (2022)
5. Rita, M.R., et al.: Simulação numérica de viga de concreto protendido em duplo balanço utilizando um modelo probabilístico. In: Proceedings of the XLI Ibero-Latin-American Congress on Computational Methods in Engineering. ABMEC, Foz do Iguaçu, Brasil (2020)
6. Rossi, P.: Fissuration du béton: du matériau à la structure-application de la mécanique lineaire de la rupture. Ph.D. thesis, L'Ecole Nationale des Ponts et Chaussées, Paris, France (1988)
7. Rossi, P.: Size effects in cracking of concrete: Physical explanations and design consequences. Proc. IA-FRAMCOS **2**(3), 1805–1810 (1995)
8. Rossi, P., Richer, S.: Numerical modelling of concrete based on a stochastic approach. Mater. Struct. **20**, 334–337 (1987)
9. Rossi, P., Wu, X., Le Maou, F., Belloc, A.: Scale effect on concrete in tension. Mater. Struct. **27**, 437–444 (1994)
10. Tailhan, J.L., Dal Pont, S., Rossi, P.: From local to global probabilistic modeling of concrete cracking. Ann. Solid Struct. Mech. **1**, 103–115 (2010)
11. Weibull, W.: A statistical theory of the strengh of materials. In: Proceedings of the Royal Swedish Institute for Engineering Research (1939)
12. Weibull, W.: A statistical distribution function of wide applicability. J. Appl. Mech. **18**, 293–297 (1951)

Numerical and Experimental Investigation of Wall Effect in Concrete

Takwa Sayari[1,2(✉)], Tulio Honorio[1], Farid Benboudjema[1], Rita Tabchoury[1], Jean-Luc Adia[3], and Christian Clergue[4]

[1] Université Paris-Saclay, CentraleSupélec, ENS Paris-Saclay, CNRS, LMPS - Laboratoire de Mécanique Paris-Saclay, 91190 Gif-sur-Yvette, France
takwa.sayari@ens-paris-saclay.fr

[2] Centre d'Essais au Feu du CERIB (Centre d'Etude et de Recherche de l'Industrie de Béton), Epernon, France

[3] EDF Lab Les Renardières, MMC Department, Moret Sur Loing, France

[4] EIFFAGE – Génie Civil, Département Innovation Représentation Matériaux (D.I.R.M.), Vélizy Villacoublay, France

Abstract. This study presents a new numerical approach using tools developed to perform a molecular simulation to investigate the wall effect on aggregates and mortar distribution in concrete. Aggregates are represented by spheres interacting via a generalized truncated Lennard-Jones potential. This approach allows obtaining the particle profiles according to the reference frame of interest (e.g., the confined directions). Then the particle-based distributions are transformed into continuum profiles of volume fractions using a convolution. Based on volume fraction profiles, transport or mechanical properties are estimated by the Mori-Tanaka scheme from classical homogenization. Results are compared to experimental work. The numerical method could be generalized and used for other applications in different fields. In civil engineering, perspectives include using the aggregate distribution to conduct a finer analysis, and the results would be extremely relevant for the prediction of the water content profile and the evolution of pathologies such as carbonation, corrosion, ISR, etc..

Keywords: Durability · Wall effect · Concrete drying · Aggregates distribution · Transport properties · Mechanical properties · Grand Canonical Monte Carlo (GCMC) simulations · Micromechanics

1 Introduction

One of the main advantages of using concrete is that the material in its fresh state can be poured into formworks and assume the most variable shapes. Nevertheless, the casting is not without effect on concrete. The formwork perturbates the arrangement of aggregates; this effect is known as 'the wall effect' and it is generally neglected in both experiments and numerical calculations. So far, there is no validated method to quantify the gradation of transport and mechanical properties induced by the wall effect. The manifestation of the wall effects can be decomposed on the contributions of (i) the effects

P. Rossi and J.-L. Tailhan (Eds.): SSCS 2022, RILEM Bookseries 38, pp. 311–326, 2023.
https://doi.org/10.1007/978-3-031-07746-3_31

of aggregate distribution from the surface; and (ii) the interactions with the environment such as drying (which may induce cracks because of drying shrinkage gradient and strain incompatibility between cement paste and aggregates), hydration slowing down (due to rapid water evaporation on exposed surfaces), and carbonation in long-term. Altogether, these aspects lead to changes in the properties of a superficial layer, and these changes can be relevant to some applications of cement-based materials and concrete structures potentially affecting their performances [1, 2]. The skin of the concrete is in direct contact with aggressive agents such as chlorides and carbon dioxide [3]. If it has weaker properties than the bulk area of the material, this implies an increase in the risk of the appearance of pathologies or the acceleration of its appearance. In reinforced concrete, the concrete cover (generally 3 cm thick on average) is critical to prevent serious durability problem such as corrosion. A better understanding of the factors affecting the properties of this zone might contribute to improve durability prediction and strategies of design for durability.

Experimental studies have identified property gradation due the wall effects in cement-based materials [1, 2, 4]. From the early 1960s, the importance of considering the effect of the (granular) inclusion of concrete was highlighted as a factor influencing the compressive strength of concrete [5]. [4] showed that the surface of the concrete is mainly composed of cement paste, and the coarse aggregates start to appear as one moves away from the walls. Numerical approaches have also been adopted to quantify how formworks alters aggregate/cement paste distribution adjacent to the walls. Simulation of hard spheres [6] and ellipsoidal [7] particles allowed the quantification of aggregates and cement paste volume fraction profiles in simplified system. To date, no modelling work have studied how such microstructure gradation affects property on the cover zone. The goal of this work is, on the one hand is to estimate the aggregates volume fraction and concrete properties profiles and to quantify numerically the gradation induced by the formwork. On the second hand, the experimental work shows whether the skin effect modifies or not the overall behavior of the concrete (about transfer and mechanical properties).

2 Material and Methods

2.1 Materials

VeRCoRs REF concrete is used in this study for both numerical and experimental work. This concrete formula is developed at EDF's laboratories for the project aiming to investigate the ageing of containment vessels. VeRCoRs concrete is an ordinary concrete C35/40 and 0.52 water-cement ratio. Table 1 shows the composition of VeRCoRs concrete. Experiments are conducted on prismatic $7 \times 7 \times 28$ cm^3 specimens which are widely used in civil engineering laboratory testing. The specimens are manufactured and then stored under sealed conditions for 60 days. Under this condition, drying (and the resulting differential cement hydration progress) is limited but the appearance of micro-cracks induced by desiccation shrinkage might occur and second to enable the hydration reaction to progress. Otherwise, these parameters would interfere the formwork effect.

Table 1. Composition of VeRCoRs concrete.

Components	Mass ($kg.m^{-3}$)	Volume fraction (%)
Cement	316.2	13.7
Sand 0/4	820.1	35.4
Gravel 4/11	439.7	18.9
Gravel 8/16	543.5	23.4
Water	192.9	8.3
Superplasticizer	2.6	0.1

2.2 Experiments

Mass loss is monitored on 9 specimens while drying at 23 °C, 50%HR over one surface (upper, lower, and left or right surface, 3 specimens per surface state). The studied drying configurations follow the scheme depicted in Fig. 1. The aim of this experiment is to investigate the impact of concrete surface state on transport properties such as permeability.

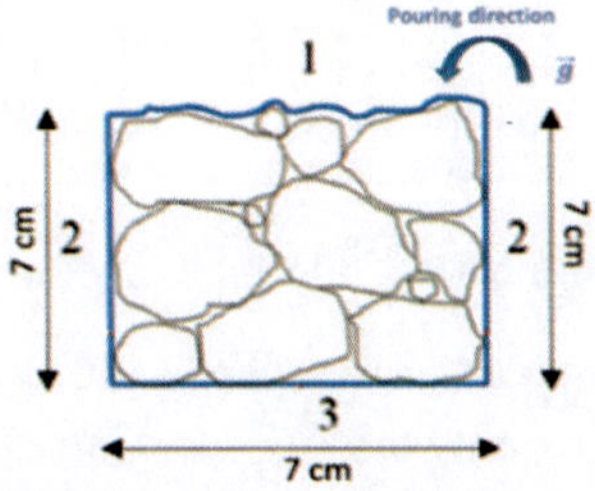

Fig. 1. The different drying configurations (1 = free non-molded surface, 2 = side surfaces in contact with the formwork and 3 = bottom surface in contact with the formwork).

The second experiment consists of conducting 3-point bending test considering the same previous configurations of specimen's surface. The idea is to investigate the effect of concrete skin state on its mechanical properties such as Young modulus and tensile strength.

Another experimental investigation of the wall effect consists of analysing the microcracking rate in molded vs non-molded concrete surfaces using a microscopic observation on a 7 × 7 × 28 cm specimen.

2.3 Simulation Granular Packing Under Confined Geometries

We propose a new approach to simulate aggregate distribution in confined geometries based on a tool developed to perform molecular simulations (using LAMMPS) [8–12]. The idea is to represent the aggregates by polydisperse spheres, adopt interactions approximating hard-sphere potential, and perform Monte Carlo simulations to fill the confined geometry. Post-processing allows computing aggregates profiles that will be used in the estimation of property gradation.

Pseudo Hard-Sphere Potential
The hard-sphere (HS) pair potential reads:

$$U_{HS}(r_{ij}) = \begin{cases} \infty, & r_{ij} < \sigma_{ij} \\ 0, & r_{ij} \geq \sigma_{ij} \end{cases} \tag{1}$$

where r_{ij} is the distance between the centers of particles i and j, and $\sigma_{ij} = \frac{\sigma_i + \sigma_j}{2}$ is the average of the diameters σ_i and σ_j of particles i and j, representing their center distance when they are in contact. The discontinuous and non-bounded nature of this potential make difficult its numerical implementation. One solution is the use of a pseudo-HS potential that is continuous. The HS potential can be approximated by a shifted and truncated generalized Lennard-Jones (or Mie) potential [13]:

$$U_{HS}^{P}(r_{ij}) = \begin{cases} \frac{\lambda_r}{\lambda_r - \lambda_a}\left(\frac{\lambda_r}{\lambda_a}\right)^{\frac{\lambda_a}{\lambda_r - \lambda_a}} \epsilon_{ij}\left[\left(\frac{\sigma_{ij}}{r_{ij}}\right)^{\lambda_r} - \left(\frac{\sigma_{ij}}{r_{ij}}\right)^{\lambda_a}\right] + \epsilon_{ij}, & r_{ij} < r_{cutoff} \\ 0, & r_{ij} \geq r_{cutoff} \end{cases} \tag{2}$$

where λ_r and λ_a are the exponents of the repulsive and attractive terms of the Mie potential, respectively; ϵ is the potential well depth; and $r_{cutoff} = \left(\frac{\lambda_r}{\lambda_a}\right)^{\frac{\lambda_a}{\lambda_r - \lambda_a}} \sigma_{ij}$ is the pair-wise cutoff distance. Therefore, only the repulsive part of the potential is effective, and the shift makes the potential smooth (class C^1) for all r_{ij}. Jover et al. [13] shows that using $\lambda_r = 50$ and $\lambda_a = 49$ lead to a good reproduction of the HS behavior. For polydisperse systems, the Lorentz-Berthelot mixing rule $\epsilon_{ij} = \sqrt{\epsilon_{ii}\epsilon_{jj}}$ and $\sigma_{ij} = (\sigma_{ii} + \sigma_{jj})/2$ can be applied to describe the interactions among the different classes of particle size.

The confining walls interaction are described by the usual 12-6 Lennard-Jones potential but also in a truncated form and with a shift:

$$U_{LJ}^{W}(r_{iw}) = \begin{cases} 4\epsilon_w\left[\left(\frac{\sigma_i}{r_{iw}}\right)^{12} - \left(\frac{\sigma_i}{r_{iw}}\right)^{6}\right] + \epsilon_w, & r_{ij} < r_{cutoff} \\ 0, & r_{ij} \geq r_{cutoff} \end{cases} \tag{3}$$

where the distance between the center of particle i and the wall, and ϵ_w is the energy well depth associated with wall interactions. The LJ wall is placed at the edge of the simulation box, and the cutoff in this case is $r_{cutoff} = \sigma_i/2$. With this assumption, the wall function as a hard wall. Figure 2 shows the LJ potential (usual 12-6 and truncated LJ).

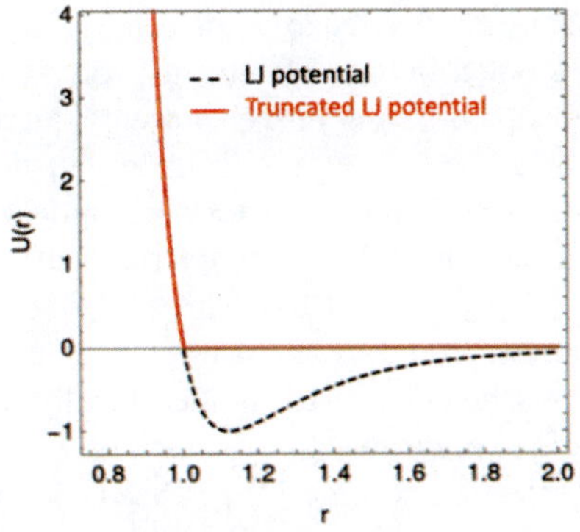

Fig. 2. Usual 12-6 Lennard-Jones potential vs truncated LJ potential.

GCMC Simulations of Concrete Granular Packing

Grand Canonical Monte Carlo (GCMC) simulations are adapted to model system with varying number of particles and is used here to simulate the granular packing in a confined system.

In a GCMC simulation a particle insertion is accepted following the probability [14].

$$acc(N \rightarrow N+1) = min\left(1, \frac{V}{\Lambda^3(N+1)} exp\left[-\frac{\Delta U - \mu}{k_b T}\right]\right) \quad (4)$$

where V is the volume, Λ is the thermal De Broglie wavelength, $\Delta U = U(N) - U(N+1)$ is the variation in potential energy due to the trial insertion, μ is the chemical potential, k_b is the Boltzmann constant, and T is the temperature. Note that in our simulations, the temperature and chemical potential do not necessarily need to have a physical meaning and can be used to enforce particle insertion and to promote a better packing if combine with dynamics. Removals can be prevented by choosing a large value of μ, which favors insertion of particles.

Aggregate Volume Fraction Profiles

The volume fraction of the aggregates can be computed directly from the particles using the convolution [15]:

$$\phi_i^s(x) = N_i(x) \otimes P_{sph}(x, R_i) = \sum_j N_i(x_j) P_{sph}(x_j - x, R_i) \quad (5)$$

$P_{sph}(x, R)$ is the projection of the sphere with the center at the point o_i and R is the radius of the sphere.

$$P_{sph}(r, R) = \begin{cases} 0, & r < -R \\ \pi(R^2 - r^2), & r < -R \\ 0, & r \geq R \end{cases} \quad (6)$$

Estimating Property Gradation Using Micromechanics

Representative elementary volume (REV) and microstructure gradation.

Homogenization schemes can be used to estimate the effective properties of heterogenous materials in which a representative elementary volume (REV) can be defined. In a REV, all the heterogeneities (different phases present) and their spatial variability (in term of spatial distribution and orientation) should be statistically represented. A REV cannot be defined in systems with microstructure gradation, but a local approximation can be done. For the zone perpendicular to the microstructure gradation, the criteria to define a REV are respected. The local behaviour per slice can be approximated as the behaviour of a 3D medium with similar microstructure as the slice. In this framework, the usual homogenization schemes can be applied locally to obtain the evolution of properties as a function of the microstructure gradation.

Mori-Tanaka homogenization scheme [16] is used to compute properties profile. The concrete in this case is considered a two-phase material, where mortar constitute the matrix and aggregates the inclusionary phase (cf. Fig. 3). More details about the used homogenization method are described in the next section.

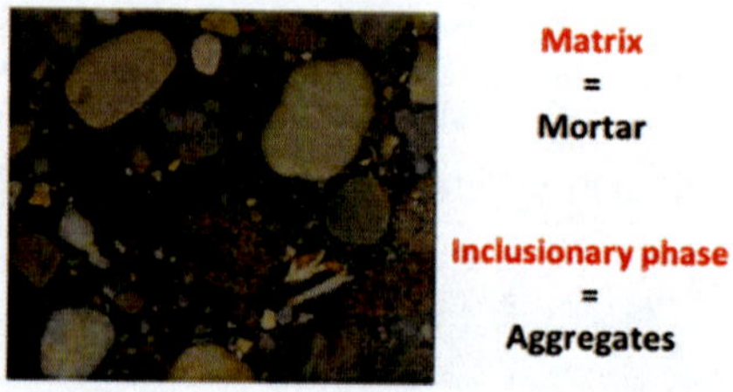

Fig. 3. Matrix and inclusionary phases of Mori-Tanaka homogenization used to estimate properties gradation.

Estimates of Elastic Properties

For an (N + 1)-phase heterogeneous material with a matrix/inclusion morphology in which N isotropic spherical inclusions are randomly placed in a host medium (percolating matrix phase), the Mori-Tanaka estimate of the effective bulk and shear moduli are, respectively [16]:

$$\frac{K^{MT} - K_0}{K^{MT} + \frac{4}{3}G_0} = \sum_{i=1}^{N} f_i \frac{K_i - K_0}{K_i + \frac{4}{3}G_0} \quad (7)$$

$$\frac{G^{MT} - G_0}{G^{MT} + H_0} = \sum_{i=1}^{N} f_i \frac{G_i - G_0}{G_i + H_0} \quad (8)$$

with

$$H_0 = \frac{G_0\left(\frac{3}{2}K_0 + \frac{4}{3}G_0\right)}{K_0 + 2G_0} \quad (9)$$

where f_i is the volume fraction of the phase i. In this expression, the subscript 0 stands for the matrix phase and the subscripts $i \in [1, N]$ for the inclusions.

The local effective young modulus E^{MT} and Poisson ratio ν^{MT} can be computed from the bulk and shear moduli using

$$E^{MT} = 9K^{MT}G^{MT}/(3K^{MT} + G^{MT}) \tag{10}$$

$$\nu^{MT} = (3K^{MT} - 2G^{MT})/[2\left(3K^{MT} + G^{MT}\right)] \tag{11}$$

Estimates of Conductivity-Like Properties

By mathematical analogy, the upscaling of the thermal conductivity, electrical conductivity, dielectric permittivity, and diffusivity are equivalent [17]. The expression of Mori-Tanaka (or Maxwell-Garnet) scheme for conductivity-like properties for an (N + 1)-phase heterogeneous material with a matrix/inclusion morphology in which N isotropic spherical inclusions are randomly placed in a host medium (percolating matrix phase) is given by [16, 18]:

$$\frac{\lambda^{MT} - \lambda_0}{\lambda^{MT} + 2\lambda_0} = \sum_{i=1}^{N} f_i \frac{\lambda_i - \lambda_0}{\lambda_i + 2\lambda_0} \tag{12}$$

where again the subscript 0 stands for the matrix phase and the subscripts $i \in [1, N]$ for the inclusions. Table 2 reports the properties of each phase. Bulk and shear moduli are taken from [3]. The permeability of aggregates is very low compared to mortars permeability. In this study, aggregates permeability is considered null.

Table 2. Properties of phases.

Properties	Aggregates	Cement matrix
Bulk modulus (GPa)	67	13
Shear modulus (GPa)	30	9
Permeability (-)	0	1

3 Experimental Results

3.1 Mass Loss Monitoring

Figure 4 reports the mean mass loss measurements on 3 prismatic specimens for each drying configuration. Mass loss is very similar at the beginning of the drying process. Then it becomes higher for the non-molded surface drying. Samples drying over the bottom and the side surfaces (which were in contact with the formwork) show a similar drying behavior. The reported difference can be explained by the roughness of the unmolded surface and/or by the rate of micro-cracking compared to a molded surface.

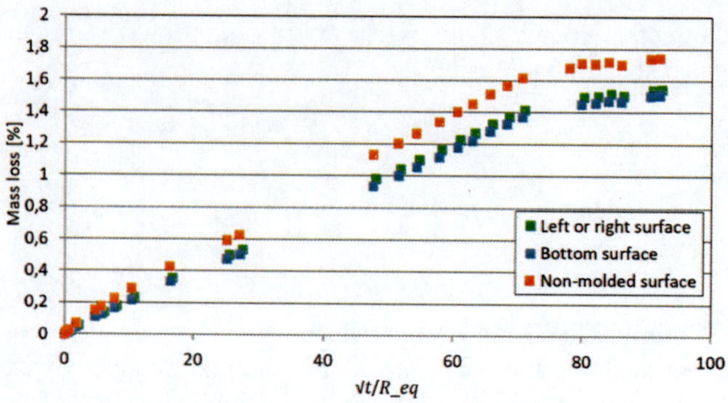

Fig. 4. Mass loss in 7 × 7 × 28 cm specimens one-sided drying in different surface configurations: bottom surface in contact with the formwork, side surfaces in contact with the formwork and free non-molded surface.

3.2 3-Point Bending Test

Table 3 reports the results of the 3-point bending test. Young modulus in the first test configuration is lower than test conducted on molded surfaces as bearing. The roughness of the non-molded surface is the main reason for the decrease in the Young modulus value. Results in the two other test configurations are almost similar.

Table 3. Young modulus measured on 7 × 7 × 28 cm specimens considering different bearing surface condition (molded and non-molded).

Condition of the bearing surface	Specimens	Young modulus (GPa)
Free non-molded surface	1	22.78
	2	23.87
	3	22.65
Free non-molded surface	4	32.04
	5	31.99
	6	31.91
Free non-molded surface	7	32.34
	8	32.34
	9	32.23

3.3 Microcracking Rate in Molded vs Non-molded Concrete Surfaces

Microscopic visualization of molded and non-molded surfaces of 7 × 7 × 28 cm concrete specimens shows a higher rate and more opened microcracks. Figure 5 shows an illustration of the visualized microcracks.

a. Non-molded surface *b. Molded surface*

Fig. 5. Concrete surface visualization using optical microscope.

4 Numerical Results: Polydisperse Systems

4.1 1D 'Confinement'

A calculation of the granular distribution in the concrete is carried out once the method is validated for a monodisperse distribution case. The application consists in simulating the granular distribution of a concrete formula starting from the granulometric curves. The simulated concrete is EDF's VeRCoRs concrete with two aggregate classes G4/11 and G8/16 Fig. 6. We consider 11 different diameters with a total volume fraction of 40% and a maximum diameter of Dmax = 25 mm.

1D and 3D confinement are studied. Confinement along the x-direction. Y independent configurations are tested to evaluate the filling variability, the effect of the confined width, and then the effect of the maximum aggregate diameter (by comparing the results obtained with a calculation based on a different aggregate distribution). The second part consists of performing a calculation considering a confinement configuration on a 7 × 7 × 28 cm VeRCoRs concrete specimen in a configuration similar to the one considered in the experimental part of this work., i.e. formwork on all surfaces + a free non-molded surface.

G4/11

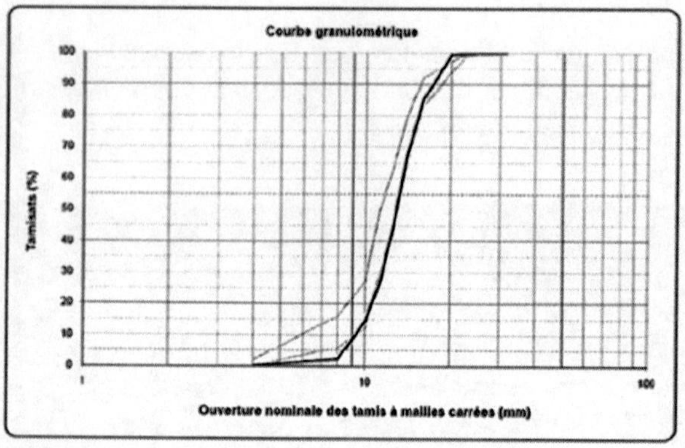

Tamis d (mm)	125	80	63	50	40	31,5	25	20	16
Passants (%)						100	100	99	85
Tamis d (mm)	14	12,5	11,2	10	8	6,3	5	4	2
Passants (%)	68	45	26	14	2			0	
Tamis d (mm)	1	0,500	0,315	0,250	0,125	0,063			
Passants (%)									

G8/16

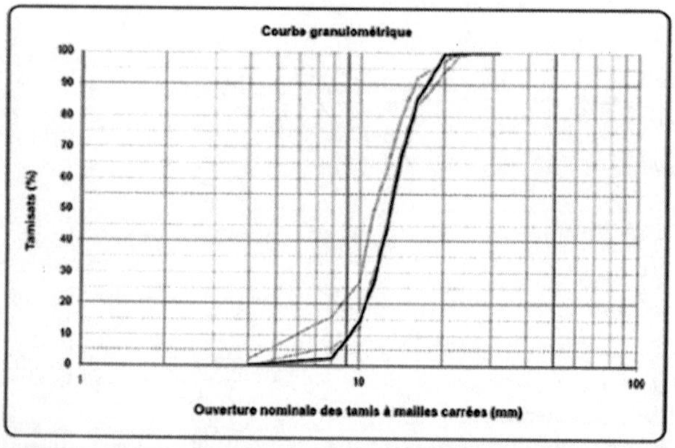

Tamis d (mm)	125	80	63	50	40	31,5	25	20	16
Passants (%)						100	100	99	85
Tamis d (mm)	14	12,5	11,2	10	8	6,3	5	4	2
Passants (%)	68	45	26	14	2			0	
Tamis d (mm)	1	0,500	0,315	0,250	0,125	0,063			
Passants (%)									

Fig. 6. Granulometric curves of VeRCoRs concrete aggregates

The configurational packing variability is evaluated by conducting 5 independent simulations in which the seed for the Monte Carlo simulation is changed. Figure 7 reports the mean value and standard deviation of the aggregates volume fraction, and the estimated profile of the effective properties. The resulting variability is low on the aggregates volume fraction (maximum standard deviation is equal to 0,015) and very much lower on the effective properties.

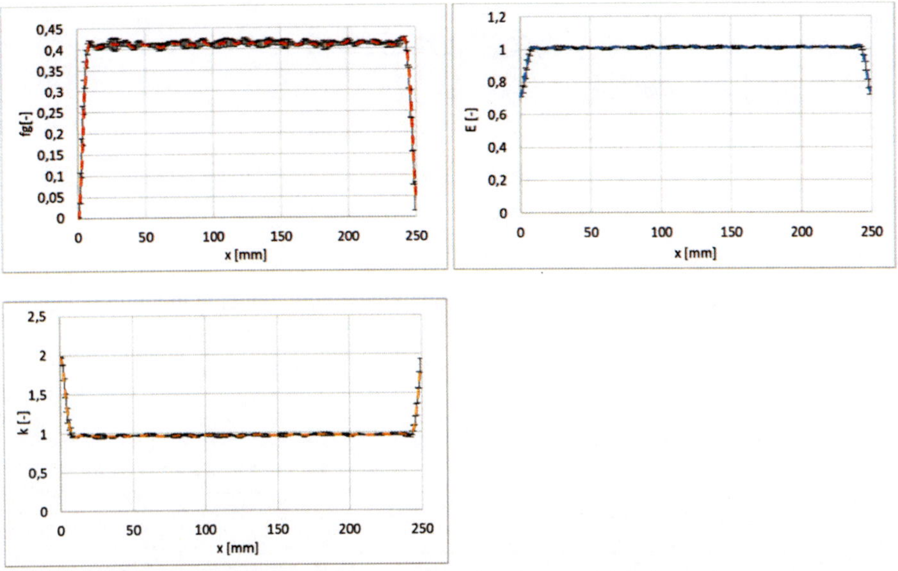

Fig. 7. Aggregates packing-related variability respectively on volume fraction profile fg, on Young modulus E, and on permeability k.

To evaluate the thickness of the zone perturbed by the wall, in Fig. 8 we compute the profile of the aggregate volume fraction over the first 50 mm of the confined direction for various confinement widths: 3Dmax, 4Dmax, 5Dmax, and 10Dmax. In the most confined system (3Dmax) the effects of the wall opposed walls superposes leading to an augmentation of the average volume fraction of aggregates even in the sample core.

Confinement width does not have a major effect on the thickness disturbed by the wall: in all cases the distance perturbed by the wall is roughly Dmax/2. This observation corroborates previous studies indicating that the thickness disturbed by the formwork depends directly on the maximum aggregate diameter [19]. We verify the validity of this observation for other aggregates granulometries in Fig. 9, in which aggregates used in other concrete mix designs are considered: F1 VeRCoRs concrete with Dmax = 25 mm, F2 CETU concrete with Dmax = 25 mm and a different aggregates distribution, and F3 the B11 concrete with Dmax = 12.5 mm (the total volume fraction is the same and is equal to 40%). The formulation with the lower Dmax exhibited a smaller thickness perturbed by the wall as expected. Thus, the disturbed thickness depends strongly the maximum aggregate diameter. It is quite inferior to half the maximum aggregate diameter. [19] shows the gradation in the wall zone where the distribution of cement pastes and aggregates would be constant from a distance R (greater than the maximum aggregate diameter) away from the wall as reported in Fig. 10. The estimated permeability profile is compared with the results obtained by Kreijger about the estimation of the concrete properties over the first 5 mm. The results are comparable in mean value as reported in Fig. 11. However, the modulus is lower on the first millimetre according to the study [1].

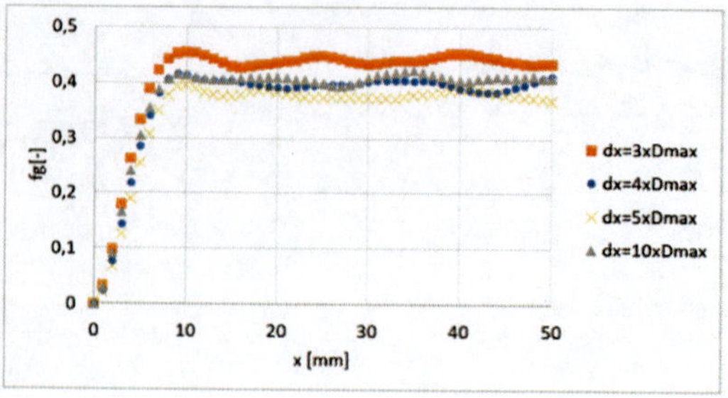

Fig. 8. Evolution of the aggregate volume fraction for different values of the confined width x over the first 50 mm - variable simulation box size dx, dy = dz = 20 × Dmax.

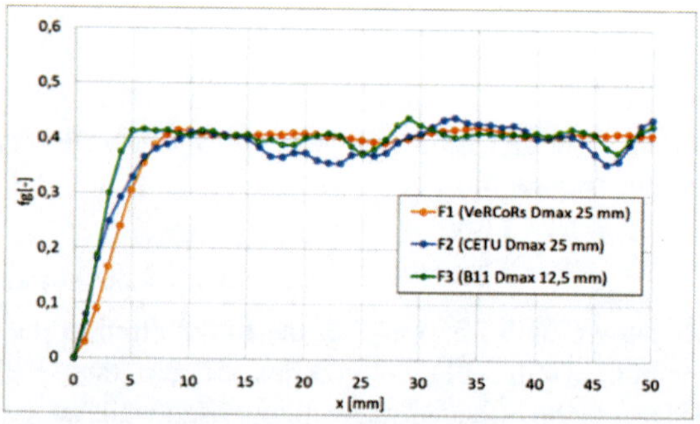

Fig. 9. Aggregate volume fraction in 3 concrete formulas: F1 VeRCoRs concrete with Dmax = 25 mm, F2 CETU concrete with Dmax = 25 mm and a different aggregates distribution, and F3 the B11 concrete with Dmax = 12,5 mm (the total volume fraction is the same and is equal to 40%).

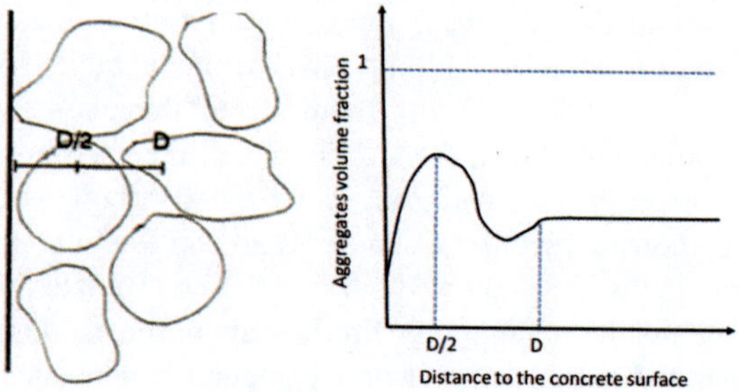

Fig. 10. Distribution of aggregates in a concrete element [19]. D is the maximum aggregate size.

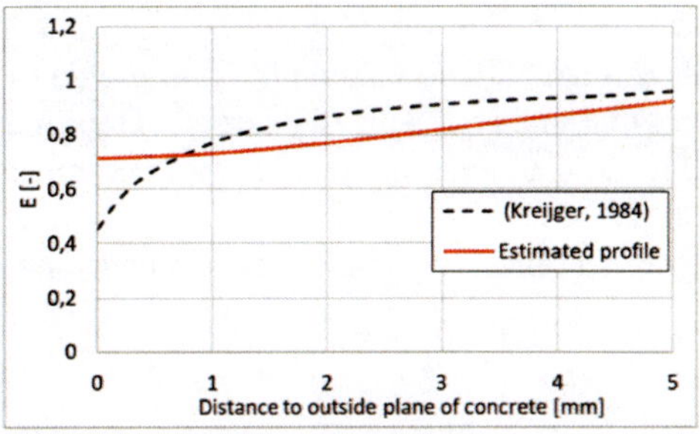

Fig. 11. Comparison of Young's modulus profile estimated over the first 5 mm with the profile obtained by Kreijger in 1984 [1].

4.2 3D 'Confinement'

We also perform simulation with 3D confinement using the same dimensions $7 \times 7 \times 28\ cm^3$ of the specimen evaluated in the experimental part of this work. The results are presented in Fig. 12 and Fig. 13. Depending on the width according to x, the thickness disturbed by the formwork at $x = 0$ and $x = 7$ cm is equal to about 8 mm, which is 1/3 of the maximum aggregate diameter in this case. Between the two walls, there is some fluctuation around the homogeneous value of about ±5% on the volume fraction profile. Regarding the profile of the estimated effective properties, from the surface of the material to 5 mm in depth, the permeability on the surface is twice that of the bulk zone. Therefore, drying and transfer in general are faster in the disturbed zone than in the bulk zone. The Young's modulus in the disturbed zone loses 10% on average compared to its value in the bulk. Since the elastic constants correlate with the mechanical strength, it is expected that the skin zone is more prone to cracking than the bulk zone.

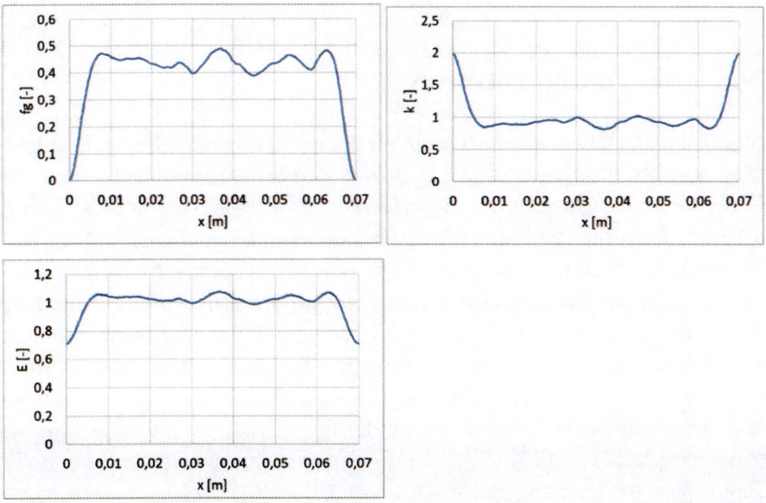

Fig. 12. Evolution of the aggregate volume fraction, permeability and Young modulus in 7 × 7 × 28 cm specimen according to x.

In the y-direction, y = 0 cm corresponds to the free unmolded surface; y = 7cm corresponds to a molded surface. The volume fraction profile of the aggregates shows that there are aggregates that stand out at the free surface. The same observations are made about the disturbed thickness and the interference on the estimated effective properties These results highlight the effect of the formwork on the concrete, which is expressed by a higher mortar content in a zone whose thickness depends on the maximum diameter of the aggregates.

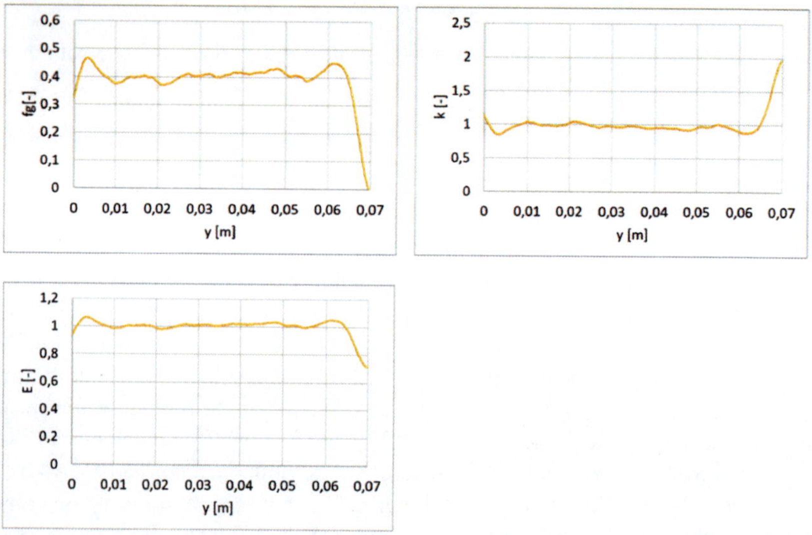

Fig. 13. Evolution of the aggregate volume fraction, permeability and Young modulus in 7 × 7 × 28 cm specimen according to y (y = 0 presents the non-molded side of the specimen).

5 Conclusion and Perspectives

We combine experiments and simulation strategies to quantify the wall effects and the repercussion in the properties of the skin of concrete. We proposed a new simulation strategy based on tool developed to performed molecular simulations. This strategy enables generating simplified microstructures in confined geometries (1-, 2- or 3D) for polydisperse particles and the computation of aggregates volume fraction profiles. These profiles can be then used as input for multiscale approaches to estimate property profiles.

In agreement with previous studies, we observe that the disturbance induced by the formwork only depends on the maximum size of the aggregates and its thickness is slightly lower than the diameter of the largest aggregate.

Properties profiles obtained from the modelling approach show good agreement with the experimental results of the literature [1], despite the simplifying assumption adopted (regarding sphericity of aggregates).

The experimental work shows that the effect of the formwork and the condition of the concrete surface have a very low impact on the overall behavior and on the global properties of the material. The effect of the formwork is counterbalanced by the effect of micro-cracking and the roughness of the non-formed surface.

In view of this study, the wall effect on concrete drying process will be evaluated by exploiting the profiles of the granular distribution and the estimated effective properties i.e., the permeability profile. The overall effect on durability could be thus predicted.

References

1. Kreijger, P.C.: The skin of concrete composition and properties. Mat. Constr. **17**(4), 275–283 (1984). https://doi.org/10.1007/BF02479083
2. Bissonnette, B., Courard, L., Garbacz, A.: Concrete Surface Engineering. CRC Press, Boca Raton (2018)
3. Bentz, D.P., et al.: Influence of aggregate characteristics on concrete performance. National Institute of Standards and Technology, Gaithersburg, May 2017. NIST TN 1963. https://doi.org/10.6028/NIST.TN.1963
4. Liu, P., Chen, Y., Sha, F., Yu, Z., Shao, G.: Study on microstructure and composition distribution of concrete surface zone based on fractal theory and XCT technology. Constr. Build. Mater. **263**, 120209 (2020). https://doi.org/10.1016/j.conbuildmat.2020.120209
5. de Larrard, F.: Structures granulaires (1999)
6. Zheng, J.J., Li, C.Q., Zhao, L.Y.: Simulation of two-dimensional aggregate distribution with wall effect. J. Mater. Civ. Eng. **15**(5), 506–510 (2003). https://doi.org/10.1061/(ASCE)0899-1561(2003)15:5(506)
7. Xu, W.X., Lv, Z., Chen, H.S.: Effects of particle size distribution, shape and volume fraction of aggregates on the wall effect of concrete via random sequential packing of polydispersed ellipsoidal particles. Phys. A **392**(3), 416–426 (2013). https://doi.org/10.1016/j.physa.2012.09.014
8. Plimpton, S.: Fast parallel algorithms for short-range molecular dynamics. J. Comput. Phys. **117**(1), 1–19 (1995). https://doi.org/10.1006/jcph.1995.1039
9. Plimpton, S., Hendrickson, B.: A new parallel method for molecular dynamics simulation of macromolecular systems. J. Comput. Chem. **17**(3), 326–337 (1996). https://doi.org/10.1002/(SICI)1096-987X(199602)17:3%3c326::AID-JCC7%3e3.0.CO;2-X
10. Clark, T.W., McCammon, J.A., Scott, L.R.: Parallel molecular dynamics. In: Proceedings of the Fifth SIAM Conference on Parallel Processing for Scientific Computing, USA, pp. 338–344, March 1991
11. Pinches, M.R.S., Tildesley, D.J., Smith, W.: Large scale molecular dynamics on parallel computers using the link-cell algorithm. Mol. Simul. **6**(1–3), 51–87 (1991). https://doi.org/10.1080/08927029108022139
12. Plimpton, S.J., Wolf, E.D.: Effect of interatomic potential on simulated grain-boundary and bulk diffusion: a molecular-dynamics study. Phys. Rev. B **41**(5), 2712–2721 (1990). https://doi.org/10.1103/PhysRevB.41.2712
13. Jover, J., Haslam, A.J., Galindo, A., Jackson, G., Müller, E.A.: Pseudo hard-sphere potential for use in continuous molecular-dynamics simulation of spherical and chain molecules. J. Chem. Phys. **14**, 144505 (2012)
14. Frenkel, D., Smit, B.: Understanding Molecular Simulation. 2nd edn. Academic Press, San diego (2002). https://www.elsevier.com/books/understanding-molecular-simulation/frenkel/978-0-12-267351-1. Accessed 11 Oct 2021
15. Nygård, K., et al.: Anisotropic pair correlations and structure factors of confined hard-sphere fluids: an experimental and theoretical study. Phys. Rev. Lett. **108**(3), 037802 (2012). https://doi.org/10.1103/PhysRevLett.108.037802

16. Mori, T., Tanaka, K.: Average stress in matrix and average elastic energy of materials with misfitting inclusions. Acta Metall. **21**(5), 571–574 (1973). https://doi.org/10.1016/0001-6160(73)90064-3
17. Torquato, S.: Random Heterogeneous Materials: Microstructure and Macroscopic Properties. Springer, New York (2002). https://doi.org/10.1007/978-1-4757-6355-3
18. Hashin, Z., Shtrikman, S.: A variational approach to the theory of the elastic behaviour of multiphase materials. J. Mech. Phys. Solids **11**(2), 127–140 (1963). https://doi.org/10.1016/0022-5096(63)90060-7
19. Granger, L.: Comportement différé du béton dans les enceintes de centrales nucléaires: analyse et modélisation (1995). https://pastel.archives-ouvertes.fr/tel-00520675. Accessed 13 Nov 2019

The Influence of Autogenous Shrinkage and Creep on the Risk of Early Age Cracking

Erik Schlangen(✉), Minfei Liang, and Branko Savija

Faculty of CEG, Microlab, Delft University of Technology, Delft, The Netherlands
erik.schlangen@tudelft.nl

Abstract. The study aims to investigate the mechanism of early-age cracks in different massive concrete structures (i.e. tunnels, bridge foundations and underground parking garages), with the objective of answering the following three specific questions:

1) How does the parameters of concrete proportion mix (e.g. w/c ratio, cementitious materials, aggregates, etc.) influence the formation of autogenous shrinkage and creep, especially at early age with the focus on the first 24 h?
2) How to build theoretical and numerical model for the process of early-age crack formation and then quantify the damage status of concrete materials?
3) How to link the results derived from material scale to structural scale, and provide useful reference for practical engineering projects?

To study the basic mechanism a research program is performed in which different mixes are tested in a Temperature and Stress Testing Machine (TSTM). Furthermore autogenous shrinkage is measured in different ways. Modeling with a FEM-tool is used to predict the risk of early age cracking. The results indicate that the combined shrinkage (or expansion and the relaxation (or creep) during the first hours of hydration have a huge influence on the stresses that develop later and with that are important to determine the risk of cracking in massive concrete structures. Since investigating the stresses that built up in the first hours after casting in such a TSTM is rather difficult, we designed a new version of the TSTM machine in which dog bone specimens are tested vertically in a Universal Testing Machine (Instron).

Keywords: Early age cracking · Autogenous shrinkage · Creep · Concrete

1 Introduction

1.1 Problem Description

Early age cracking in concrete is a well-known problem. Imposed deformation due to heat of hydration and autogenous shrinkage develops during the hydration phase of the concrete. The mechanical properties like stiffness, strength and visco-elastic behavior is gradually increasing with hydration. When the imposed deformation is restrained stresses in the material will develop and if too high cracks will occur. In Fig. 1 a typical

P. Rossi and J.-L. Tailhan (Eds.): SSCS 2022, RILEM Bookseries 38, pp. 327–334, 2023.
https://doi.org/10.1007/978-3-031-07746-3_32

example of vertical early age cracks in a wall-slab connection are shown. These cracks typically start a short distance above the floor and propagate a few meters up. These are cracks through the complete cross section and when it is a water retaining structure, leakage is often the consequence. The cracks occur typically one or two weeks after casting when the structure is cooling down and shrinking.

Fig. 1. Water leakage through vertical early age cracks in a wall-slab connection.

The cracking is influenced by many parameters like mix-design and early age properties, temperature of the fresh mix, environmental conditions, dimensions of the structure and restraining conditions. A lot of research has been performed on this topic already and a very good summary is given in a state of the art report by Rilem [1]. Furthermore various simulation tools are available (also described in [1]) that are used for research purposes but also for engineering and designing measures to prevent this kind of cracking in practice. The success of these simulation and design tools depends to a large extent as always on the input parameters. Many properties of the material are needed and these properties are influenced by the condition the material is in. Generally all the properties are related to the degree of hydration of the material or a maturity concept is used in which the time is influenced by temperature.

1.2 Initial Lab Research

In the work of Van Bokhorst [2] an initial study of combined experimental and numerical research is performed to investigate especially the influence of autogenous shrinkage and creep during the first day on the risk of early age cracking. For a concrete mix that was going to be used for tunnel segments various properties were tested. The compressive strength and splitting tensile strength were determined at 1, 2, 5, 8, 14 and 28 days. Next to that the adiabatic heat development and the autogenous shrinkage was measured and a test was performed in the TSTM-machine at TU-Delft, see Fig. 2. In this test a dog-bone sample is cast in the machine with the chosen concrete mix. A temperature regime is applied which is expected to happen in the application in the tunnel. For this water with the desired temperature is circulated through the mold. The deformation of the dog-bone specimen is restrained by the machine by measuring with LVDT's the length (750 mm) of the straight part the dog-bone and keeping that zero by continuous adjusting the load in a closed loop system.

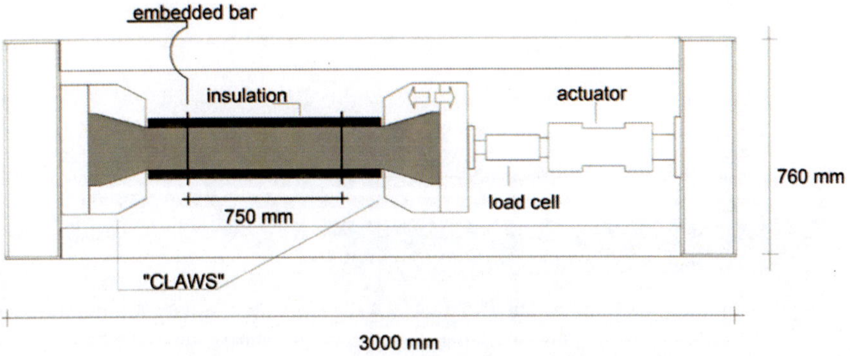

Fig. 2. A sketch of the TSTM-machine at TU-Delft.

1.3 Modelling with FEM-Tool

Modelling of the tunnel segment (wall-slab connection) was performed with MLS of femmasse [3]. This is a EM-based software tool specially designed to perform simulations to determine the risk of cracking. It is based on the state parameter concept in which maturity is used to determine all the properties of the elements in the model, depending on time, temperature and location.

First the TSTM-experiment as discussed above is simulated. As input for the model the properties that are measured of the concrete are used. For properties that are not measured values recommended in Eurocode are adopted. Especially the early age creep or relaxation is an unknown parameter. By using the standard values for creep the results that were obtained from the simulations were very different from the experimental outcome. Then it was decided to fit the input for the creep (modelled with Maxwell chains) in such a way that a good match between the simulations and the TSTM test was obtained.

With these values the tunnel segment was simulated. The results are shown in Fig. 3. The cross section of the structure is given in which the stresses perpendicular to the cross section are shown that occur after cooling down of the wall. In the graph the (tensile) strength development is given as well as a line for 0.85 and 0. Times the strength. Values that are often used in design. The development of the tensile stress is given for the situation in which the creep is according to the Eurocode (grey line) and for the fitted creep data from the TSTM-test (black line). It can be clearly seen that with the fitted data a much higher tensile stress will occur in the cross section with a much higher risk of cracking.

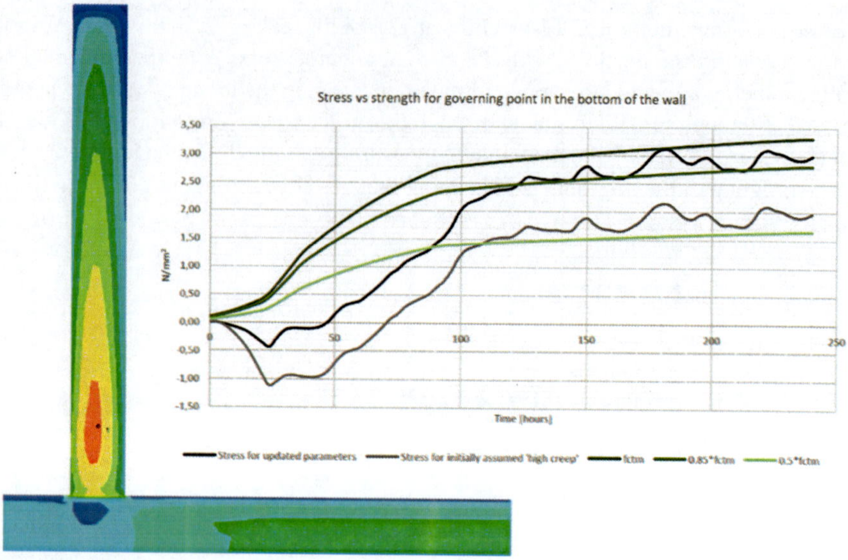

Fig. 3. Simulation of a wall to slab connection in a tunnel section. Stress profile after cooling down of the structure and stress and strength development in the core of the wall where the highest stress occurs.

1.4 Research Objective

The conclusion from this initial study [2] is that especially the early age creep or relaxation combined with the autogenous shrinkage in the first hours or day of hydration is extremely important for the stresses that will occur such structures. In contrary the measurement of creep and the autogenous shrinkage in the first hours or day is very difficult and results depend a lot on the set-up and conditions that are used for the measurements. Therefore it was decided to start a more detailed study to investigate these parameters.

2 Extended Lab Research

In the extended lab research we started with testing four different concrete mixes as explained in Table 1. The mixes are composed of CEMM III/B cement with different water to cement ratio (0.5, 0.42 and 0.35) and one mix with CEM I with w/c of 0.42.

These mixes (Table 1) were tested in the TSTM machine at 20 C and with restrained deformation and in the ADTM machine in which a prism is casted and the autogenous deformation is measured. Furthermore the strength development is followed (see Fig. 4) to have the basic properties. In the ADTM and TSTM tests the measurements and control cannot be started immediately after the concrete is casted, since a steel pin is placed inside the concrete to which the LVDT's are attached. For this the concrete needs some stiffness. The measurement of autogenous deformation and the control of deformation in the TSTM test therefore starts after 7 h. To check the autogenous shrinkage of the cement paste mixtures that are used in the concretes, the autogenous deformation is measured in standard corrugated tubes test measurements [4].

Table 1. Concrete mixes used

	W/C	Cement	Water	Sand (0–4 mm)	Gravel (4–16 mm)	SP
Mix 1	**0.50**	320 (CEM III/B)	160	811.8	1032	0
Mix 2	**0.42**	320 (CEM III/B)	134.4	811.8	1032	0.475
Mix 3	**0.35**	320 (CEM III/B)	112	811.8	1032	1.9
Mix 4	**0.42**	320 (CEM I)	134.4	811.8	1032	0

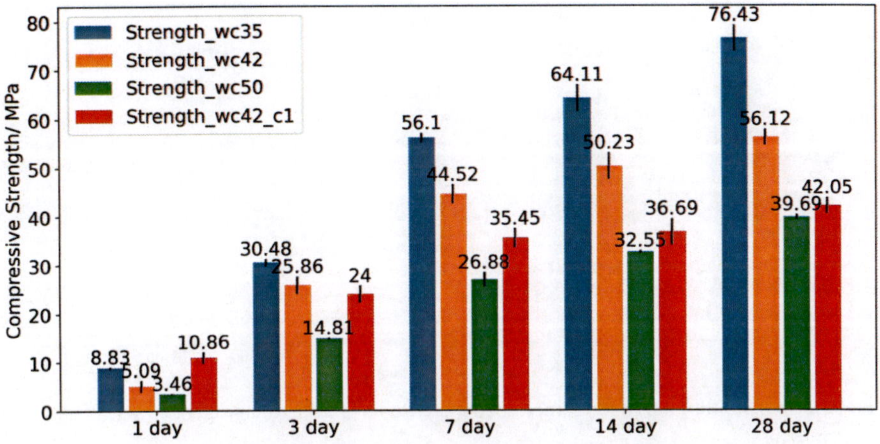

Fig. 4. Strength development of 4 concrete mixes.

3 Obtained Results

The results obtained from the ADTM tests for measuring the autogenous shrinkage on the 4 concrete mixtures is shown in Fig. 5. The temperature in all specimens is kept to 20 C all the time. However some variation in the beginning of max ±.5 C is observed due to hydration heat because the control with running water through the moulds is not that fast. Note that all mixes first show and autogenous expansion after which a shrinkage starts. The results seems not very consistent. Only the amount of shrinkage in the material with the lowest w/c is the highest, which is expected.

In Fig. 6 the stress that develops in the TSTM machine in which the deformation of the dogbone specimens are restrained is plotted for the 4 mixes. The tensile stress development of the mix with w/c of 0.35 is clearly the highest. In Fig. 7 the aoutonous shrinkage measured on the pastes is shown. These results are consistent with the measurements on the concrete, except that the expansion part is far smaller.

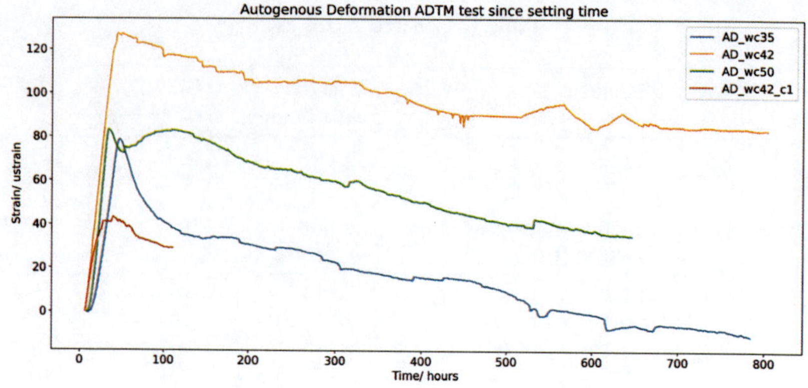

Fig. 5. Autogenous shrinkage of 4 concrete mixes.

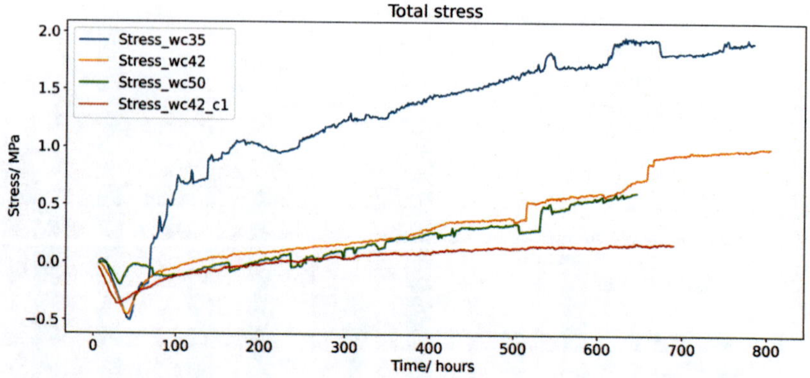

Fig. 6. Stress building up in restrained specimens in TSTM due to autogenous shrinkage for the 4 concrete mixes.

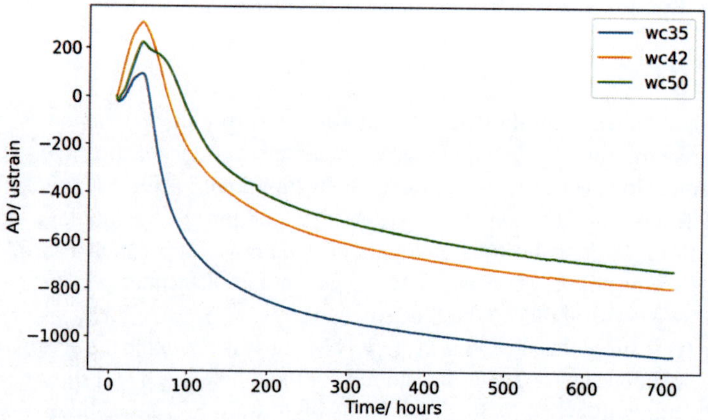

Fig. 7. Autogenous shrinkage measured in corrugated tube tests for 3 mixes with CEM III/B cement.

It is however difficult to dervive strong conclusions from these measurements. It has to be kept in mind that the measurements in ADTM and TSTM do not start immediately after casting. A second point is the temperature control in the mould, which is not extremely accurate. And finally there will be some friction between mould and concrete specimen which will influence deformation of the specimen and also the stresses that develop.

4 Design of New Set-Up

In order to obtain more accurate measurements of the shrinkage and stresses that develop in the material, especially at early age a new TSTM setup was designed. In this setup the specimen will be vertical to reduce frictional effects and the temperature control will be much faster. Furthermore the specimens will be smaller which also enables faster control of the surface temperature and reduction of the gradients. The setup will be placed inside a Universal Testing Machine (Instron) and active control of deformations can be done from time zero, immediately after casting, since contactless measurements of deformations will be done.

At the moment of writing this article the set-up is still in testing phase. The first results look promising (Fig. 8).

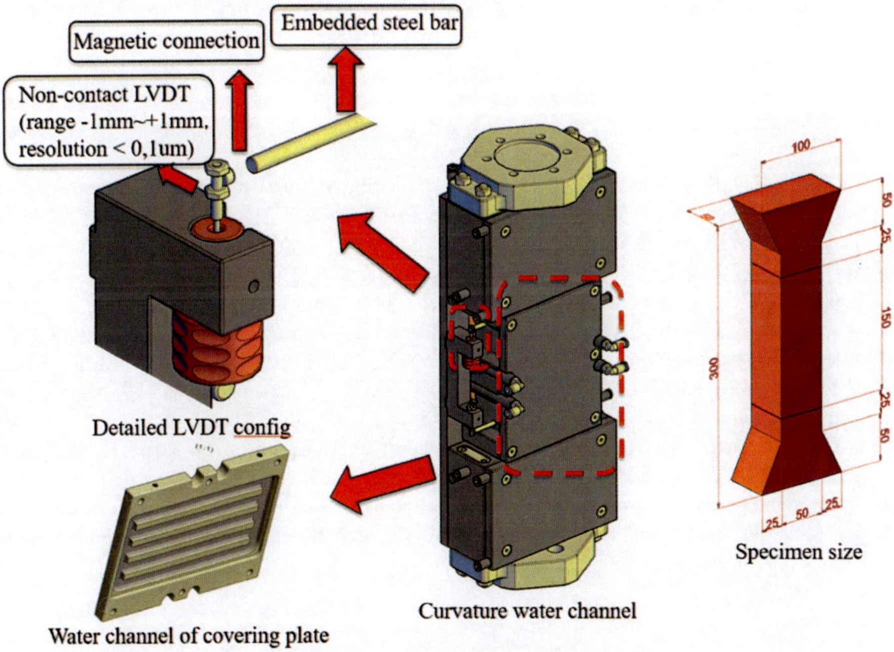

Fig. 8. New design of TSTM set-up.

5 Conclusions

This paper presents a combination of experimental and numerical research to study the effect of autogenous deformation on the stresses that develop in a restrained situation in early age concrete.

The results show that assumed (or measured) autogenous shrinkage and creep (or relaxation), especially during first day have an enormous effect on the stresses that develop in a real structure.

For regular concretes we can still rely to some extent on experience for estimating the material properties. However, we are moving more towards alterrnative concretes like alkali activated materials using waste materials and also the use recycled aggregates in concrete becomes a standard. These materials are known to have autogenous shrinkage and also the visco-elastic behaviour is different from what we are used to with regular concrete. Having accurate material properties to feed modelling will then be crucial to predict the risk of cracking.

The measurement of the shrinkage and creep is however quite complicated due to various effects: exact temperature control, time to start the measurement, and friction between mould and specimen.

A new TSTM setup is designed which could prevent the measurement errors.

In the continuation of this research project the new set-up will be tested and results will be used to predict stresses in early age concrete structures. Furthermore the input will be used in combination with ongoing machine learning approaches [4, 5] to better understand the influence of shrinkage and creep effects at early ages.

References

1. Fairbairn, E.M.R., Azenha, M. (eds.): Thermal Cracking of Massive Concrete Structures. RSR, vol. 27. Springer, Cham (2019). https://doi.org/10.1007/978-3-319-76617-1
2. Van Bokhorst, J.R.: Early-age cracking of concrete. M.Sc. thesis, Delft University of Technology (2020)
3. Femmasse, B.V.: User manual MLS version 8.5 (2006). www.femmasse.com
4. Wyrzykowski, M., Hu, Z., Ghourchian, S., Scrivener, K., Lura, P.: Corrugated tube protocol for autogenous shrinkage measurements: review and statistical assessment. Mater. Struct. **50**(1), 1–14 (2016). https://doi.org/10.1617/s11527-016-0933-2
5. Liang, M., Schlangen, E., Savija, B.: Bayesian inverse modelling of early-age stress evolution in GGBFS concrete due to autogenous deformation and aging creep. In: Rossi, P., Tailhan, J. (eds.) SSCS 2022. RILEM, vol. 38, pp. xx–yy. Springer, Cham (2022)
6. Liang, M., Gan, Y., Chang, Z., Wan, Z., Schlangen, E., Šavija, B.: Microstructure-informed deep convolutional neural network for predicting short-term creep modulus of cement paste. Cem. Concr. Res. **152**, 106681 (2022)

Relevance of Behavior Laws of Homogenized Reinforced Concrete in the Context of Finite Elements of Different Sizes

Alain Sellier[1](✉) and Alain Millard[2]

[1] LMDC, Université de Toulouse, INSA/UPS Génie Civil, 135 Avenue de Rangueil, 31077 Toulouse Cedex 04, France
alain.sellier@insa-toulouse.fr

[2] Formerly CEA DEN/DANS/DM2S/SEMT/LM2S, Bâtiment 607 – CEN Saclay, 91191 Gif sur Yvette Cedex, France

Abstract. Despite the increase in computational power, the accurate modelling of crack openings in reinforced concrete remains an open problem for structural elements of complex shapes. Since the national CEOS.fr project dedicated to the control of cracking for large structures, it is accepted that the statistical scaling effect on the tensile strength of concrete and the realism of the concrete steel slip law are two essential ingredients to achieve a good level of accuracy in the prediction of crack spacing and openings. The implementation of these two aspects in finite element codes faces two problems: on the one hand, the statistical scaling effect depends on the dimension of the tensioned zone, which varies with cracking, and on the other hand, the steel-concrete slip is generally longer than the size of the finite elements. In this context, proposing homogenized reinforced concrete finite elements is problematic. The most common methods consist in using random mechanical property fields to integrate the statistical aspect of the tensile strength, and the slip is explicitly modelled by joint elements, preventing the use of homogenized elements. However, it is possible to avoid both random draws and explicit modelling of concrete-steel interfaces. The proposed method generalizes the notion of phase fields to address both problems simultaneously. A first phase field integrates the scale effect, and another one the steel-concrete slip. The FE code Castem (CEA) has been modified to allow such generalizations. After having given the theoretical bases of the methods used, we will comment two implementations.

Keywords: Finite element · Phase fields · Reinforced concrete · Statistical scale effect · Rebars sliding

1 Introduction

The modeling of large reinforced concrete structures requires on the one hand the consideration of the impact of defects in the homogeneity of concrete, leading to the appearance of the first cracks located in the areas of weakness of the concrete, and on the other hand the realistic consideration of the concrete steel sliding which helps to reduce crack

P. Rossi and J.-L. Tailhan (Eds.): SSCS 2022, RILEM Bookseries 38, pp. 335–344, 2023.
https://doi.org/10.1007/978-3-031-07746-3_33

openings by relocating the deformation of steels on either side of the cracks. The heterogeneity of concrete concerns scales of lengths of the order of the size of the stretched area, a priori these lengths are much larger than the meshes of the discretization by finite elements.

The concrete steel slide is held on decimetric scales. This scale is often included in the largest meshes, in this case the sliding must be integrated into the behavior law of reinforced concrete, but in other circumstances the meshes may be less than the length concerned by the slip. It is then tempting to resort to an explicit mesh of the steel bars and concrete steel interfaces, but this leads to refining the meshes in a way that is not very relevant to the calculation times.

Thus, both the statistical scale effect and the concrete-steel sliding have the particularity of involving radii of action greater than the size of the meshes. We will see how to express these problems in order to obtain behavior laws of "homogenized reinforced concrete" integrating these different radii of action. We will then see that there are several possibilities to integrate these differential equations into finite element codes, and that one of them has advantages in terms of ease of implementation and precision, without degrading performance in terms of volume and calculation time. It is a implicit "segregated method". This approach has been implemented in the CEA's Castem software.

2 Helmholtz's Formulation of the Weibull Scale Effect

2.1 The Statistical Scale Effect

The national project CEOS.fr [1] confirmed that the gradual appearance of cracks in a long reinforced concrete tie was explained by Weibull's theory: The first crack appears in the weakest area of the tie, the next appears for higher strength and so on until the re-anchoring of the reinforcements is no longer sufficient to create more cracks. For a 5 m long reinforced tie such as the one studied by Mivelaz in [2], the resistance to 1st cracking is twice as low as that measured on a small sample using a splitting test. At the end of the tensile test of the reinforced tie, the number of cracks stabilizes at 4 with a cracking resistance 30% higher for the last crack than for the 1st. Although Weibull's theory is quite capable of describing this effect, it has been found to lead to underestimating the 1st crack resistance of very long ties. The reason is explained in [3] by the existence on site and in prefabrication plants, of control procedures that lead to the rejection of concrete of too poor quality. As in Weibull's theory the increase in the tensile volume decreases the strength, it is necessary to artificially limit the volumes considered in Weibull's theory so that the lowest strength does not fall below the value leading to the "inadmissibility of concrete on the construction site". The work carried out in the CEOS.fr project made it possible to estimate that the volume of concrete to be investigated to find the weakest link generally does not exceed 2 cubic meters, beyond this value, an underestimation of the resistance of 1st cracking was possible. To automatically consider this maximum investigated size in Weibull's theory, we proposed in [3] a probabilistic weight all the lower as the volume elements considered by Weibull's theory are far from the Gauss point for which the most probable tensile strength is seek. Thus, the theory of the weakest link whose original expression is given by Eq. (1) has been replaced by Eq. (3). In expressions (1) and (3) Pf_v is the probability of cracking on a volume structure V, Pf_{ref} the probability on a reference volume V_{ref}.

$$\ln(1 - Pf_v) = \int_V \frac{\ln(1 - Pf_{ref})}{V_{ref}} dV \tag{1}$$

The probability of cracking can be related to the mean strength and standard deviation of the resistance by the Weibull probability distribution (2).

$$Pf_{ref} = 1 - \exp\left(-\left(\frac{\sigma}{\lambda_{ref}}\right)^m\right) \tag{2}$$

In (2) λ_{ref} is close to the average resistance and m is related to the coefficient of variation of the resistance (for a coefficient of variation of 12% m is 10 for example).

$$\ln(1 - Pf_{V/M}) = \int_V \frac{\ln(1 - Pf_{ref})}{V_{ref}} \psi(l/lc) dV \tag{3}$$

In expression (3) the function $\psi(l/lc)$ is a weighting function, decreasing as a function of the distance l between the point at which the most probable resistance is sought and any point of integration of the structure. lc is a characteristic length related to the volume investigated by the Eq. (3). For a volume of the order of 2 cubic meters and ψ a Gaussian form, lc is therefore close to 0.5 m.

$$\int_V \psi(l/lc) dV \rightarrow (lc\sqrt{2\pi})^3 \tag{4}$$

Once the integral (3) has been calculated, it is possible to define the tensile equivalent volume (5). After repeating this operation on all the Gauss points of the model, we have a field of "equivalent volumes" $V_{eq(M)}$.

$$V_{eq(M)} = \frac{\int_V \left(\frac{\sigma}{R_{t/Vref}}\right)^m \psi(l/l_c) dV}{\left(\frac{\sigma_{Max}}{R_{t/Vref}}\right)^m} \tag{5}$$

The field $V_{eq}(M)$ allows to calculate the field of $Rt(M)$ using (6).

$$Rt(M) = Rt(Vref)\left(\frac{V_{ref}}{V_{eq}(M)}\right)^{1/m} \tag{6}$$

The result of this type of calculation is consistent with the experiment as illustrated by the Fig. 1. Thanks to this modified Weibull method, it is no longer necessary to carry out random draws of tensile strength to consider the structural effects due to the dispersion of resistance, on the other hand it is necessary to reassess the field $Rt(M)$ as the structure becomes damaged, but the calculation uses the integral (3) that would have to be redone at each Gauss point. This is a similar problem, in terms of method, to that of the non-local integral often used to regulate the problem of crack localization [4].

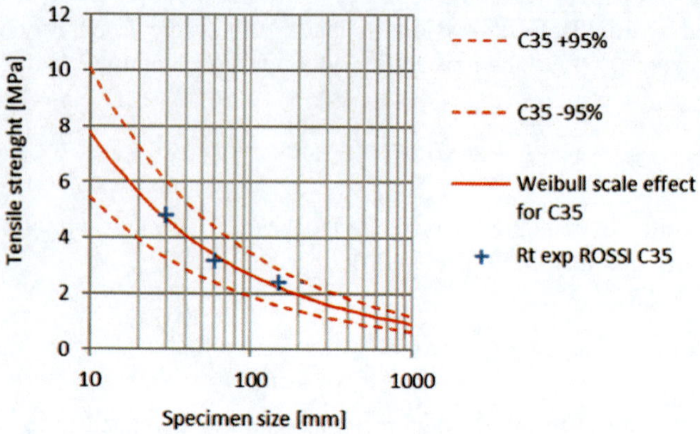

Fig. 1. Weibull scale effect, experimental results for a C35 concrete (from Rossi [5])

2.2 Differential Formulation of the Statistical Scale Effect

To remedy the problem of calculating non-local integrals, De Borst and Pamin [6] used an approximation based on a series development at order 2 of the function to be integrated; this approximation allowed them to replace the multiple non-local integral calculations by solving a single differential equation. We transposed their approach to the case of the non-local Weibull integral (3): By posing (7), the calculation of the non-local Weibull integral becomes (8).

$$\alpha = \int_V \left(\frac{\sigma}{R_{t/Vref}}\right)^m \psi(l/l_c)dV \tag{7}$$

$$\alpha - \left(\frac{lc^2}{2}\right)\Delta\alpha = V_{Max}\left(\frac{\sigma}{R_{t/Vref}}\right)^m \text{ with } V_{Max} = \int_V \psi(l/lc)dV \tag{8}$$

In (8) Δ is the Laplacian of α. The implementation in the Castem software of this equation made it possible to find the statistical scale effects provided by [7] (Fig. 2).

Fig. 2. Self-determination of the most likely local value of Rt in a 3-points bending beam during cracking (thanks to Eq. (8), Rt is known only in non-blue areas, the color scale represents the normalized tensile strength. At the end of the analysis, the most likely tensile strength is known only in cracked or micro-cracked zones).

We therefore have at this stage a differential Eq. (8) capable of directly estimating the most likely tensile strength to be used for structural calculation (i.e. representative of a set of random structures). It is interesting to note that a single calculation of the structure is enough to know the most likely behaviour, the use of random fields is therefore no longer necessary, and that it is sufficient to solve Eq. (8) in parallel with the mechanical equilibrium calculation. In the classification of partial differential equations, Eq. (8) belongs to the category of "Helmholtz" equations.

3 Helmholtz Formulation of Concrete Steel Sliding

3.1 Concrete Steel Sliding

Another well-known problem can also be reduced to a Helmholtz equation, it is the concrete steel sliding. The system of equations handling the problem is detailed in [8]. It contains the 4 equations of the differential system (9): the axial equilibrium of the bar, the behavior law of steel, the shear-slip behavior law of the concrete steel interface, and the kinematic relationship connecting the deformations of steel and concrete to the sliding at the interface. In (9) the behavior law of steel was derived twice with respect to the axis of the bar to, subsequently, be introduced into the equilibrium equation, likewise the behavior law of the interface is derived once, and therefore the kinematic relation as well. These derivations make it possible to combine the 4 equations into a single one presented in the following paragraph (Fig. 3).

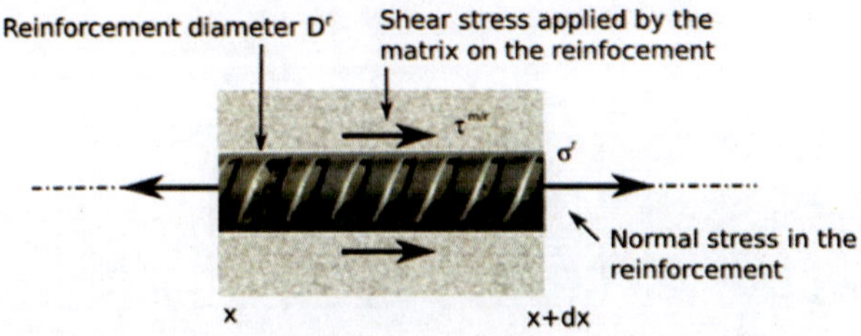

Fig. 3. Axial and shear stresses applied to a reinforcement

$$\begin{cases} \dfrac{\partial^2 \sigma^r}{\partial x^2}\dfrac{D^r}{4} + \dfrac{\partial \tau^{m/r}}{\partial x} = 0 \\ \dfrac{\partial^2 \sigma^r}{\partial x^2} = E^r \dfrac{\partial^2 \epsilon^{re}}{\partial x^2} \\ \dfrac{\partial \tau^{m/r}}{\partial x} = H^i \dfrac{\partial g^{m/r}}{\partial x} \\ \dfrac{\partial g^{m/r}}{\partial x} = \epsilon^m - (\epsilon^{re} + \epsilon^{ra}) \end{cases} \tag{9}$$

3.2 The Helmholtz Formulation of Concrete Steel Sliding

The combination of system Eq. (9) leads to Eq. (10). This last equation is also of the Helmholtz type, it has, as its main variable, the stress of the reinforcement σ^r; the second member of the equation consists of the combination of the deformation of the matrix ϵ^m and plastic deformation of steel ϵ^{ra}, they too in the direction of reinforcement. The "diffusion" term of Eq. (10) contains the steel module E^r, the diameter of the steel bar D^r and the tangent stiffness H^i of the shear-slip law of the interface.

$$\sigma^r - \frac{E^r D^r}{4H^i}\frac{\partial^2 \sigma^r}{\partial x^2} = E^r\left(\epsilon^m - \epsilon^{ra}\right) \tag{10}$$

It is remarkable that it is not necessary to mesh the reinforcements if Eq. (10) is available in the finite element code, because all the variables or parameters of the equation can be represented by bi- or three-dimensional continuous fields. The combination of stresses in steel and concrete is then made by a trivial mixing (11), in which the ρ^n are the density of the different steel categories, N^r the total number of steel rebar types. Each type of steel rebar is managed by a type Eq. (10); σ^m is the stress in the concrete, assessed with a nonlinear model, and σ^{rn} the stress in type steel rebar n. All the details to implement this method are available in ref [8].

$$\sigma_{ij} = (1 - \sum_{n=1}^{N^r} \rho^n)\sigma_{ij}^m + \sum_{n=1}^{N^r} \rho^n \sigma_{ij}^{rn} \tag{11}$$

The Fig. 4 depicts a reinforced concrete tie rod with axial reinforcement, a mid-length attenuation of the tensile strength leads to the localization of the damage.

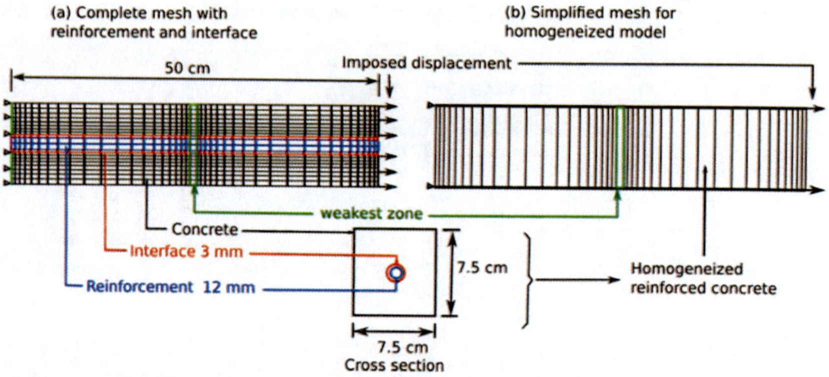

Fig. 4. Reinforced concrete tie rod from ref [8], left reference mesh with steel, concrete and mesh interfaces, right homogenized reinforced concrete element using Eq. (10),

In a first calculation the concrete interface and steel are meshed, and each area of the mesh follows its own behavior law, leading to the so-called reference solution. In a second step the steel is not meshed, nor the interface, only the reinforced concrete is meshed, it contains an adequate longitudinal reinforcement density whose behavior is managed by fields supported by the same mesh than for RC but with the Eq. (10). In this context the nodes of the mesh have each one 4 variables: three displacements and σ^r.

The comparison, in Fig. 5, shows that the two methods lead to almost the same result in terms of force-displacement as well as stress profile in steel. The model using Eq. (10), however, is significantly more economical in terms of the number of nodes.

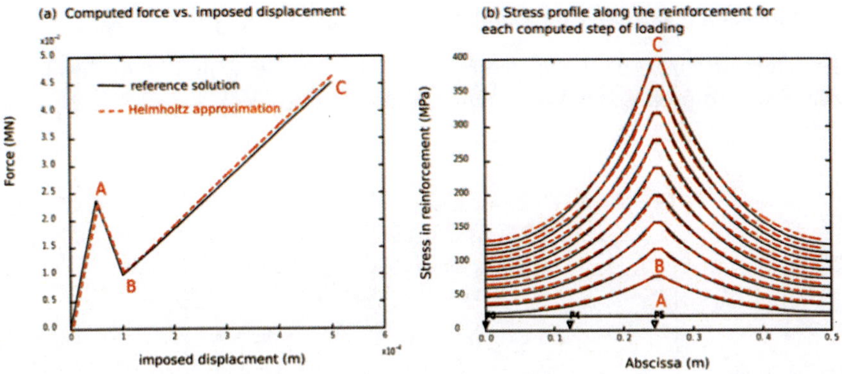

Fig. 5. Comparison of the solution obtained with the reference mesh and the solution obtained with the Helmholtz Eq. (10)

4 Coupling Phase Fields with Mechanical Formulations

The two Helmholtz equations we have described in (8) and (10) can also be called phase fields. This name has been used in particular by [9] to test various forms of non-local damage laws resolved by the second gradient method. Thus, many problems can be considered as "phase fields". It is therefore necessary to implement this method in the most efficient way possible, but at least two implementation variants can be considered, we comment on them below by applying them to the coupling of the previous set of problems. If the problem to be solved is three-dimensional with three types of reinforcement, the vector of the nodal variables contains $\left[u, v, w, \alpha, \sigma^1, \sigma^2, \sigma^3\right]$. In this vector u, v, w are the components of the displacement field, α is the modified Weibull integral solved by (8), $\sigma^1, \sigma^2, \sigma^3$ are the stresses in the three types of steel bars solved by (10). Because of the mixing law (11) on the one hand, and the behavior law of the concrete matrix on the other hand, the variables $\alpha, \sigma^1, \sigma^2, \sigma^3$ all intervene in the equilibrium equations to find u, v, w. Conversely, the right-hand terms of Eqs. (8) and (11) involve stress in concrete (for 8) and deformation of concrete (for 10). The problem is therefore strongly coupled.

4.1 Dimension of the Digital Problem in the Case of a Classic Resolution

Considering the strong coupling between the equilibrium equations of reinforced concrete, Weibull's law and the behavior of steel rebars leads to the formulation of a finite element whose stiffness matrix will be proportional to 7^2 since there are 7 scalar equations to solve. In the case where the mesh has N nodes, the global matrix will be of the order of magnitude of $(7N)^2$. If to cancel the overall residue of the equations I iterations are needed, the numerical cost of each converged step will be $I(7N)^2$. Let's compare, below, this value with another implementation method.

4.2 Dimension of the Digital Problem in the Case of a Segregated Resolution

Segregation of our strongly coupled problem can be done using the two-level convergence flowchart proposed below.

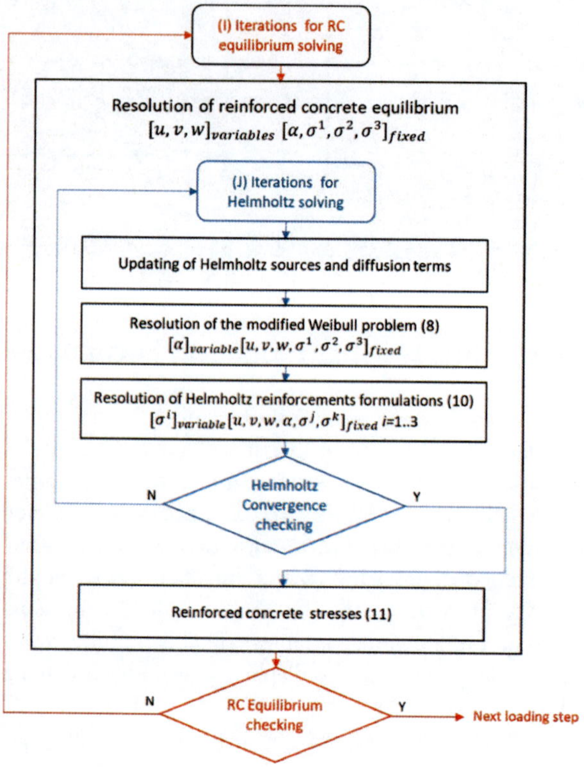

Fig. 6. Two convergence levels Chart flow for segregated solving of strongly coupled equations

Internal convergence (in bleu on the Fig. 6) allows to satisfy the equations of type Helmholtz (Phase fields), this area of the algorithm uses a number of calculations whose order of magnitude is $4JN^2$ since each segregated problem has a single unknown variable. Once convergence is reached at this level, the variables are returned to the upper level to be used in the balance of reinforced concrete. The order of magnitude of the total number of calculations to be carried out to achieve the convergence of the upper level is therefore $I((3N)^2 + 4JN^2)$. In the implementation we have achieved in Castem, a same type of convergence accelerator [10] is used at each level, the same level of convergence difficulty would therefore lead significantly to $I \cong J$ and therefore the total time for the segregated algorithm should be of the order of magnitude $I(3N)^2 + 4(IN)^2$. This order of magnitude can be compared to that resulting from a classical resolution (above). The equality between the orders of magnitude of the number of calculations to be made to have convergence on a step is done for a number of iterations $I = 10$ whatever the number of nodes N of the problem. This means that if at each time step the convergence requires less than 10 iterations per level, the segregated method is always the fastest, regardless of the size of the problem. Another significant advantage is that the segregated method does not require the formulation of a new finite element with each addition of "phase fields".

5 Conclusion

We have seen that problems as varied as Weibull's weakest link theory, or concrete steel sliding in the context of homogenized reinforced concrete can be approached by differential equations of the Helmholtz type. These equations are also used in the literature on non-local calculations where they are also called "phase fields". We have implemented in the CEA's Castem software a solving method that can be used to solve "phase field" problems strongly coupled to equilibrium equations. This method, unlike the classical method of strong coupling, does not require the creation of a new finite element for each new strong coupling: it uses only the matrices of the sub-problems, but two levels of convergence: a higher level for the equilibrium of homogenized reinforced concrete, and a lower level for the phase fields. We have also shown that the so-called "segregated" method is always the fastest as long as the number of iterations for each time step per nonlinear problem remains less than 10. Further study of the convergence conditions of each of the levels should be done to verify whether there are no beneficial effects induced by the segregation of phase fields. Indeed, the resolution of the mechanical problem of the top level uses only licit fields of phase fields; the upper level could then converge faster than if we had used a conventional method mixing all non-zero residues on the same level.

Thus, it is now possible, thanks to Helmholtz's formulations of the Weibull scale effect and steel-concrete sliding, to consider these phenomena in a homogenized model for reinforced concrete. Helmholtz's equations are indeed very well adapted tools to consider, in the behavior law of reinforced concrete, phenomena that occur at scales larger than the meshes. From this point of view, the behavior law of reinforced concrete can only be "non-local", not necessarily in the sense of the regularization of damage, but rather because it is the best way not to mesh steels and not to make random draws.

References

1. Barre, F., et al.: Control of Cracking in Reinforced Concrete Structures. Wiley, Hoboken (2016)
2. Mivelaz, P.: Etanchéité des structures planes en béton armé-fuites au travers d'un élément fissuré. Ecolue Polytechnique Fédérale de Lausanne (1996)
3. Sellier, A., Millard, A.: Weakest link and localisation WL 2: a method to conciliate probabilistic and energetic scale effects in numerical models. Eur. J. Environ. Civ. Eng. **18**(10), 1177–1191 (2014). https://doi.org/10.1080/19648189.2014.906368
4. Pijaudier-Cabot, G., Bazant, Z.: Nonlocal damage theory. J. Eng. Mech. **113**, 1512–1533 (1987). https://doi.org/10.1061/(ASCE)0733-9399(1987)113%3A10(1512)
5. Rossi, P., Wu, X., Maou, F., Belloc, A.: Scale effect on concrete in tension. Mater. Struct. **27**(8), 437–444 (1994). https://doi.org/10.1007/BF02473447

6. De Borst, R., Pamin, J.: Gradient plasticity in numerical simulation of concrete cracking. Eur. J. Mech. - A/Solids **15**(2), 295–320 (1996)
7. Syroka-Korol, E., Tejchman, J., Mróz, Z.: FE calculations of a deterministic and statistical size effect in concrete under bending within stochastic elasto-plasticity and non-local softening. Eng. Struct. **48**, 205–219 (2013). https://doi.org/10.1016/j.engstruct.2012.09.013
8. Sellier, A., Millard, A.: A homogenized formulation to account for sliding of non-meshed reinforcements during the cracking of brittle matrix composites: Application to reinforced concrete. Eng. Fract. Mech. **213**, 182–196 (2019). https://doi.org/10.1016/j.engfracmech.2019.04.008
9. Miehe, C., Hofacker, M., Welschinger, F.: A phase field model for rate-independent crack propagation: robust algorithmic implementation based on operator splits. Comput. Methods Appl. Mech. Eng. **199**(45–48), 2765–2778 (2010). https://doi.org/10.1016/J.CMA.2010.04.011
10. Chow, Y.K., Kay, S.: On the aitken acceleration method for nonlinear problems. Comput. Struct. **19**(5–6), 757–761 (1984). https://doi.org/10.1016/0045-7949(84)90175-5

Design Optimization of Concrete Railway Tracks by Using Non-linear Finite Element Analysis

Jean-Louis Tailhan[1](✉), Pierre Rossi[1], and D. Salin[2]

[1] MAST-EMGCU, Univ Gustave Eiffel, 77447 Marne-la-Vallée, France
jean-louis.tailhan@univ-eiffel.fr
[2] Hexagon, 91320 Wissous, France

Abstract. Between 2007 and 2014, UGE, Alstom and other industrials partners developed a new concept of railways track called New Ballastless Track (NBT). The concept was validated under 10 million fatigue cycles on a real-size mock-up at UGE. A first numerical study using a non-linear model was performed to evaluate the possibility of replacing the original reinforced concrete layer of the track slab with Steel Fibre Reinforced Concrete. The objective is to simplify the NBT track's construction and take advantage of the redistribution of mechanical stresses on a hyper-static structure. This study led to the conclusion that this replacement was very relevant.

This paper is on optimizing this Fibre Reinforced Concrete (FRC) solution by using the same non-linear numerical model. It is shown that this optimization procedure leads to a significant reduction of CO2 emissions compared with the initial one.

Keywords: Railway track · Slab · FRC · Numerical model · Cracking · Design optimisation

1 Introduction and Objectives of the Study

Fibre Reinforced Concrete (FRC) is increasingly used in structural applications. One of the principal reasons for its popularity is the recent emergence of national and international recommendations efficient, for instance, to designing simply supported structures in bending using this type of material. However, they do not possess an excellent physical base to propose relevant solutions for more complex structures such as these statically indeterminate. Hence, it is claimed that the most efficient approach for designing such structures for both safety and sustainable development is to use non-linear finite element analysis.

A probabilistic discrete cracking model, developed since 1985, to simulate the cracking process of concrete has been extended to analyse reinforced concrete and fibre reinforced concrete structures. All these numerical models have been fully and deeply validated [1–9].

Between 2007 and 2014, UGE, Alstom and other industrials partners developed a new concept of railways track called New Ballastless Track (NBT) [10]. The concept is

P. Rossi and J.-L. Tailhan (Eds.): SSCS 2022, RILEM Bookseries 38, pp. 345–354, 2023.
https://doi.org/10.1007/978-3-031-07746-3_34

based on two superimposed independent layers of concrete slabs (i.e., foundation and track slabs) and was designed to achieve a 100 years life span under a mixed high speed and freight traffic. It was first validated through a FEM model and then under 10 million fatigue cycles on a real-size mock-up at UGE [10–12]. A trial section of 1 km was then built by Alstom on the French railway's network [10] and opened to regular speed traffic by the end of 2013. It was monitored for two years. and after five years of service. it behaves well. confirming the concept's relevance.

Figure 1 presents the original concept and the geometry of the experimental mock-up. In this figure, the upper track layer is reinforced concrete (C35/45) called BC5, while the foundation slab is plain concrete (C25/30) called BC3. In addition, an elastic layer is used to reproduce the mechanical reaction of the ground simulating the soil bearing capacity.

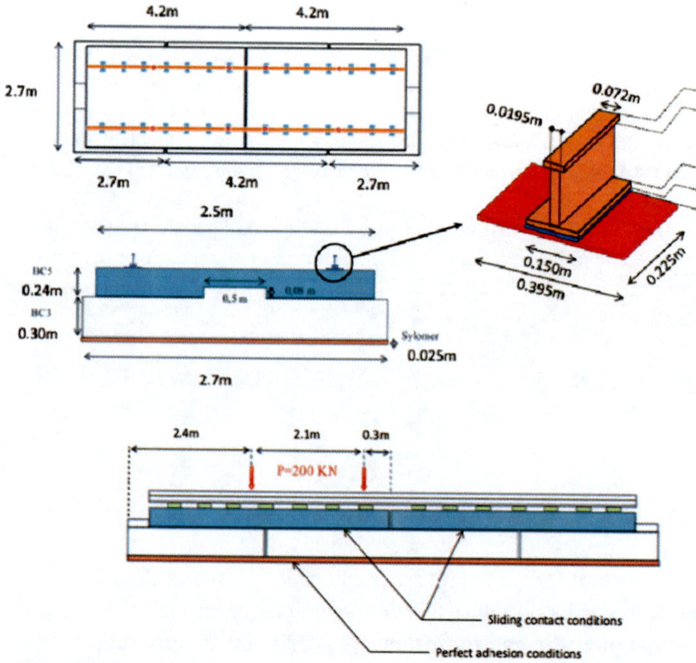

Fig. 1. Geometry and dimensions of the UGE Railway Track Mock-up.

A first numerical study [13] evaluated the possibility of replacing the original reinforced concrete layer BC5 of the track slab with a Steel Fibre Reinforced Concrete (using 78 kg/m^3 of fibre) to simplify the NBT track's construction and take advantage of the redistribution of mechanical stresses on a hyper-static structure. This numerical study was performed using the probabilistic discrete cracking model evocated above to analyse and compare their respective cracking process. The principal conclusion was that the FRC solution is mechanically more efficient than the one with a usual ratio of rebars. Nevertheless, it is preferable with such a solution to use some local rebars in zones of high concentration of tensile stresses[1], to ensure a necessary level of safety for the structure.

The objective of this second study on the same topic is to optimize the steel fibre reinforced railway track (still by using the probabilistic discrete cracking model).

For this objective. the following strategy of optimization has been followed:

- Step 1: First level of optimization: different dosages of fibres from 40 to 78 kg/m^3, different sectional percentages of bottom rebars from 0 to 8.8% and different thicknesses of the upper track layer (B3) from 15 to 24 cm are considered. The rebars, with an arbitrarily length of 40 cm, are placed in zones of high concentration of tensile stresses as in the first study [13]. Acceptable solutions are those leading to maximal crack openings smaller than 100 μm (maximal value allowed by the industrial for this application.
- Step 2: Second level of optimization: complementary numerical simulations are carried out to enrich the base of knowledge around the previous acceptable solutions, in the space of the tested parameters.
- Step 3: Carbon footprint evaluations: carbon footprint evaluation is performed for all technical solutions resulting from step 2.

2 Numerical Models

2.1 Probabilistic Explicit Cracking Model of Concrete

The model was first developed by Rossi et al. [14–16]. It describes the behaviour of concrete via its two significant characteristics: heterogeneity and sensitivity to scale effects [18]. The physical basis of the model has been already presented in detail in [14–16]). The model is called explicit in the manner he considers cracks. Each massive finite element represents au given volume of uncracked heterogeneous material. They remain linear elastic. Cracks are then explicitly represented by 2D or 3D quadratic interface elements of zeros thickness, and Rankin and Tresca failure criteria are used in tension or shear respectively. If stresses remain lower than their critical values, the interface element ensures the continuity of displacements between the nodes of its two neighboring volume elements. If stresses exceed their critical values, the interface element opens creating an elementary crack. The tensile and shear strengths and the normal and tangential stiffness values related to this interface element become equal to zero [14–16]. In the case of crack re-closure, the interface element recovers its normal stiffness and follows a classical Coulomb's law [14–16]. Tensile and shear strengths are randomly distributed over the mesh before any simulation. The tensile strength's mean value and standard deviation depend on the heterogeneity degree of the material and the quality of the cement paste. The mean value of the shear strength is assumed constant and equal to the half of the average compressive strength. Its standard deviation depends on the element size, and is assumed to be the same (for elements of same size) as that of the tensile strength.

[1] i.e., in the upper track layer. B3. at the vertical of the joints of the foundation slab. B5.

This numerical model is summarized in Fig. 2.

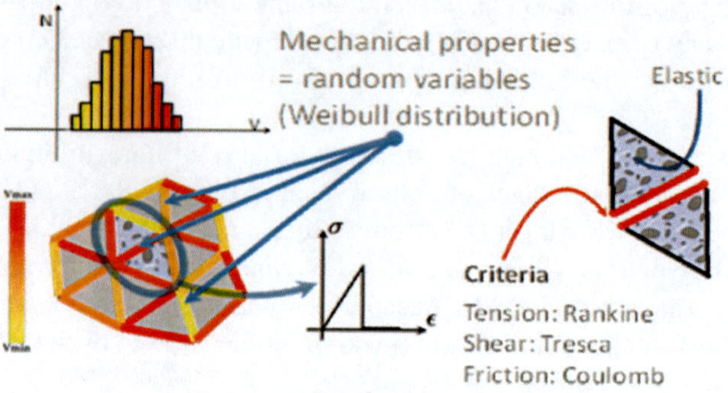

Fig. 2. Probabilistic concrete cracking model – explicit approach

2.2 Probabilistic Explicit Cracking Model of Fibre Reinforced Concrete (FRC)

The creation of cracks in the concrete matrix is represented by an elastic perfectly brittle behaviour. When the interface element is opened, the fibres' bridging effect is represented with an elastic damage behaviour. $K_n^{'}$ and $K_t^{'}$, normal and tangential rigidity values respectively, are different since the fibres' rigidity is physically more critical in tension than in shear. (From a threshold displacement, ζ_0, the mechanical behaviour of the interface element changes, and the normal stress is considered linearly decreasing with the normal displacement). It represents both the damage of the concrete/fibre bond and the fibre pullout. Finally, the interface element is considered definitively broken when the displacement reaches a maximal value, ζ_c (Fig. 3). This value corresponds to the state where the effect of fibres is considered negligible. It is determined from a uniaxial tensile test and characterises the FRC. At this point, normal and tangential rigidities are set to zero.

The post-cracking energy dissipated by the bridging effect of the fibres is considered randomly distributed over the mesh elements, and following a log-normal distribution function with a mean value independent of the mesh elements' size [19] and a standard deviation due to the heterogeneity of the material, increasing as the mesh elements' size decreases. All these values can be determined from experimental tests and/or inverse analysis.

Figure 3 presents the numerical mechanical behaviour adopted to represent the experimental post-cracking behaviour. Again, only the normal stress-normal displacement curve is considered in this figure.

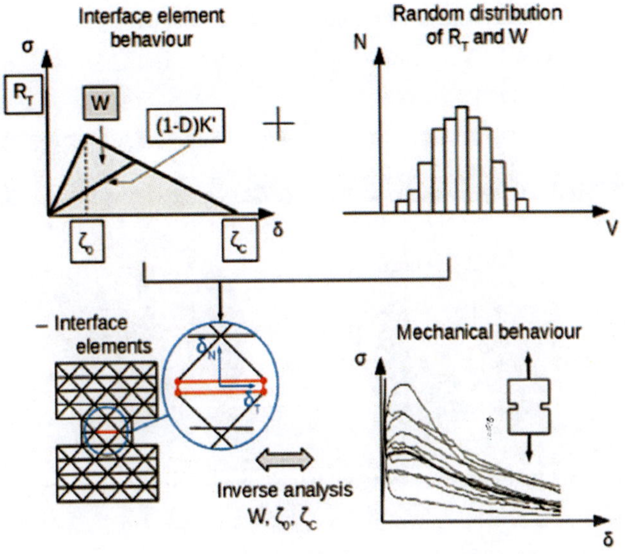

Fig. 3. Probabilistic FRC model

2.3 Rebars and Concrete/Steel Bond Modelling

In the 2D modelling approach, an equivalent plate mixes concrete and steel rebars. The height of the plate is taken equal to the greater diameter of the local rebars. The plates' equivalent Young moduli is determined using a classical mixture rule taking into account the steel reinforcement ratio (t_a). Here again, interface elements are also used to simulate the concrete cracking through the reinforcement. Their strengths are similar to those used previously and follow the same opening criteria. After the opening of the interface element, the residual stiffness corresponds to the one of the steel reinforcements only (a hypothesis of a total rupture of adherence along the two volume elements concerned by the interface element is assumed).

3 Numerical Simulations

3.1 Finite Element Mesh

Figure 4 presents an example of finite element mesh used in numerical simulations. This mesh is 2D, although the mechanical problem is a 3D one. Since non-linear 3D simulations could be extensively time consuming, and regarding that the model used considers volume effects, the 2D plane stress condition has been retained. The width (i.e., the length measured in the direction perpendicular to the plane of the figure) of the mock-up has therefore to be considered in the calculation. For a reason related to volume effects, considering the total width of the mock-up will cause the models to produce a poor description of the cracking processes. Nevertheless, a more detailed description can be achieved by choosing to perform simulations with a smaller value of this width. The width of the cast iron plates under the rail, equal to 0.395 m (see Fig. 1), is then considered in the present 2D numerical analysis. This choice has been justified in previous studies [13].

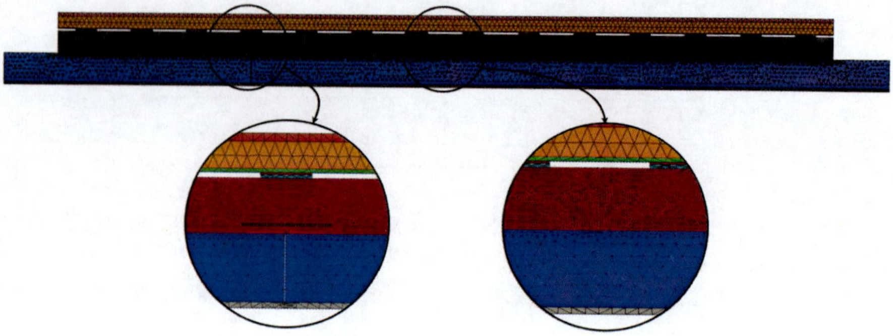

Fig. 4. 2D finite element mesh related to the numerical simulations

3.2 Non-linear Finite Element Simulations

The parameter values of the different FRCs (different fibre dosages) considered in this study are determined from the FRC used in [13]. The fibre content was 78 kg/m^3.

Then, based on some knowledge concerning that type of FRCs [20], it is assumed that the post-cracking dissipated energy is linearly proportional to the fibre content in a range between 40 and 80 kg/m^3. It is also assumed that the standard deviation related to this post-cracking energy is kept identical. Consequently, the coefficient of dispersion related to the FRC containing fewer fibres is higher than that of the FRC containing more fibres. This assumption is reasonable because it is well known that the mechanical degree of heterogeneity decreases when the percentage of fibres augments, this latter staying below the fibres' saturation percentage [20].

All the technical solutions simulated during the first and second steps of optimization are presented in Table 1. Table 2 presents the results obtained with the different technical solutions. They are expressed in terms of crack openings (mean and dispersion) and number of cracks for the three zones (left, central and right) of the loaded BC5 slab. The central one corresponds to the zone that includes the steel reinforcement (left circle in Fig. 4), the other two zones being on either side. Technical solutions leading to average crack openings less than 100 μm, everywhere in the three zones, are considered mechanically acceptable. They are underlined in dark grey in Table 2. Note that these results are still an underestimation or the reality, since the synergic effect between fibres and steel reinforcement is not well represented in this modelling.

Table 1. Technical solutions studied

	First step													Second step						
Technical solution n°	1	2	3	4	5	6	7	8	9	10	11	12	13	14	15	16	17	18	19	20
Concrete slab height (cm)	15	15	17	19.5	19.5	19.5	22	24	24	17	17	19.5	22	22	22	22	19.5	19.5	24	15
Steel reinforcement (%)	2.20	6.61	5.29	0.00	2.20	8.81	7.93	4.41	4.41	4.41	2.64	7.05	6.17	5.29	5.29	7.93	5.29	7.93	8.81	17.6
Fibre content (kg/m^3)	72	80	56	56	48	40	76	48	64	68	64	60	56	64	76	64	76	64	80	72

Table 2. Obtained results for the three zones (left, central and right) of the loaded BC5 slab.

	Left part			Central part			Right part		
Tech.	Crack opening		Nb of	Crack opening		Nb of	Crack opening		Nb of
sol. n°	Mean	Std dev.	cracks	Mean	Std dev.	cracks	Mean	Std dev.	cracks
1	13.4	2.4	6.90	138.2	245.7	0.40	0.0	0.0	0.00
2	12.9	2.0	7.00	152.5	188.9	0.50	0.0	0.0	0.00
3	382.3	289.9	1.60	133.0	160.9	0.90	21.4	3.5	0.40
4	196.8	287.0	3.20	127.7	178.8	1.00	24.7	4.6	1.00
5	125.8	232.1	4.25	42.7	31.9	1.00	34.2	8.0	0.63
6	674.3	402.0	1.40	382.7	708.2	1.10	32.1	9.6	1.70
7	16.7	3.4	8.50	31.6	9.3	0.90	21.7	5.4	2.80
8	20.2	6.5	7.30	144.4	304.2	1.70	58.9	10.3	1.00
9	17.9	4.8	7.50	44.5	27.5	1.80	44.2	7.1	1.00
10	99.9	163.7	3.80	19.8	6.1	0.60	0.0	0.0	0.0
11	137.9	189.5	3.50	286.8	392.1	0.90	0.0	0.0	0.00
12	120.3	215.9	3.70	101.9	294,4	1.00	19.8	4.8	1.10
13	18.0	4.8	4.80	150.5	300.1	1.00	23.4	9.2	2.30
14	17.6	5.2	3.40	56.3	16.2	1.10	21.6	8.2	2.60
15	15.5	4.4	5.80	34.3	8.6	1.10	18.9	7.8	2.80
16	17.1	4.5	4.57	49.8	13.7	1.00	22.9	8.4	2.29
17	62.1	98.0	4.50	22.5	6.7	1.00	18.4	3.4	0.60
18	129.5	187.1	3.80	15.3	4.0	0.90	18.7	5.4	0.70
19	16.0	3.9	3.00	45.8	13.7	1.20	27.1	7.1	1.40
20	13.5	2.4	6.40	16.0	2.6	0.30	0.0	0.0	0.0

4 Carbon Footprint Evaluations

In this chapter, comparisons of carbon footprint evaluation concerning two families of technical solutions are proposed: the classical solution of railway tracks reinforced only with rebars [8] and that of railway tracks reinforced with a mix of steel fibres and rebars. These comparisons are based on some data from literature [20] and given Table 3.

Figure 5 shows an example of a crack pattern obtained with one technical solution studied (n°15, Table 1 and Table 2).

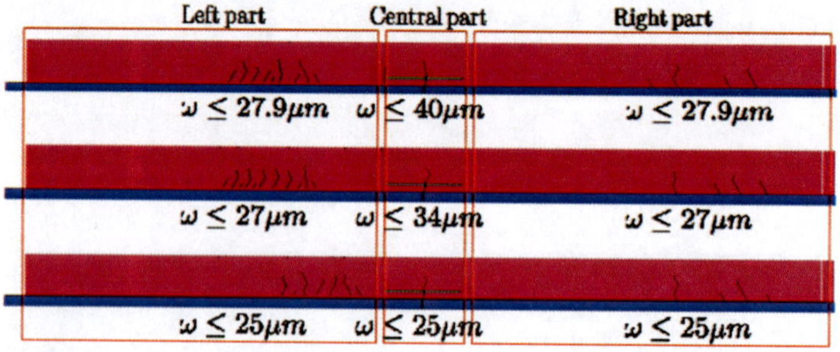

Fig. 5. Example of cracking patterns related to technical solution (n°15) – 3 simulations.

Table 3. CO_2 emissions per slab's constituent. (*) As an order of magnitude, CO_2 emissions per concrete m^3 are between 200 and 300kg/m^3. (**) Values from [21].

Concrete	Fibres	Steel reinforcement
250 kg/m^3(*)	2.425 kg/t(**)	1.932 kg/t(**)

4.1 Classical Solution

The reinforcement ratio at the top of the track slab is different from that at its bottom [8]. They are, respectively, $1.5310^{-3}m^2$ (7 rebars of diameter 14 mm +4 rebars of diameter 12 mm) and $3.3810^{-3}m^2$ (14 rebars of diameter 16 mm +5 rebars of diameter 12 mm). In this work, the rebars' length is 4.2m. Consequently, according to values given in Table 3 and quantities of concrete and steel for this solution, the amount of CO_2 emitted is 63 t per slab.

4.2 Mixed Solutions

Table 4 summarizes the carbon footprint evaluations of mechanically acceptable technical solutions. CO_2 amounts are calculated for one slab using values given in Table 3 and according to their respective geometries (see Fig. 1), fibre contents and steel reinforcements.

Table 4. CO_2 emissions per technical solutions. The gain, in CO_2, emitted is expressed here by comparison with the classical solution. The minus sign denotes a reduction in CO_2 emission.

Tech. sol n°	7	9	14	15	16	19	20
CO_2 (t)	57.8	63.0	57.7	57.8	57.7	63.0	39.4
Gain (%)	-8	0	-8	-8	-8	0	-37

Results show that the most significant part of the emission gain is related to the reduction in concrete's volume. From Table 1, Table 2 and Table 4, one can note that a simple reduction in slab's height of 2 cm (22 cm instead of 24 cm) leads to a non-neglectable reduction of quite 10% of CO_2 per slab. It represents in fact around 5.2 t of CO_2, an amount that can become important for the entire railway line. Moreover, this solution is immediately technically viable.

Solution number 20 also appears very interesting, since it allows a much larger reduction of CO_2 emission (−37% compared to the classical solution). However, it remains to be verified that the placement of a greater quantity of steel in the central section of the slab is technically acceptable, as well as a greater deformability and possibly increased stresses in the foundation. Nevertheless, a more substantial optimization in this sense remains a lead to pursue.

5 Conclusions

This study concerns the use of numerical models to analyse and compare the cracking process of two types of Railway Tracks: one classical in reinforced concrete and the second in FRC with local rebars (mixed solution). Probabilistic Explicit Cracking models, developed at UGE and fully validated in the framework of previous works, are used.

The main result obtained can be summarized as follows:

- The mixed solution is, at least, as mechanically efficient as the classical one to control cracks openings.
- The mixed solution emits more than 10% of CO_2 less than the classical one.

The results presented will also serve as a database for automatic learning tools used in an optimization process in order to arrive at one (or more) solutions that are truly optimized both mechanically and with regard to its environmental impact.

References

1. Phan, T.S., Tailhan, J.-L., Rossi, P.: 3D numerical modelling of concrete structural element reinforced with ribbed flat steel rebars. Struct. Concr. **14**(4), 378–388 (2013)
2. Tailhan, J.-L., Rossi, P., Daviau-Desnoyers, D.: Numerical modelling of cracking in steel fibre reinforced concrete (SFRC) structures. Cement Concr. Compos. **55**, 315–321 (2015)
3. Rossi, P., Daviau-Desnoyers, D., Tailhan, J.L.: Analysis of cracking in steel fibre reinforced concrete (SFRC) structures in bending using probabilistic modelling. Struct. Concr. **16**(3), 381–388 (2015)

4. Rastiello, G., et al.: Macroscopic probabilistic cracking approach for the numerical modelling of fluid leakage in concrete. Ann. Solid Struct. Mech. **7**(1–2), 1–16 (2015)
5. Rossi. P., et al.: Numerical models for designing steel fiber reinforced concrete structures: why and which ones? In: Massicotte. B., Charron. J.-P., Plizzari. G., Mobasher. B. (eds.) FRC 2014: ACI-fib International Workshop. FIB Bulletin 79 - ACI SP-310. Pp. 289–300 (2016)
6. Rossi, P., Tailhan, J.-L.: Numerical modelling of the cracking behaviour of steel fibre reinforced concrete (SFRC) beam on grade. Struct. Concr. **18**(4), 571–576 (2017)
7. Rossi. P., Daviau-Desnoyers D., Tailhan. J.-L., Probabilistic numerical model of ultra-high performance fiber reinforced concrete (UHPRFC) cracking process. Cement Concrete Comp. **90**, 119–125 (2018)
8. Nader, C., Rossi, P., Tailhan, J.-L.: Numerical strategy for developing a probabilistic model for elements of reinforced concrete. Struct. Concr. **18**(6), 883–892 (2017)
9. Nader. C., Rossi. P., Tailhan J.-L., Multi-scale strategy for modelling macrocracks propagation in reinforced concrete structures. Cement Concrete Comp. **99** (2018). https://doi.org/10.1016/j.cemconcomp.2018.04.012
10. Robertson, I., et al.: Advantages of a new ballastless trackform. Constr. Build. Mater. **92** 16–22 (2015)
11. Chapeleau, X., et al.: Study of ballastless track structure monitoring by distributed optical fiber sensors on a real-scale mockup in laboratory. Eng. Struct. **56**, 1751–1757 (2013)
12. Sedran, T., et al.: Development of a new concrete slabs track (NBT). In: 12th International Symposium on Concrete Road. Prague. Czech Republic. 23–26 September 2014
13. Tailhan, J.L., Rossi, P., Sedran, T.: Comparison between the cracking process of reinforced concrete and fibres reinforced concrete railway tracks by using non-linear finite elements analysis. In: Grassl, P.C., Bayonne, L. (eds.) 10th International Conference on Fracture Mechanics of Concrete and Concrete Structures. FRAMCOS 10, France. 23–26 June 2019
14. Rossi, P., Richer, S.: Numerical modelling of concrete cracking based on a stochastic approach. Mater. Struct. **20**, 334–337 (1987)
15. Rossi, P., Wu, X.: Probabilistic model for material behaviour analysis and appraisement of concrete structures. Mag. Concr. Res. **44**(161), 271–280 (1992)
16. Rossi, P., Ulm, F.-J., Hachi, F.: Compressive behaviour of concrete: physical mechanisms and modeling. J. Eng. Mech. **122**(11), 1038–1043 (1996)
17. Rossi, P., Wu, X., Le Maou, F., Belloc, A.: Scale effect on concrete in tension. Mater. Struct. **27**, 437–444 (1994)
18. Rossi, P.: Experimental study of scaling effect related to post-cracking behaviours of metal fibres reinforced (MFRC). Eur. J. Environ. Civ. Eng. **16**(10), 1261–1268 (2012)
19. Rossi, P.: Les bétons de fibres métalliques. Edité par les Presses de l'Ecole Nationale des Ponts et Chaussées.1998 (in french)
20. ITAtech Guidance for Precast Fibre Reinforced Concrete Segments 2016, vol. 1, Design Aspects ISBN: 978-2-9701013-2-1

Numerical and Experimental Evaluation of Multi-ion Kinetics Associated with Macro-cell Corrosion by Pseudo-concrete

Zhao Wang, Hiroki Takeda, and Koichi Maekawa(✉)

Yokohama National University, Yokohama, Kanagawa, Japan
maekawa-koichi-tn@ynu.ac.jp

Abstract. Macro-cell corrosion, as one of the typical deterioration factors to structural concrete, has drawn enormous engineering interests in the past years. Steel corrosion accompanied with macro-cell circuit is related to both electric fields and chemical substances of polarized metals and electrolytes. The dynamically-equilibrated various substances in the pore solutions of concrete act an important role in terms of electron carrying as well as interaction with the porous skeleton of cementitious material. Therefore, it's meaningful to investigate such multi-ion kinetics associated with macro-cell corrosion to further study the deterioration of reinforcement as well as concrete. This paper contains a numerical and experimental evaluation of the multi-ion kinetics with pseudo concrete electrolyte. As a key species of cementing material, the authors focus on the profiles of Ca^{2+} concentrations as a major cation of concrete around the positive and negative electrodes, which are significantly affected by the initial saturation of calcium hydroxide and the supply of carbon dioxide. Three predominant mechanisms are summarized depending on the $Ca(OH)_2$ and CO_2 condition. Experiments and numerical simulation applied to pseudo concrete show satisfactory correlation to reveal the governing mechanism based on polarization and the Nernst-Plank theory with multi-ion mass equation.

Keywords: Macro-cell corrosion · Multi-ion kinetics · Pseudo-concrete · Pore solution

1 Introduction

As one of the deterioration factors to structural concrete, steel corrosion causes loss of reinforcement and cracking of concrete. Usually, there are two types of corrosion: micro-cell and macro-cell one, depending on whether the anodic and cathodic reactions happen at same place or not [1]. When the targeted domain has uniform corrosion potential, micro-cell corrosion takes place without accompanying an external circuit. While macro-cell corrosion takes place when two half-cells with different corrosion potentials are separated but electrically connected. Latter case is associated with an external circuit where the electrons move along reinforcement bars and are conveyed by migrating ions within the electrolytes of concrete. Besides, these ions originated from cement hydration

P. Rossi and J.-L. Tailhan (Eds.): SSCS 2022, RILEM Bookseries 38, pp. 355–364, 2023.
https://doi.org/10.1007/978-3-031-07746-3_35

also affect the corrosion potential and therefore affect the polarization. Never the less, the mutual interaction between various types of ions, i.e., ionization, has made the issue even more complicated. In other words, under macro-cell corrosion conditions, the ions migrate to form a circuit in electrolyte, which is related to several factors including the electric field, ionic concentration gradient, ionization, and polarization. It's of significant importance to investigate such multi-ion kinetics associated with macro-cell corrosion to further study the deterioration of reinforcement as well as concrete.

Some studies have been conducted to investigate the multi-ion profiles with corrosion damage numerically and experimentally. Bazant [2] proposed a physical-mathematical model of corrosion including the diffusion of oxygen and chloride ion as well as the consumption of oxygen and the production of rust by polarization near the electrodes. With the development of a finite element method, Liu et al. [3] proposed multi-phase modelling of ionic transport in concrete subjected to an impressed charge by solving the mass conservation equation and Poisson's equation. Following this model and integrating the polarization kinetics, Xia et al. [4] successfully applied the simulation to evaluate steel corrosion in concrete. These works adopt the simplification to neglect the geochemical events, such as the mutual reaction between different ions in porous media and the interaction between pore solution (liquid) and skeleton (solid).

The authors present an integrated simulation platform to cover the coupled electrical and chemical fields as well as the geochemical balance in structural concrete. This platform can handle the dynamic multi-ion kinetics during macro-cell corrosion. As the first step, instead of applying it to real concrete material, an independent porous medium is assumed for simulation, which has similar pore structure and pore solution compared to real concrete [5]. Similarly for the experimental validation, pseudo concrete materials (transparent polymer, starch slurry, glass balls and silica sand) are adopted, which allows the direct, dynamic and synchronous sampling of small pieces of electrolyte to measure the ion profiles with almost no destruction to specimens. In addition, the pseudo concrete has the advantages to identify the anodic and cathodic reaction by naked eyes. Beyond the satisfied agreement between simulation and experiment, both results yield that the ionic kinetics of cation Ca^{2+} is closely related to the supply of calcium hydroxide and carbon dioxide in the atmosphere. Accordingly, three predominant mechanisms are summarized and proposed.

2 Numerical Analysis

2.1 Simulation Platform

In the integrated simulation platform, the spatial coupled chemical and electric fields are solved following the governing equations of continuity formulas where the flux terms of ion and electron are calculated by the Nernst-Plank-Poisson equation [6], see Fig. 1.

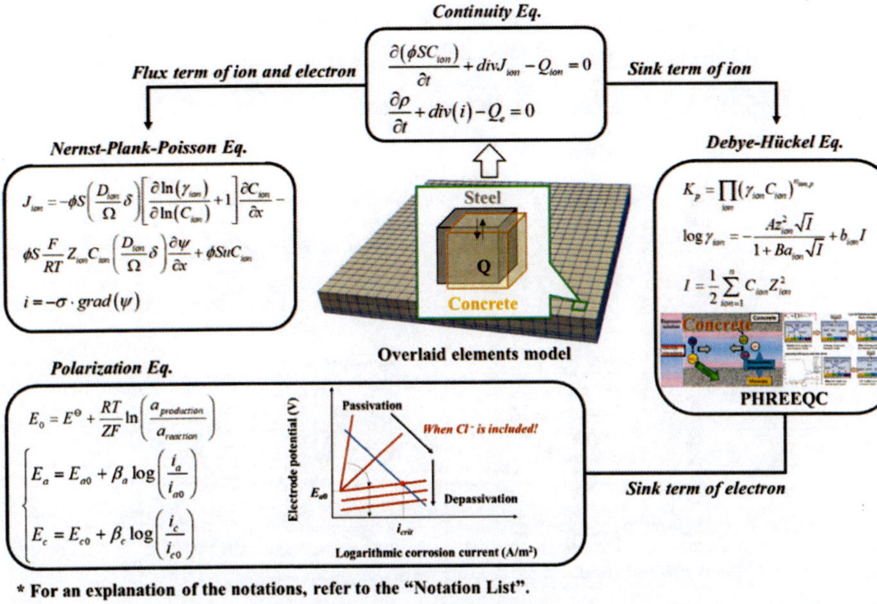

Fig. 1. Governing theories of integrated simulation platform

In the meantime, PHREEQC, known as a geochemical program that deals with the equilibrium among chemical species, is also integrated as the solution of sink term of ion [7]. Finally, the sink term of electron (anodic/cathodic polarization) was solved by the electro-chemical dynamics with overlaid domain approach [8]. Unlike applying a 2D interface element between concrete and rebar to account for the ion-electron transfer, the 3D overlaid elements model uses smeared concrete-steel elements with shared DOFs (degree of freedom) to calculate the polarization, which allows the analytical domain to be members and even structures. Moreover, such consistent conversion from 2D to 3D continuum modeling allows coupling with 3D smeared crack models of structure concrete for further mechanical analysis. In summary, combining the Nernst-Plank-Poisson equation, Debye-Hückel equation and polarization dynamics, the continuity equation could be solved in terms of ion concentration and electric potential, taking into consideration the multi-ion kinetics with macro-cell corrosion. As the first step, the current simulation ignores the time-dependent change of migration factors of concrete due to the varied pore structures by interaction between pore solution and solid skeleton. Instead, an independent porous material with similar pore framework and pore solution are adopted, which is consistent with the experiment described in next chapter.

2.2 Simulation Models

Four models are prepared in terms of (1) reinforcement (without: NP; with: P, S-NaCl, G-NaCl) (2) saline inclusion (without: NP, P; with: S-NaCl, G-NaCl) (3) saturation of $Ca(OH)_2$ (over saturation of 3 kg/m^3: NP, P, S-NaCl; under saturation of 0.3 kg/m^3: G-NaCl), as shown in Fig. 2. The dimension of electrolyte, reinforcement and meshing as well as the charging voltage are also indicated in Fig. 2. After charging for certain time, the ion profiles are outputted and analyzed.

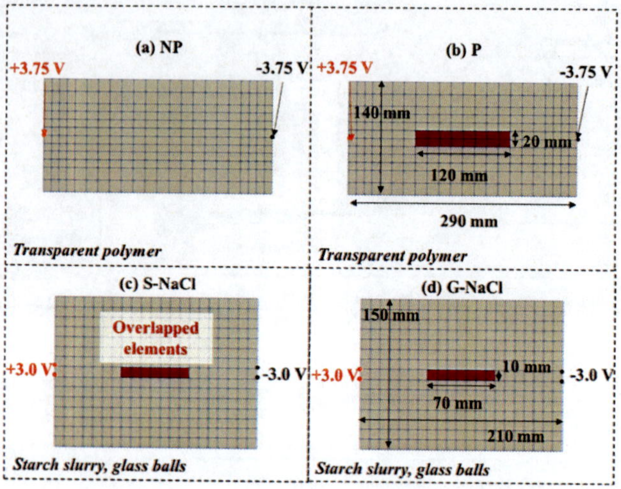

Fig. 2. Simulation models of NP, P, S-NaCl, and G-NaCl

3 Experiment

3.1 Specimen Preparation

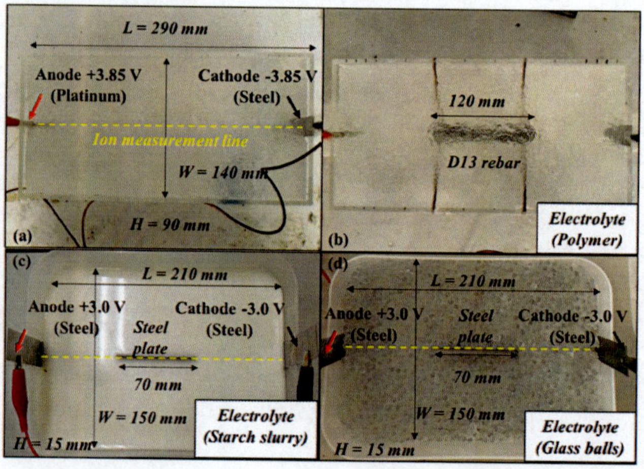

Fig. 3. Experiment specimens of NP, P, S-NaCl, and G-NaCl

To validate the simulation results, experiment is conducted with pseudo concrete including sodium poly-acrylate superabsorbent polymer, starch slurry and glass balls of different grain sizes. In accordance with the simulation models, four specimens (NP, P, S-NaCl,

G-NaCl) are prepared which exactly follows the dimension and set-ups of the analytical models as mentioned in last chapter, see Fig. 3. By observation of steel and electrolyte as time passes, the anodic and cathodic regions can be clearly captures through bubbles or brownish rust by using such pseudo concrete materials. Besides, the dynamic ion profiles are easily measured by taking samples of pore solution at different locations.

3.2 Results and Discussions

The ion profiles of hydroxyl (OH^-) for NP, P and S-NaCl, and sodium (Na^+) for S-NaCl are plotted in Fig. 4 with showing the location of rebar if any. Dashed lines indicate the simulation results while solid lines represent the experimental result. By comparison, it is found the simulation platform can provide the rational multi-ion kinetics associated with macro-cell circuit. If comparing the hydroxyl concentration around middle reinforcement between P and S-NaCl in Fig. 4(b) and (c), the presence of chloride ion is proved to accelerate the macro-cell corrosion, which is due to the breaking of passive film of steel by Cl^-.

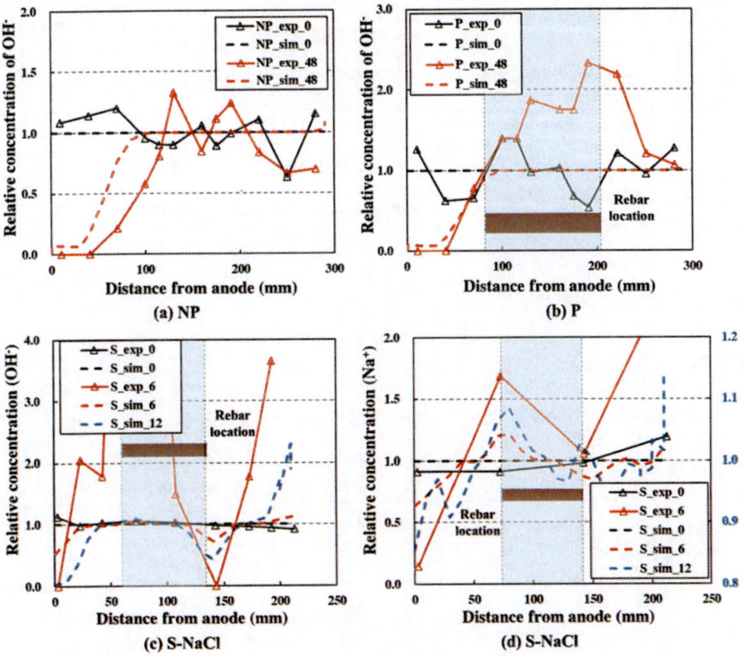

Fig. 4. Ion profiles of NP, P and S-NaCl

Furthermore, the ion profiles of calcium (Ca^{2+}) for S-NaCl and G-NaCl are given in Fig. 5, where good agreement could only be found for former case (S-NaCl) with over-saturated calcium hydroxide solution. For S-NaCl, unlike the common sense where cation moves towards cathode and achieve an increasing concentration there, Ca^{2+} is gathered around anode as captured by both simulation and experiment. It could be explained by the ionization of $Ca(OH)_2$, in which the reaction $Ca(OH)_2 = Ca^{2+} + 2OH^-$

will tend to create more Ca^{2+} with huge consumption of OH^- at anode, accompanied with the undissolved $Ca(OH)_2$ solid. However, it is contradictory that experimental data shows the same tendency for under-saturated calcium hydroxide case (G-NaCl) as well, unlike simulation results. For investigation, the fact is recalled that the boundary condition in the experiment is different from the simulation. As shown in Fig. 3, the specimen in the experiment was open to air containing carbon dioxide gas, which might react with the electrolyte of calcium hydroxide. But in the simulation, the model was equivalent to the sealed condition where nothing was supplied at boundary (Fig. 2). Such inconsistency was believed to cause the contradictory phenomenon between experiment and simulation. Therefore, to further investigate such uncorrelation due to the supply of carbon dioxide in the air, parametric study is conducted.

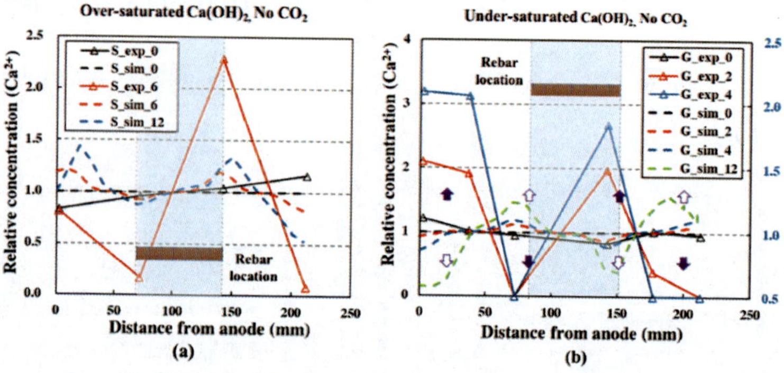

Fig. 5. Ca^{2+} ion profiles of S-NaCl and G-NaCl

4 Parametric Study

4.1 Summary of Cases

Table 1. Mix proportions and initial components of the pore solution.

Specimen	$Ca(OH)_2$ (kg/m^3)	CO_2 (kg/m^3)	Initial components
U-0.0	1.0	0.0	Ca^{2+}, OH^-
U-0.3	1.0	0.3	Ca^{2+}, OH^-, $CaCO_3$
U-0.6	1.0	0.6	$CaCO_3$
O-0.0	3.0	0.0	Ca^{2+}, OH^-, $Ca(OH)_2$
O-0.5	3.0	0.5	Ca^{2+}, OH^-, $Ca(OH)_2$, $CaCO_3$
O-1.5	3.0	1.5	Ca^{2+}, OH^-, $CaCO_3$
O-1.8	3.0	1.8	$CaCO_3$

Seven cases are prepared with respected to the initial saturation degree of calcium hydroxide and supply of carbon dioxide, as listed in Table 1. Both simulation and experiment are conducted, where silica sand is adopted as the pseudo concrete since the observation of anodic/cathodic reaction is less important compared with the quantitative measurement of Ca^{2+} ion profiles. The dimension and charging voltage of models/specimens in parametric study are same as NP/P.

4.2 Results and Discussions

The ion profiles of Ca^{2+} after charging for 24 h of all the seven cases by both simulation and experiment are shown in Fig. 6, 7, 8. The calcium ion has a different profile for each case depending on the initial pore solution of electrolyte and supply of carbon dioxide. Generally, the seven cases can be roughly divided into three genres depending on the phenomenon and mechanisms:

(1) Genre 1 including Case 1 (U-0.0), Case 2 (U-0.3) and Case 6 (O-1.5): As shown in Fig. 6, the calcium ion had the same tendency to increase near the cathode and decrease near the anode. The initial pore solution contained un-saturated Ca^{2+}, OH^- ions and a certain amount of $CaCO_3$ solids (nil for U-0.0). The presence of these ions ensured the mobility and conductivity of electrolyte. Thus, cations moved towards the cathode while anions moved towards the anode based on the Nernst-Plank equation. Although for Case 2 and Case 6, Ca^{2+} was also created at the anode from the dissolution of the $CaCO_3$ solids due to consumption of OH^- (acid environment), but the speed was much less compared with the Ca^{2+} migration towards the cathode. As a result, an obvious Ca^{2+} concentration rise at the cathode and drop of the concentration at the anode are captured by both numerical simulation and experiment with pseudo-concrete of silica sand. This case shown in Fig. 6(a) represents current simulation result of G-NaCl in Fig. 5(b) where CO_2 has not yet been included in the boundary condition.

(2) Genre 2 including Case 3 (U-0.6) and Case 7 (O-1.8): As shown in Fig. 7, the calcium ion increased near the anode while it decreased near the cathode electrode, which is opposite to the evidence of Genre 1. In this genre, the initial pore solution contained almost no Ca^{2+}, but only a certain amount of $CaCO_3$ solids. Therefore, the ion mobility and conductivity of electrolyte were significantly reduced. When polarization happened with further consumption of the OH^- at the anode, the calcite dissolved in the acid environment, generating ionized Ca^{2+}. Such ionization effect explains why the concentration of the cation calcium ion increased near the anode. Moreover, as the initial pore solution was close to neutral, the ion profile of OH^- varied drastically in logarithmic scale compared to Genre 1 and 3, which are characterized by the alkaline state. This case shown in Fig. 7(a) is also supposed to be the real condition for G-NaCl in Fig. 5(b) where CO_2 should be taken into consideration.

(3) Genre 3 including Case 4 (O-0.0) and Case 5 (O-0.5): As shown in Fig. 8, the calcium ion concentration was almost unchanged near the cathode, while it increased around anode or at a certain distance from the anode. In this genre, the initial pore solution contained saturated Ca^{2+}, OH^-, and a certain amount of $Ca(OH)_2$ solids and $CaCO_3$ solids (nil for O-0.0). Under these conditions, the ion mobility was large

since sufficient ion existed in the electrolyte. While calcium moved towards cathode, its concentration could not increase anymore since calcium hydroxide was already fully-saturated. The only result was to participate more $Ca(OH)_2$ solids near the cathode. Therefore, the migration of calcium ion towards the cathode was also slowing down. Meanwhile, associated with huge consumption of hydroxide at anode, $Ca(OH)_2$ solids started to dissolve and release calcium ion here. As a result, the calcium ion near anode showed increase.

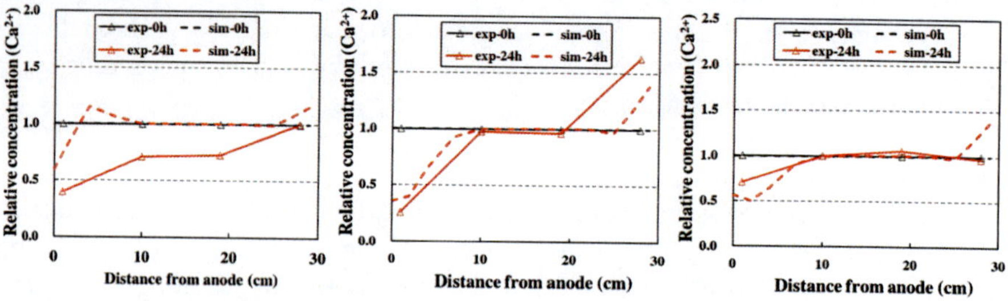

Fig. 6. Ca^{2+} ion profiles of U-0.0, U-0.3 and O-1.5

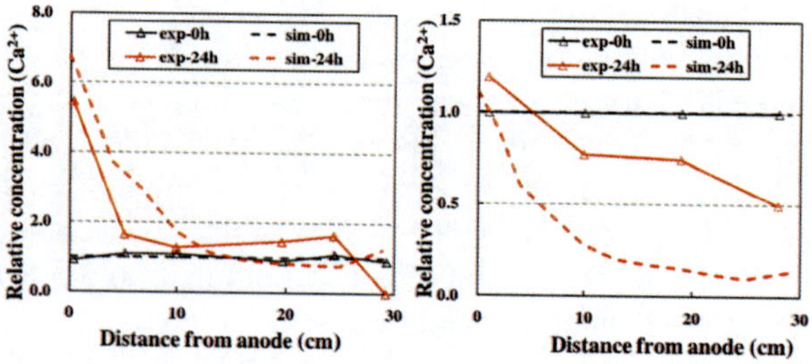

Fig. 7. Ca^{2+} ion profiles of U-0.6 and O-1.8

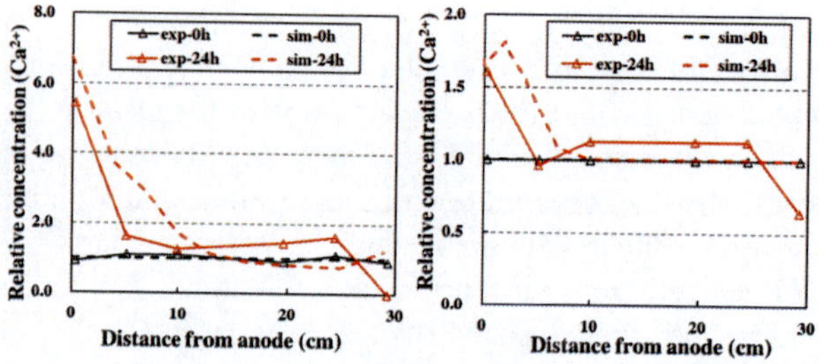

Fig. 8. Ca^{2+} ion profiles of O-0.0 and O-0.5

4.3 Re-analysis of G-NaCl and Summary

The analysis of G-NaCl is re-conducted with coupling the carbon dioxide after the above-mentioned parametric study. The ion profiles of Ca^{2+} are shown in Fig. 9, where (a) is without considering the effect of CO2, while (b) is considering the effect of CO_2. From Fig. 9, it indicates that agreement is obtained by attaching CO_2, where Ca^{2+} increases near the anode, while it decays near the cathode. Following the parametric study, three governing mechanisms are summarized in Fig. 10.

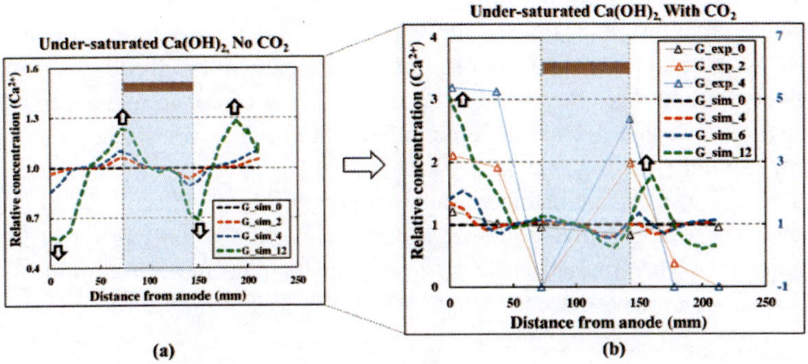

Fig. 9. Computed Ca^{2+} profiles of G-NaCl (a) without CO2 and (b) with CO2

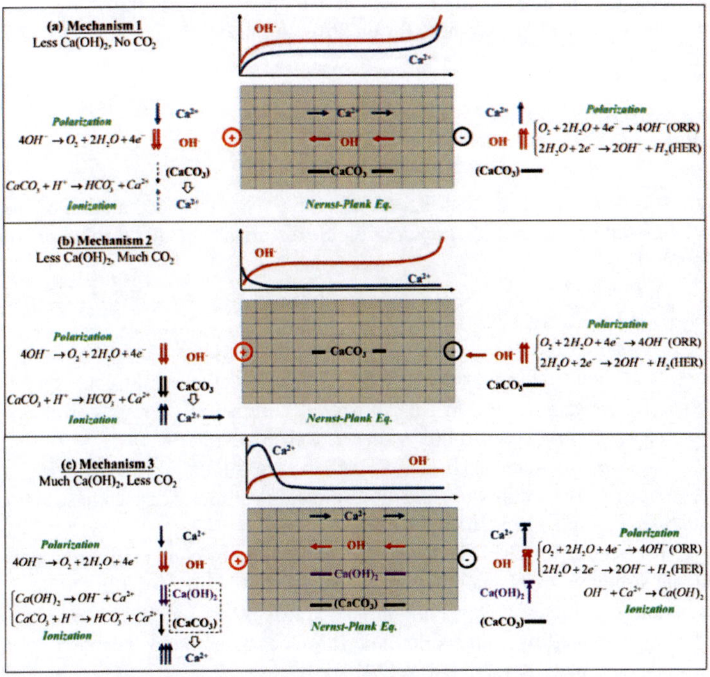

Fig. 10. Summary of three governing mechanisms

5 Conclusions

(1) Multi-ion kinetics with macro-cell corrosion is simulated and verified by using pseudo concrete materials whose characteristics as electrolyte are similar to those of the pore solution of real concrete, and the corresponding corrosion of reinforcing bars and accompanying dynamics of cation and anion substances are measured.
(2) Provided that less solid Portlandite is included in the mixture, the concentration of positive calcium ion was found to rise at the negative electrode. On the contrary, it is also found that calcium ion as a major cation of concrete does not decay but gathers on the positive anodic electrode unlike other cations when over-saturated Portlandite is dosed in the matrix.
(3) Furthermore, CO_2 inclusion is found to change the profile of ion concentration again. These apparently complex kinetics of calcium ion can be quantitatively predicted by the simulation platform to consider the multi-ion equilibrium with precipitated solids of hardened cement paste in hydration.

Acknowledgments. The first author would like to express his sincere thanks to the Japan Society for the Promotion of Science (JSPS) for providing a fellowship (20F20367) to his postdoctoral study. In addition, the authors would like to express their sincere gratitude for a Grant-in-Aid for Scientific Research (A) of Japan Society for the Promotion of Science (20H00260).

References

1. Maruya, T., Takeda, H., Horiguchi, K., Koyama, S., Hsu, K.L.: Simulation of steel corrosion in concrete based on the model of macro-cell corrosion circuit. J. Adv. Concr. Technol. **5**(3), 343–362 (2007)
2. Bazant, Z.P.: Physical model for steel corrosion in concrete sea structures – theory. J. Struct. Div. **105**(6), 1137–1153 (1979)
3. Liu, Q.F., Li, L.Y., Easterbrook, D., Yang, J.: Multi-phase modelling of ionic transport in concrete when subjected to an externally applied electric field. Eng. Struct. **42**, 201–213 (2012)
4. Xia, J., Li, T., Fang, J.X., Jin, W.L.: Numerical simulation of steel corrosion in chloride contaminated concrete. Constr. Build. Mater. **228**, 116745 (2019)
5. Wang, Z., Maekawa, K., Takeda, H., Gong, F.: Numerical simulation and experiment on the coupled effects of macro-cell corrosion and multi-ion equilibrium with pseudo structural concrete. Cement Concr. Compos. **123**, 104181 (2021)
6. Maekawa, K., Ishida, T., Kishi, T.: Multi-scale modeling of concrete performance integrated material and structural mechanics. J. Adv. Concr. Technol. **1**(2), 91–126 (2003)
7. Elakneswaran, Y., Ishida, T.: Development and verification of an integrated physicochemical and geochemical modelling framework for performance assessment of cement-based materials. J. Adv. Concr. Technol. **12**(4), 111–126 (2014)
8. Maekawa, K., Okano, Y., Gong, F.: Space-averaged non-local analysis of electric potential for polarization reactions of reinforcing bars in electrolytes. J. Adv. Concr. Technol. **17**(11), 616–627 (2019)

Discrete Models of Structural Concrete: Discretization Strategies

Qiwei Zhang and John E. Bolander(✉)

University of California, Davis, CA 95616, USA
{qvvzhang,jebolander}@ucdavis.edu

Abstract. Discrete models, which are based on two-node elements or particle assemblies, offer advantages in modeling cracking and other discontinuous phenomena. Within the community of developers of discrete models, there are two main veins of thought: 1) the discrete nature of the model represents physical features of the material; or 2) the discretization strategy should not have a dominant influence on the analysis results. Advocates of the first vein of thought can argue, with some merit, that the discrete approaches effectively capture the effects of large-scale heterogeneity of concrete. From another viewpoint, however, it is attractive for the models to retain selected qualities of the continuum approaches including, for example, elastic homogeneity under uniform straining. Similar qualities are attractive in scalar field analyses (e.g., of moisture or heat transport) that couple with mechanical analyses for simulating the durability of concrete. The research presented herein resides in the second vein. One goal is to provide an unbiased framework without artificial heterogeneity, such that the presence of actual heterogeneity (in the form of phase fractions, interfaces, reinforcing bars, short fibers, etc.) can be explicitly represented.

Keywords: Structural concrete · Discrete models · Fracture · Reinforcement

1 Introduction

Cracking of concrete can adversely affect its service life performance. The questions regarding whether structural concrete cracks, and then the consequences of such cracking, are difficult to resolve from laboratory studies alone. Computational modeling plays a key role in this respect, enabling full-field parametric evaluations that are beyond the reach of laboratory studies.

Models of structural concrete cracking, as well as other forms of damage, are based on continuum or discrete formulations. The continuum approaches, including the finite element method, have a long history of usage and have led to numerous advancements in concrete modeling. Discrete approaches have been motivated by, in large part, their natural abilities to represent cracking and other forms of displacement discontinuity that occur in structural concrete.

P. Rossi and J.-L. Tailhan (Eds.): SSCS 2022, RILEM Bookseries 38, pp. 365–372, 2023.
https://doi.org/10.1007/978-3-031-07746-3_36

One of the key aspects of discrete modeling of concrete materials and structures is the discretization strategy. It is well known that cracking patterns can be affected by the choice of regular versus irregular forms of discretization [10,14]. For the former case, regular grid patterns can strongly bias the potential directions of crack propagation. Irregular grids induce less bias on cracking directions and are therefore preferred. Based on random (or semi-random) placements of nodal points, Voronoi/Delaunay tessellations have emerged as effective means for domain discretization. With this choice as a starting point, this paper looks at another important aspect of discrete modeling: role of the discretization strategy. In broad terms, discretization is either (1) non-physical or (2) based on some physical units of the material structure [3]. The non-physical form acts as a pure discretization technique, such that the model response should be sufficiently independent of the size or shape of the discrete elements. In contrast, physical discretizations represent, either directly or indirectly, relevant features of the material. They have been effective in simulating the behavior of concrete materials and structures [1,6–8].

This paper investigates some of the qualities of non-physical forms of discretization. This is done through a benchmark example involving a panel structure loaded in pure shear. Comparisons are made between non-physical and physical forms of discretization, both of which are based on the rigid-body-spring concept of Kawai [11]. The models differ in their approaches to simulating the Poisson effect.

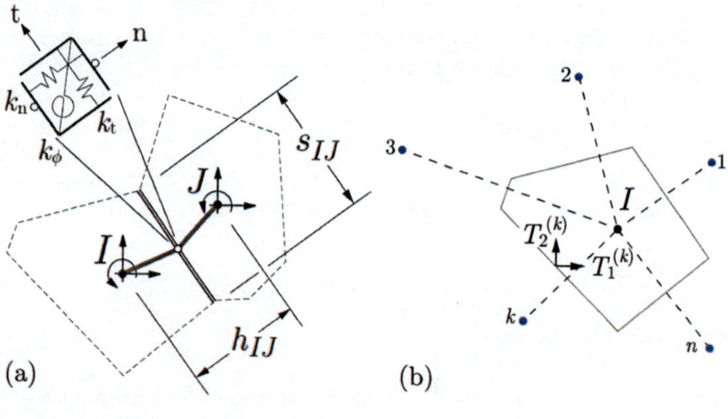

Fig. 1. (a) Rigid-body-spring element; and (b) spring-set force components for nodal stress calculation

2 Nodal Stress Calculation

Nodal stress calculation is a means for studying the character of the discretization schemes. Each rigid-body-spring element is composed of a zero-size spring set that is connected to the element nodes via rigid-body constraints, as shown in Fig. 1a where k_n, k_t and k_ϕ represent the normal, tangential, and rotational spring stiffness coefficients, respectively. The collection of these simple, two-node elements fits well within the family of lattice models [9,14]. At any stage of loading, the forces in the springs of the rigid-body-spring elements framing into a given node are known. Referring to Fig. 1b, tensorial representations of nodal stress are then calculated using [5]

$$\sigma_{ij} = \frac{1}{V} \sum_{k=1}^{n} x_i^{(k)} T_j^{(k)} \tag{1}$$

where k is the number of elements connected to node I; T_j are the spring-set forces, resolved into components aligned with the global coordinate axes; x_i are the positions of the spring sets relative to node I; and V is the volume of the Voronoi cell. The two-dimensional depictions in Fig. 1 extend naturally to three dimensions.

3 Elastic Shear Panel Simulation

3.1 Boundary Conditions

A uniform state of strain (or stress) can be introduced in a homogeneous plate by prescribing displacements along its boundaries according to

$$\begin{aligned} u(x,y) &= c_1 x + c_2 y + c_3 \\ v(x,y) &= c_4 x + c_5 y + c_6 \\ w(x,y) &= (\partial u/\partial y - \partial v/\partial x)/2 \end{aligned} \tag{2}$$

where u and v are displacements in the $x-$ and $y-$directions, respectively; w is rotation; and c_i are constant coefficients. To introduce a state of pure shear into the panel structure, as represented by the rigid-body-spring network shown in Fig. 2, the boundary nodes are constrained to follow

$$\begin{aligned} u(x,y) &= cy \\ v(x,y) &= cx \\ w(x,y) &= 0 \end{aligned} \tag{3}$$

where constant c determines the magnitude of the shear strain.

3.2 Introducing the Poisson Effect

Discrete modeling approaches based on 1D elements, or particle-particle interactions, do not have a direct means for representing both elastic constants of a homogeneous material. For rigid-body-spring networks based on a Voronoi partitioning of the material domain, and setting $k_n = k_t$, the resulting model is elastically homogeneous under uniform straining [4]. Young's modulus E is precisely represented, albeit with Poisson's ratio $\nu = 0$.

The conventional approach for simulating the Poisson effect is to set the spring stiffness coefficients using either a mechanics of materials formulation or a calibration process. As an example of the former option, the coefficients have been defined by [11]:

$$k_n = \frac{1-\nu}{(1+\nu)(1-2\nu)} \frac{EA}{h} \tag{4}$$

and

$$k_t = \frac{1}{1+\nu} \frac{EA}{h} \tag{5}$$

where A is the area of the Voronoi facet that defines the element cross-section. For two-dimensional models, as shown in Fig. 1, $A = st$ where t is the thickness dimension. These approaches approximate the Poisson effect macroscopically, but not in a local sense.

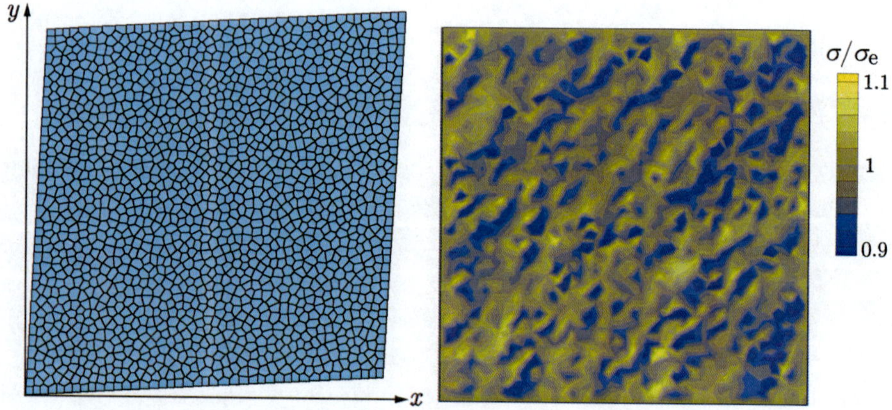

Fig. 2. Shear panel model in deformed configuration (left); and contours of major principal stress for the case of $k_n \neq k_t$ (right)

Schlangen and Garboczi [13] developed a method for assigning the cross-section properties of frame-type elements to achieve an elastically uniform lattice for prescribed values of E and ν. Alternatively, Asahina et al. [2] begin with a rigid-body-spring network based on a Voronoi partitioning of the domain and $k_t = k_n = AE/h$, which results in $\nu = 0$. The Poisson effect is then introduced iteratively using the concept of auxiliary stress. In the examples that follow, comparisons are made between this auxiliary stress approach and a conventional approach (based on Eqs. 4 and 5) for introducing the Poisson effect.

3.3 Simulation Results

For the panel structure shown in Fig. 2, which is subjected for pure shear, L_2 error norms of the nodal displacements and stresses are given in Table 1.

Poisson's ratio does not influence the displacement field for this problem, but it does affect the stress values. Whereas conventional approaches simulate the elastic behavior in a macroscopic sense, the relatively large error in the nodal displacements reflects the presence of artificial heterogeneity. In contrast, the auxiliary stress approach reproduces the displacement field exactly (i.e., to within machine precision).

Table 1. Relative error in nodal displacements $\mathbf{u}$ and major principal stress σ_1

Formulation	Coefficient settings	Nodal quantity	L_2 value*
Conventional approach	$k_t \neq k_n$	$\mathbf{u}$	2.43×10^{-3}
		σ_1	3.00×10^{-3}
Auxiliary stress approach	$k_t = k_n$	$\mathbf{u}$	3.97×10^{-16}
		σ_1	1.02×10^{-11}

* L_2 norms exclude the contributions of the boundary nodes

To visualize the presence of artificial heterogeneity induced by the conventional approach, contours of major principal stress are plotted in Fig. 2. The nodal stress values have been normalized by the magnitude of the uniform stress field, according to theory. One interesting aspect of these results is the apparent directionality of the stress distribution. That is, some of the extreme values appear to be in sequence, aligned in one of the principal stress directions. The corresponding result for the auxiliary stress approach has not been plotted, since normalized stress values match theory, as indicated by its small L_2 error norm in Table 1.

4 Reinforcing Component Discretization

Most applications of concrete materials involve reinforcing components in the form of bars or short fibers. There are several means for introducing such reinforcing components within the rigid-body-spring networks [3,12]. Herein, reinforcement is modeled using frame elements connected in series. The discretization of a bar into frame elements depends on the Voronoi cell discretization of the concrete, as shown in Fig. 3. To simulate bond behavior, link elements connect the frame element nodes to the nearest concrete nodes (which are the generator points of the Voronoi cells).

The aforementioned iterative means for introducing the Poisson effect has been validated for the case of a plain material (i.e., without reinforcing elements) [2]. To account for the presence of reinforcement, as done herein, the iterative procedure must include the effects of the frame and link elements in the internal force calculations.

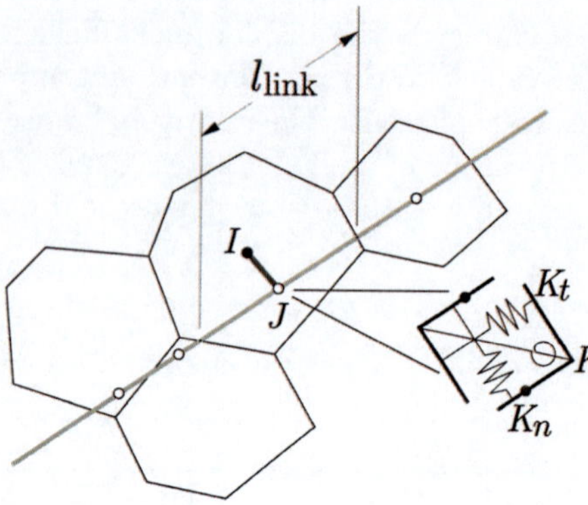

Fig. 3. Linkage between concrete node I and frame element node J representing reinforcement. Link element properties are based on tributary length l_{link}.

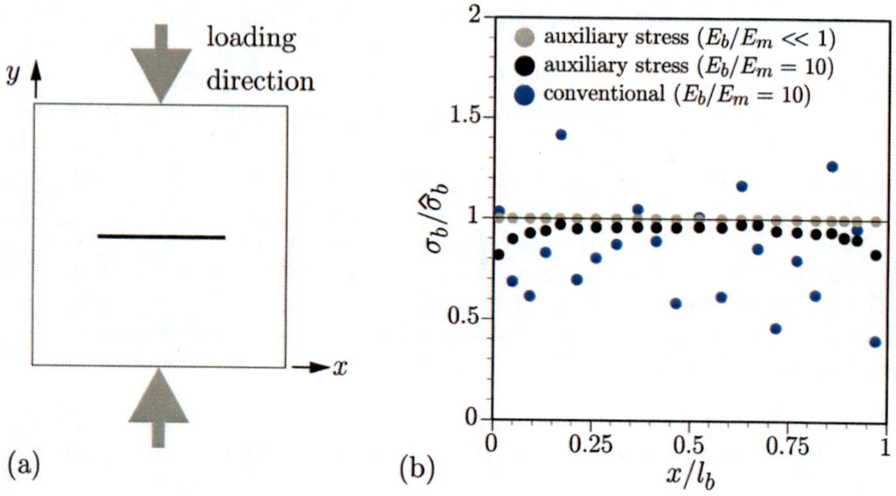

Fig. 4. (a) Bar placement within panel structure; and (b) normalized axial stress in frame elements representing the bar

Consider a bar (or fiber) placed horizontally within the panel structure, as shown in Fig. 4a. The matrix material is discretized as previously shown in Fig. 2. Imposed vertical displacements $v = -y$ along the top and bottom boundaries induce uniaxial compression. In the absence of the bar, principal strains $\epsilon_y = -1$ and $\epsilon_x = -\nu\epsilon_y$ act throughout the domain. When including a bar with negligibly small axial stiffness, the axial stress in the bar should be $\hat{\sigma}_b = -\nu\epsilon_y E_b$, where E_b is the modulus of elasticity of the bar material. The auxiliary stress approach [2] based on $k_t = k_n$ reproduces this theoretical result, as shown in Fig. 4b.

As the stiffness of the bar inclusion increases, the presence of the bar eventually alters the local strain field and thus the stress conditions (Fig. 4b). For a modular ratio of $E_b/E_m = 10$ and a bar diameter of $t/10$ (where E_m amd t represent the elastic modulus and thickness of the panel material, respectively), several observations can be made: 1) bar stresses are reduced; 2) there is a shear lag effect at the bar ends; and 3) the stress values do not form a clean trend line due to, at least in part, the eccentricity of the connections made between the bar and the concrete (Fig. 3).

Although conventional approaches based on $k_t \neq k_n$ may simulate the Poisson effect in a global sense, they do not do so in a local sense. This is evident when plotting axial stress in the bar for this case. The values exhibit scatter about the theoretical result (Fig. 4b). Prior to cracking, bars typically play a secondary role in concrete structures, such that this scatter might be of secondary importance.

5 Conclusions

The use of physical or non-physical forms of discretization depends on the objectives of the analysis. Physical models can be used to effectively simulate the presence of heterogeneity. Non-physical discretizations are devoid of artificial heterogeneity and therefore can serve as an unbiased framework for explicitly representing reinforcing bars or other forms of inclusions. The following conclusions can be made:

- Using the auxiliary stress approach developed by Asahina et al. [2], rigid-body-spring networks can be rendered elastically uniform for general values of Young's modulus and Poisson's ratio. For the case of a panel structure loaded in pure shear, as demonstrated herein, the nodal displacement and stress values match theory.
- As expected, the use of $k_t = \alpha k_n$ (for $\alpha \neq 1$) to simulate the Poisson effect introduces a form of heterogeneity into the discrete model. The heterogeneity appears to induce directionality in the contours of principal stress.
- The auxiliary stress approach developed for plain materials can be extended to the case where reinforcing components are present. When the stiffening effects of the reinforcement are small, the axial stress in the reinforcing components matches theory. When the stiffening effects of the reinforcement modify the strain field of the concrete, which occurs in practice, axial stress varies along the bar length.

The examples presented herein have been confined to elastic behavior. The elastic properties of these lattice networks should be understood on route toward simulating concrete cracking and other forms of nonlinear behavior. Current efforts include the simulation of reinforced concrete shear panels under monotonic and cyclic loading.

References

1. Alnaggar, M., Cusatis, G., Di Luzio, G.: Lattice Discrete Particle Modeling (LDPM) of Alkali Silica Reaction (ASR) deterioration of concrete structures. Cement Concrete Comp. **41**, 45–59 (2013)

2. Asahina, D., Aoyagi, K., Kim, K., Birkholzer, J.T., Bolander, J.E.: Elastically-homogeneous lattice models of damage in geomaterials. Comput. Geotech. **81**, 195–206 (2017)
3. Bolander, J.E., Eliáš, J., Cusatis, G., Nagai, K.: Discrete models of concrete fracture. Eng. Fract. Mech. **257** (2021)
4. Bolander, J.E., Saito, S.: Fracture analyses using spring networks with random geometry. Eng. Fract. Mech. **61**(5–6), 569–591 (1998)
5. Christoffersen, J., Mehrabadi, M.M., Nemat-Nasser, S.: A micromechanical description of granular material behavior. J. Appl. Mech. **48**(2), 339–344 (1981)
6. Eddy, L., Nagai, K.: Numerical simulation of beam-column knee joints with mechanical anchorages by 3D rigid body spring model. Eng. Struct. **126**, 547–558 (2016)
7. Eliáš, J., Vořechovský, M., Skoček, J., Bažant, Z.P.: Stochastic discrete meso-scale simulations of concrete fracture: comparison to experimental data. Eng. Fract. Mech. **135**, 1–16 (2015)
8. Fascetti, A., Ichimaru, S., Bolander, J.E.: Stochastic lattice discrete particle modeling of pervious concrete. Comput. Aided Civil Infrastr. Eng. (2022). https://doi.org/10.1111/mice.12816
9. Herrmann, H.J., Roux, S.: Statistical Models for the Fracture of Disordered Media. Elsevier, Amsterdam (1990)
10. Jirásek, M., Bažant, Z.P.: Particle model for quasibrittle fracture and application to sea ice. J. Eng. Mech. ASCE **121**(9), 1016–1025 (1995)
11. Kawai, T.: New discrete models and their application to seismic response analysis of structures. Nucl. Eng. Des. **48**(1), 207–229 (1978)
12. Pan, Y., Kang, J., Ichimaru, S., Bolander, J.E.: Multi-field models of fiber reinforced concrete for structural applications. Appl.. Sci. **11**, 184 (2021)
13. Schlangen, E., Garboczi, E.: New method for simulating fracture using an elastically uniform random geometry lattice. Int. J. Eng. Sci. **34**(10), 1131–1144 (1996)
14. Schlangen, E., Garboczi, E.J.: Fracture simulations of concrete using lattice models: computational aspects. Eng. Fract. Mech. **57**(2–3), 319–332 (1997)

Author Index

P. Rossi and J.-L. Tailhan (Eds.): SSCS 2022, RILEM Bookseries 38, pp. 373–374, 2023.
https://doi.org/10.1007/978-3-031-07746-3